FUEL EFFICIENT POWER TRAINS AND VEHICLES

VECON '84

Conference Planning Panel

D Broome, MA, CEng, FIMechE, MSAE,
  MIRTE (Chairman)
Ricardo Consulting Engineers plc
Shoreham-by-Sea
Sussex

M Barnard, BA
BL Technology Limited
Lighthorne Heath
Warwickshire

W T Birge (SAE Representative)
The Bendix Corporation
Southfield
Michigan
USA

M R Dunn, BSc, MEng, MIMechE
Rolls–Royce Motors
Crewe
Cheshire

K Parmee, CEng, FIMechE
Eaton Limited
Manchester

J D Savage, CEng, FIMechE, HonFIRTE,
  MSAE
BP Oil International Limited (Consultant)
London

INTERNATIONAL CONFERENCE ON

FUEL EFFICIENT POWER TRAINS AND VEHICLES

VECON '84

IMechE CONFERENCE PUBLICATIONS 1984–14

Organized jointly by
The Automobile Division of
The Institution of Mechanical Engineers and
The Society of Automotive Engineers
and held under the patronage of
Fédération Internationale des Sociétés d'Ingénieurs
des Techniques de l'Automobile (FISITA)

22–24 October 1984
1 Birdcage Walk, Westminster, London SW1

Published for
The Institution of Mechanical Engineers
by Mechanical Engineering Publications Limited
LONDON

First published 1984

This publication is copyright under the Berne Convention and the International Copyright Convention. Apart from any fair dealing for the purpose of private study, research, criticism or review, as permitted under the Copyright Act 1956, no part may be reproduced, stored in a retrieval system, or transmitted in any form or by any means, electronic, electrical, chemical, mechanical, photocopying, recording or otherwise, without the prior permission of the copyright owners. Inquiries should be addressed to: The Managing Editor, Mechanical Engineering Publications Limited, PO Box 24, Northgate Avenue, Bury St Edmunds, Suffolk IP32 6BW

© The Institution of Mechanical Engineers 1984

ISBN 0 85298 546 0

The Publishers are not responsible for any statement made in this publication. Data, discussion and conclusions developed by authors are for information only and are not intended for use without independent substantiating investigation on the part of potential users.

Printed by Waveney Print Services Ltd, Beccles, Suffolk

CONTENTS

C451/84 SAE 841312	Total vehicle economy—the challenge (Europe) *F W Lohr*	i
C452/84 SAE 841313	Total vehicle economy—the challenge (United States of America) *C L Knighton*	xv
C453/84 SAE 841314	Total vehicle economy—the challenge (Japan) *K Matsumoto*	xxi
C454/84 SAE 841315	Energy and the vehicle *D A G Simon*	xxv
C422/84 SAE 841281	The interaction of fuel economy and emission control in Europe—a literature study *C D de Boer and J A Jeyes*	1
C441/84 SAE 841282	U.S. passenger car fuel economy *B H Simpson*	13
C429/84 SAE 841283	Design of a high-efficiency, high-BMEP, small car powerplant *M A Pulick, G F Leydorf Jr, R C Innes, R A Stein and B A S Shannon*	31
C443/84 SAE 841284	Torque characteristics and fuel efficiency of various gasoline engine concepts *D Gruden and H Richter*	41
C425/84 SAE 841285	Comparison of a turbocharger to a supercharger on a spark ignited engine *T G Adams*	49
C432/84 SAE 841286	Fuel economy opportunities with an uncooled DI diesel engine *W R Wade, P H Havstad, E J Ounsted, F H Trinker and I J Garwin*	59
C440/84 SAE 841287	An appraisal of advanced engine concepts using second law analysis techniques *R J Primus, K L Hoag, P F Flynn and M C Brands*	73
C421/84 SAE 841288	Fuel economy, emissions and noise of multi-spray light duty DI diesels—current status and development *K P Mayer*	89
C434/84 SAE 841289	Knock protection—future fuels and engines *D L Sutton and D Williams*	99
C424/84 SAE 841290	A cold-start track test procedure for evaluating fuel-efficient oils *G B Toft*	113
C430/84 SAE 841291	Impact of fuel composition on octane requirement *F J Marsee, J P Sunne and W E Adams*	125
C433/84 SAE 841292	New temperature control criteria for more efficient gasoline engines *H-P Willumeit, P Steinberg, H Ötting, B Scheibner and W Lee*	135
C446/84 SAE 841293	Austin-Rover Montego programmed ignition system *E W Meyer, R Green, R Woodhouse and M H Cops*	141
C427/84 SAE 841294	A study on dual circuit cooling for higher compression ratio *H Kobayashi, K Yoshimura and T Hirayama*	151
C426/84 SAE 841295	A single cylinder engine for crankshaft bearings and piston friction losses measurement *R Cerrato, R Gozzelino and R Ricci*	159

C449/84 SAE 841296	Piston and ring mechanical losses *L Feuga and C Bury*	173
C444/84 SAE 841297	Adaptive air/fuel control applied to a single point injection system for SI engines *E M Gamberg and T A Huls*	181
C436/84 SAE 841298	Development of a high powered, economical four-valve engine Toyota 4A-GE *Y Kimbara, M Konishi, N Nakamura and S Kobuki*	189
C450/84 SAE 841299	Friction in the piston group and new ideas for piston design *H Oetting, D Pundt and W Ebbinghaus*	197
C435/84 SAE 841300	Three-cylinder, 1.0 litre diesel engine mounted on fuel-efficient vehicle of New Charade *Y Ino*	203
C428/84 SAE 841301	Passenger car diesel engines charged by different systems for improved fuel economy *P W Manz and K-D Emmenthal*	211
C447/84 SAE 841302	Control strategies for a chain drive CVT *U Eggert and H D Schneider*	219
C431/84 SAE 841303	Fuel economy with small automatic transmissions *R P Jarvis*	227
C420/84 SAE 841304	Mercedes–Benz automatic transmissions for passenger cars *H J Förster*	235
C445/84 SAE 841305	The Ford research dual mode continuously variable transmission *T R Stockton*	247
C439/84 SAE 841306	The design of an engine-flywheel hybrid drive system for a passenger car *N A Schilke, A O DeHart, L O Hewko, C C Matthews, D J Pozniak and S M Rohde*	255
C423/84 SAE 841307	The Unomatic transmission *G Falzoni*	277
C442/84 SAE 841308	Field experience with a fleet of test cars equipped with Comprex supercharged engines *P Rebling and F Jaussi*	283
C437/84 SAE 841309	Development of Mazda fuel-efficient concept car *K Yokooku, A Nagao and H Oda*	295
C438/84 SAE 841310	Methods of reaching vehicle fuel consumption and weight reduction *H Lagrange and P Baudoin*	309
C448/84 SAE 841311	Fuel economy and performance—the case for lightweight vehicles *A Ancliff*	321

C451/84

SAE 841312

Total vehicle economy—the challenge (Europe)

F W LOHR
Adam Opel AG, Russelsheim, Federal Republic of Germany

Thank you very much.

It is an honor to be with you today.

I would like to thank the organizers of this conference for inviting me to be one of your keynote speakers. The agenda you have set forth -- to explore the future of fuel efficiency in power trains and vehicles -- is a most important one. And I am delighted to share some of my thoughts with you regarding the challenge of vehicle economy from a European perspective.

Over the next three days, you will hear from many knowledgeable people in the world automotive industry. They will be discussing the progress of fuel efficiency from every angle, from the standpoint of virtually every vehicle system and component.

In my remarks this morning, however, I would like to examine potential economy gains of the total vehicle -- the powertrain and the body, as well as the importance of aerodynamics and electronics. And I would like to point out some of the special challenges and opportunities that automotive engineers will face in the years ahead.

But first, let me turn briefly to the European auto industry and focus on where it is today and where we believe it is heading in the future.

Overall -- given a healthy economic climate -- the European new car market will probably climb from its current level of about 10.5 million annual sales to some 12.5 million units by 1990.

That is the potential ... the anticipated demand.

But the market is intensely competitive.

This competition, generated by consumer needs and expectations, will be the driving force behind future technical innovation. But market forces will also demand higher levels of quality and reliability than ever before.

Thus, I believe that technical progress in overall vehicle economy will be evolutionary rather than revolutionary. Change will come at a rapid pace. But it will be change that has proven itself in the laboratories, on the test stands and on the proving grounds.

Certainly this has always been the case in our industry. For example, the weight-saving unitized body, which was first introduced in 1935, required more than 40 years of development before it was used on a volume basis all over the world.

Other technical improvements, because of high development costs, have been introduced first on up-scale models that are less price sensitive. This was the case with anti-lock brake systems. They were tested as early as 1972 but had to wait for advancements in electronics before gaining acceptance on a broad scale.

In the same way, fuel injection first appeared in the more expensive models before expanding to other vehicle lines.

Clearly, it made good business sense to introduce our technical achievements this way in the past. For the future, however, a more sophisticated car-buying public will require innovations to be launched in our most popular models.

I see this as a challenge to our profession. New developments must be tested as rigorously and as completely as ever before -- but we must find better ways to get them to our customers faster, in lower priced models as well as up-scale models.

There is no doubt that much of the new customer sophistication I have mentioned was spawned by the growing awareness of the need for vehicle economy.

No one in this room needs to be reminded of how two oil shocks in the 1970's changed our industry. It caused manufacturers all over the world to put new emphasis on fuel consumption, an

C451/84 © IMechE/SAE 1984

objective that had always been important to European manufacturers. This is shown here by the example of the German car manufacturers.

Almost at the same time, new efforts were undertaken to drop noise levels and create more effective protection of the environment by lowering exhaust emissions.

Even as the European passenger car market recovered from both oil shocks and appears to have a strong future, the combined effect of these recent developments resulted in a new set of ground rules for our industry.

It meant that automotive engineers would have to work besides cost against a backdrop of the following main objectives: they would have to develop cars of economy ... cars that maintained the performance and excitement demanded by our customers ... and cars that met the standards and regulations issued by governments of several European countries.

The supply of energy itself has fluctuated greatly in recent years, making this job even more difficult. Today, supplies of oil are plentiful. But we should always be mindful that the quantity of oil in the world is finite. The mineral oil used by the transportation sector because of its high energy content, easy handling and cost effectiveness can only last so long. The world's supply of oil is diminishing ... and our industry has a responsibility to develop vehicles that use it sparingly.

I believe our achievements have been outstanding in this area. An analysis published last year by a major oil company, for example, showed that in the Federal Republic of Germany fuel consumption for the automobile sector reached 26 million tons in 1980 with a car population of 23.0 million. Since that time, the number of cars has gone up to nearly 24.6 million while the total fuel consumption has remained about the same.

Looking ahead, we see continued improvement. The estimate for the year 2000 is 30 million cars as against 21 million tons of fuel.

There is a long-term potential to reduce oil usage even further if alternative fuels like liquified gas, methanol, and mixtures of gasoline and alcohol can be consumed on a larger scale. Nationwide tests on gasoline mixed with 15% methanol have already been carried out in Germany with cooperation of Opel, and in Sweden, resulting in the compliance of all fuel systems in Opel cars to gasoline mixtures. Alternative fuels, however, currently have several disadvantages. They have a lower energy rate ... they require more costly vehicle fuel and storage systems ... and in some instances additional energy is required to produce them.

This being the case, it is expected that by the year 2000, oil-derived fuels will still account for about 97 percent of the total fuel consumed by cars, compared with today's 99 percent.

We believe that energy will continue to be a concern and that the trend toward lighter and slightly smaller models will continue. Energy consciousness is established in the minds of our customers -- and we believe it will stay that way. And for our industry, the challenge will be to reduce consumption <u>throughout</u> the vehicle -- not just on any single component or system.

Let me now briefly discuss how greater economy and efficiency can be achieved in several key areas of the vehicle. I'd like to begin with the powertrain -- the engine and transmission -- because I believe it offers the greatest potential for savings.

The engine is the heart of the vehicle. And, as such, it should be the central focus of our efforts to achieve improved economy. We should keep in mind that our customers' tastes are always shifting and that styling is an ever-changing element of our business.

In other words, the fuel economy provided by slippery shapes and low air drag coefficients will do us little good if our customers want to buy jeeps or convertibles.

But the powerplant is independent of appearance and is isolated from design. It therefore offers us the greatest opportunity in our search for efficiency.

In Europe in the years between 1971 and 1983 considerable reductions in exhaust emissions have been achieved. The trend up to the year 1995 assumes that emission standards will become much more stringent and the lead content in petrol will gradually diminish.
While this is still being discussed in several European countries the German government had recently decided to adopt the US'83 emission standards, requiring catalytic converters and lead free gasoline.

Thus, our industry will have to develop engines which will offset resulting lower octane values while still maintaining power levels demanded by the customers. At the same time, these engines will have to become even more fuel efficient.

I believe these new demands upon our engines in Europe have to be realized maintaining todays displacements, but time is needed to find ways to offset the loss of performance caused by emission controls, and to develop methods to maintain fuel economy.

Meeting these new objectives for engines will call for a combination of variable technology -- such as variable valve timing and induction systems -- together with more traditional approaches such as improved fuel combustion, mass reduction, reduced friction and supercharging.

As we see it, two primary types of engines are likely to emerge: a standard engine ... and a high performance/fuel efficient engine. Both will feature good economy.

Of the two types, the standard engine will be the most balanced in terms of fuel efficiency, power and cost. And it will be the springboard for improvements to the performance engine as progress is made in lean burning, higher compression ratios, smaller size and lighter weight.

The high performance/fuel efficient engine will be distinguished by high output and response but will maintain excellent fuel economy. It will feature double overhead camshafts, four valves per cylinder, variable valve timing and induction systems and supercharging.

It is certain that electronic engine control technology will become indispensable for both types of engines. With the introduction of throttle actuators and torque sensors, electronic controls will be used to improve fuel efficiency and engine transitional response, to simplify choke systems and to gain common control of engine and transmission.

Assuming emission control technology is in hand, both lean burning and high compression ratios are major future areas regarding fuel combustion technology. Since lean burn is the most promising approach, emphasis will likely center on closed loop engine control systems combined with the linear oxygen sensor.

We know that average compression ratios dropped in Japan and the United States after unleaded gasoline was introduced. But those ratios will again become higher as improved fuel efficiency and output are required.

It is assumed that by 1990 the average ratio will be 10.0 for premium fuel engines. In addition to the mechanical technologies like reduction in flame transmission time, knock sensors will be used more commonly for achieving high compression ratios.

Developments in diesel engine technology during the 1980's will be aimed at including ceramics around the combustion chamber, such as at the piston head and the cylinder liner. In addition the utilization of direct fuel injection will be of equal importance. The combination of ceramics and direct injection will promote the application of turbochargers.
Furthermore we look for reducing costs and at improving those features in which the diesel is still inferior to the gasoline engine. These would be areas such as horsepower-to-mass ratio, size, noise and vibration, noise, starting features and acceleration.

It has been assumed thus far there will be no substantial tightening of regulations for diesels in the area of NOx emissions or particulates. If, however, stricter regulations take effect, the diesel engine for the economy segment will have to be controlled electronically. A particulate filter device will have to be used and the exhaust gas recirculation application will have to be adapted.

I believe the conventional internal-combustion engine will remain the powerplant of the future -- at least through the end of this century. It has proven itself as a durable and dependable performer, more practical than other concepts which have challenged its dominance.

It has been improved time and time again ... and it will be perfected even more in the future as we continue to adapt more sophisticated computer and software technology ... as we learn more about the use of materials ... and as we increase our understanding of internal processes.

The function of transmitting power to the wheels has been performed very well over the years by the manual transmission ... so well that there will probably be few changes in the basic design during the remainder of this decade.

Improvements will likely include friction reduction of oil seals, reduced oil viscosity and improved shift feeling.

Most manual transmissions will be five-speeds because of the outstanding benefits they provide in the way of fuel consumption, acceleration performance, and shift frequency during normal running.

On the other hand, with increased customer demand for cars that are easy to drive but still maintain good performance and fuel economy, four-speed automatic transmissions will gradually gain acceptance in Europe in the latter half of the 1980s.

The new generation automatic transmission is based on the traditional automatic but contains fundamental changes in structural design to make it smaller, lighter and cheaper.

In addition, a number of mechanisms and functions have been added to it such as a torque converter with lock-up clutch in all gears, except first and reverse, and electronic control of shifting.

A major transmission development during this decade could well be the continuously variable transmission, or CVT. It has tremendous potential for ease of operation as well as savings in fuel consumption by means of electronic controls.

The first applications of the continuously variable transmission will be limited to small cars because of lesser torque capacity needs. Later it will be used in other model lines, providing the same performance and economy levels as a 5 speed manual transmission.

Let us now shift our focus to the contributions of economy from the largest part of the car, the body.

There are essentially two ways of gaining increased fuel efficiency from the body: through lower mass and reduced aerodynamic drag.

I might add that aerodynamics is particularly significant in Europe where there are major highway networks and where speed limits are higher than in the United States and Japan.

Let's first consider the materials that will comprise our passenger car bodies of tomorrow.

The search for fuel efficiency has brought with it a drop in the use of sheet metal. There has, however, been an increase in the use of high tensile steel.

We believe the use of aluminum will increase in Europe through 1990. And, except for aluminum, the use of non-ferrous metals will decline. There will be a tendency to use less glass over the long term, but the use of plastics in Europe, we believe, will increase.

The use of plastics offers some particularly interesting possibilities. There is a good chance to produce vehicles in a limited production volume having plastic outer body panels like hood, decklid and tailgate or even the entire outer body surface in plastics similar to the GM Pontiac Fiero concept.

Taken together, this will mean that the average body mass of automobiles will drop steadily. Looking ahead to 1990, we see a reduction of average body mass by about five percent compared to 1983.

These predictions of body mass, however, do not take into consideration changing customer tastes and government regulations.

In the future there will very likely be greater demand for comfort and convenience items such as five doors, sophisticated radios, power steering and richer trims. Individually, none of these features adds very much mass to the car. Taken together, however, several options like these represent a substantial mass gain to the total vehicle.

And when you consider the added mass of equipment mandated by legal requirements for emissions control, noise reduction and safety, holding down the total weight becomes a major challenge to engineers in search of vehicle efficiency.

Generally, we believe the tendency to add more weight in equipment will be offset both by the increased use of lighter and better utilized materials and by the use of computer aided tools which enable engineers and designers to minimize the mass of each component.

Perhaps no single aspect of an automobile is more closely related to its marketability than styling. Simply stated, a car must look good to sell. We know that aerodynamics and attractive appearance can go hand-in-hand. So the role of the designer must be to develop a style that minimizes air drag and looks good as well.

The average drag area of European cars today is 0.72. We believe that through the end of this decade, drag areas will continue to drop to an average of .60. Our newly-introduced Opel Kadett GSi is a good example of this trend. It has a low drag area of 0.57, yet is attractive and functional in every manner.

Looking ahead to 1990, the drag area of extremely good aerodynamic cars could even drop to 0.55, depending on customer tastes, cost effectiveness and government regulation.

In addition to lowering aerodynamic drag, engineers will look increasingly for ways to reduce road resistance through more efficient tires. Over the years, tires that minimize rolling resistance have become available, and we expect more improvements in the future.

The challenge of developing a new tire, wheel and suspension system, that contributes to fuel efficiency, yet maintain good ride and handling characteristics is one that we believe offers great potential to our industry.

No discussion of vehicle economy would be complete without a reference to electronics. Advances in electronics

© IMechE/SAE 1984 C451/84

have changed our products, our industry, and, to a large degree, the way we do business.

Thanks to the explosion in microprocessing technology, our future cars in Europe will be ablaze with electronic information displays and control functions.

Down the road, changes to mechanical components technology will be characterized in almost every case by a joining together of this technology with electronics.

In the past, electronics were frequently added to mechanical systems, thus creating more size and mass. In the years ahead, however, systems integration will eliminate bulky mechanical forms, creating streamlined systems that are lighter, more efficient, and frequently less costly.

With electronics, we believe our opportunities are unlimited. They will provide for common and more effective management of engine and transmission ... they have already given and will give new efficiency to such features as start-stop-system, body computer, and adaptive ride control ... and they are providing driver and passengers with infinetly more comfort, convenience and entertainment features.

Just as we must seek out every contribution to fuel efficiency in the vehicle, so we must look to external factors to increase our economy wherever possible. For this reason, the vehicle manufacturers of Europe are increasingly calling on governments and institutions for parallel activities in the improvement of highway systems and traffic management.

There has already been substantial research done, to demonstrate that smooth traffic flow can make substantial contributions to the improvement of fuel efficiency as well as the reduction of pollution. This is verified by a study in a German city, where a fuel saving up to 20% and adequate reduction of exhaust emissions were achieved.

Car manufactureres have a responsibility to develop vehicles that are as efficient and as safe to operate as possible. But other major institutions of society can and should help. I believe that efficient and economical motoring, like safety, is 'everybody's business'.

In my remarks thus far, I have described the developments going on in all areas of the vehicle, which are of course strongly influenced by the behaviour and taste of the customer.

That means it is of no benefit if stream lined body shape and low aero drag coefficient are realized for passenger cars, while people demand Jeeps and convertibles.

Reduction of drag coefficient can 'only be one part' of the vehicle development in the direction to better fuel economy.

Most important tasks must remain at all progress we need to concentrate on the power source, because independent from vehicle appearance and usage the power plant will be the heart of the automobile.

For this reason I believe that a high concentration of the upcoming developments and innovations will happen in this area.

That there will be no stillstand in the achievement of total vehicle economy will also be caused by the strong competition inbetween car manufacturers in Europe and worldwide.

It must be the responsibility of all car manufacturers - in a society which is partly hostile to industry - to take care of the human beings, the environment and its resources and therefore to guarantee the proper utilization of energy and raw materials.

This is to a great deal particularly true for the industrialized countries as it is their responsibility to maintain the scarce resources so that other countries with high growth in population will be able to achieve the same mobility as their wealth in economy will improve and motor vehicles will become available for everybody.

As the automobile industry is absolutely concious to solve the remaining problems in the future I am extremely positive about the automotive engineering.

This is a dynamic and growing industry. And I can assure you we have some exciting times ahead of us.

Again, I want to thank you for inviting me here today ... and I want to extend my best wishes for a successful conference.

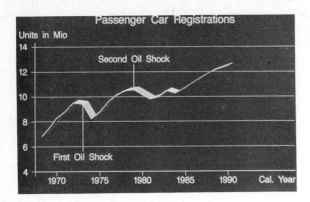

Fig 1 European new car market

Fig 2 Electronic control modules for passenger cars

Fig 3 Unitized body (1935)

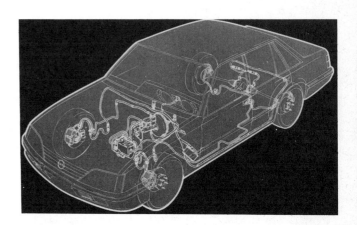

Fig 4 Anti-lock brake system

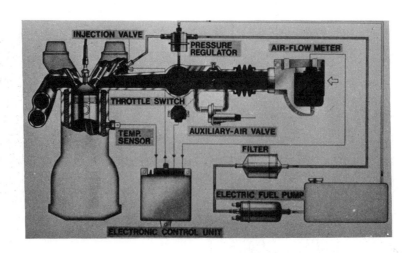

Fig 5 Electronic fuel injection

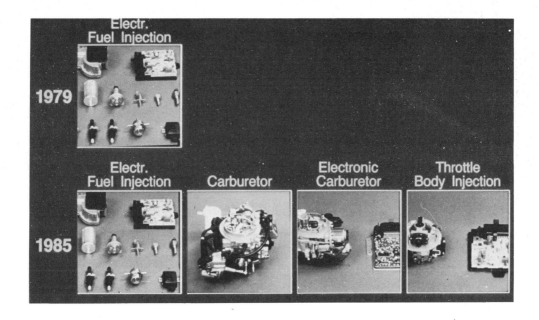

Fig 6 Deceleration fuel control

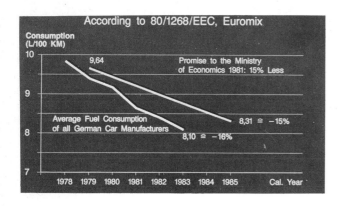

Fig 7 Fleet fuel consumption in Germany

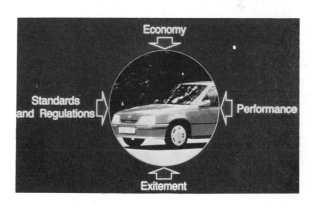

Fig 8 Central objectives

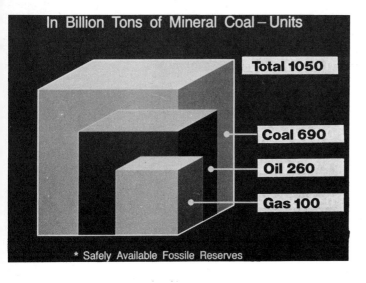

Fig 9 World energy resources*

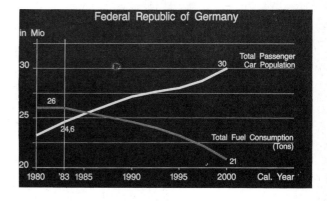

Fig 10 Fuel consumption versus passenger car population

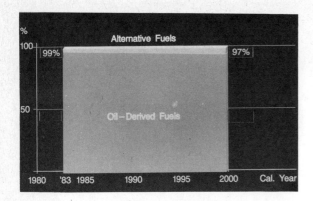

Fig 11 Use of alternative fuels

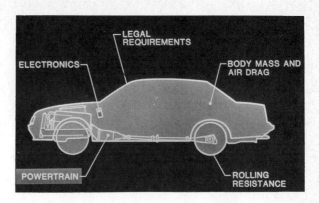

Fig 12 Key areas of vehicle economy

Fig 13 The engine — heart of the vehicle

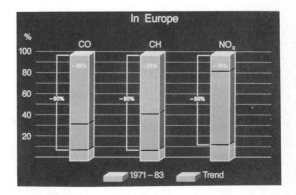

Fig 14 Trend of exhaust emissions

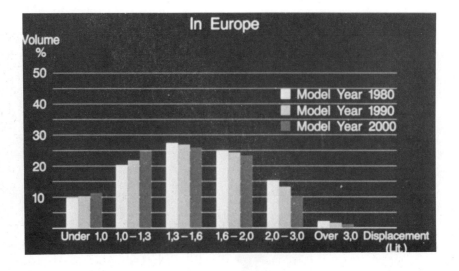

Fig 15 Trend of vehicle displacement

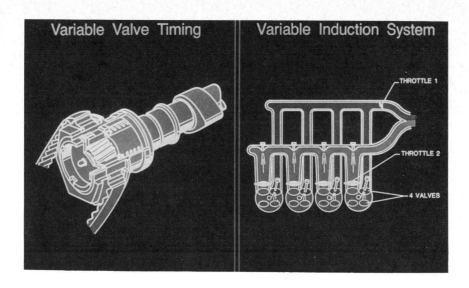

Fig 16 Variable engine technology

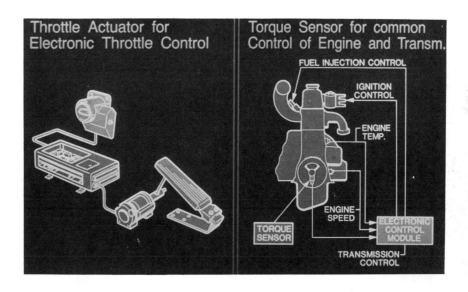

Fig 17 Throttle actuators and torque sensors

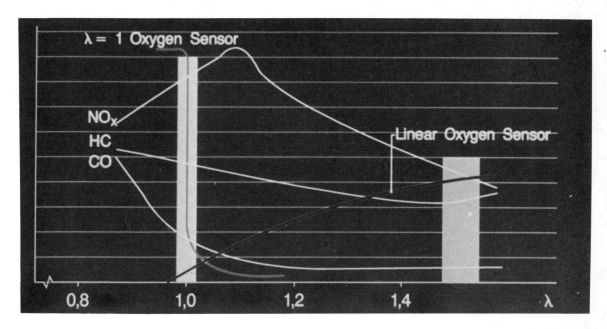

Fig 18 Closed loop engine control with oxygen sensor

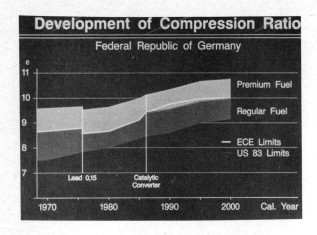

Fig 19 Development of compression ratio

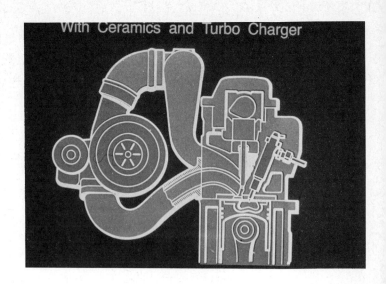

Fig 20 Direct injection diesel engine

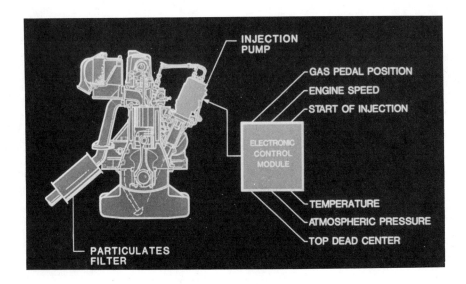

Fig 21 Electronic diesel control

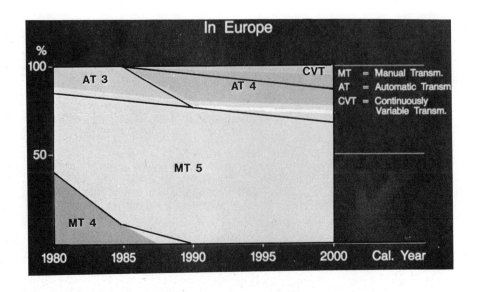

Fig 22 Trend of transmissions

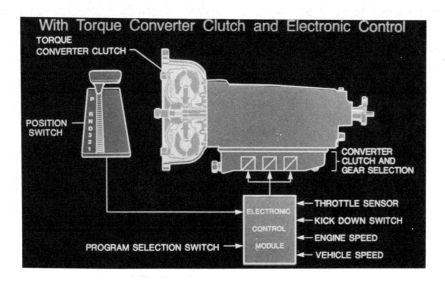

Fig 23 Four speed automatic transmission

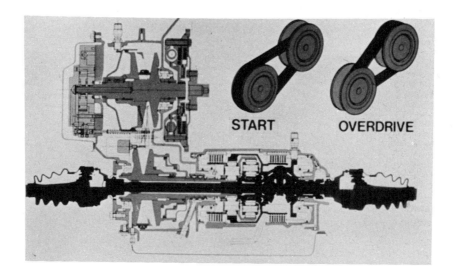

Fig 24 Continuously variable transmission

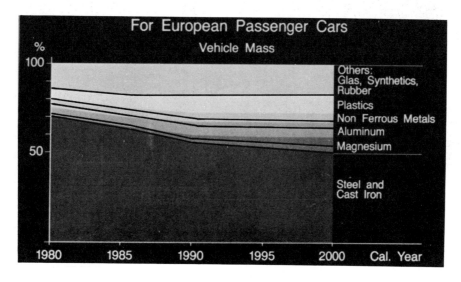

Fig 25 Trend of material usage

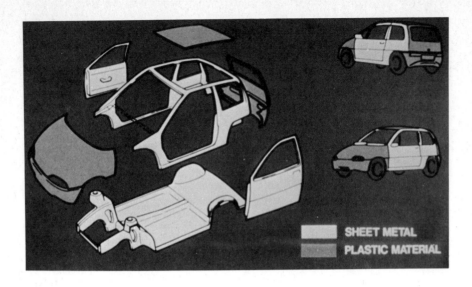

Fig 26 Light mass components

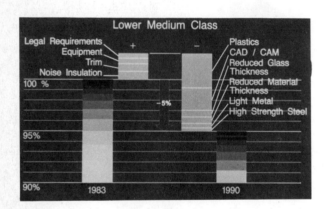

Fig 27 Trend of average vehicle body mass in Europe

Fig 28 Low air drag — good appearance

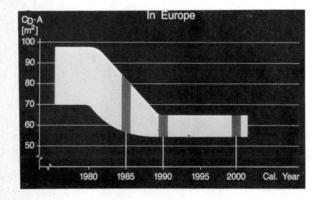

Fig 29 Drag area trend

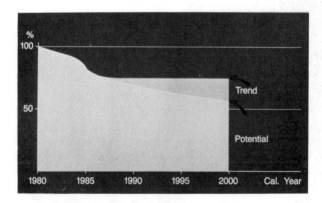

Fig 30 Rolling resistance of radial tires

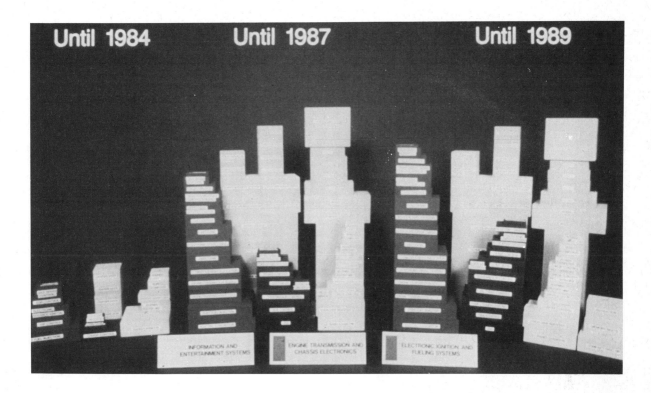

Fig 31 Increase of electronic components in car

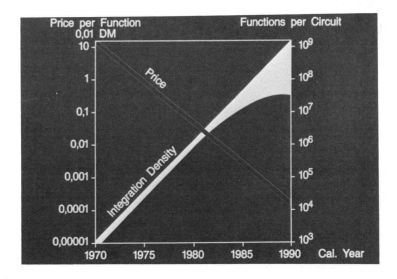

Fig 32 Development of integrated electronic circuits

Fig 33 Highway systems and traffic management

Total vehicle economy – the challenge (United States of America)

C L KNIGHTON, BSME, SAE
Ford Motor Company, Michigan, United States of America

It is a great pleasure to visit again with so many friends and business acquaintances from the European technical community. The four years I spent at Ford of Europe were absolutely invaluable to me from a professional standpoint. And from a purely personal perspective, my wife and I have many fond memories of the UK and Continent – memories we intend to freshen while we are here for this meeting.

I believe I can speak for the senior engineering management of the American automobile manufacturers when I say that conferences like Vecon '84 are very useful. We in Ford's Dearborn Technical Center are linked by computers, closed circuit television and private telephone lines to our major product centers at Dunton, England, and Merkenich, Germany, so we do have access to considerable information about the European scene. But there is much that we cannot stay abreast of and only conferences like this one can provide the full insight we need. Clearly this need will grow as American and European products and markets converge.

When I was invited to serve as one of your keynoters, I was given two guidelines: first, to address fuel economy within the context of the vehicle as a total system and, second, to approach the talk from the – and I quote – 'unique market perspective' of the United States.

That's exactly what I intend to do, with a modest modification. I would like to address the second point first, to insure that we start from the same knowledge base, and I shall confine my comments on fuel economy to vehicle factors, so that I will not duplicate the remarks on engine factors that will be covered by Herr Lohr and Mr. Matsumoto.

Let me begin by confessing something you may have suspected – fuel efficiency has not had a high priority among American automobile engineers until fairly recently – October 1973, to be exact. When I began my engineering career more than three decades ago, my first boss gave me some advice. 'Knighton,' he said, 'gray iron costs just a nickle a pound. Don't be afraid to use it.'

That was sound advice in those days. If you recall, American cars were uncomplicated, quite inexpensive in terms of United States purchasing power and their sales success made our auto industry the envy of the world.

In the process of achieving that success, American cars became unique in the world. For example, three-quarters of our cars had V-8 engines, preferred by customers because our cars typically weighed two tons, travel distances were long, roads were designed for high speeds and government policy kept the price of petrol below the world market price.

Needless to say, American manufacturers were not prepared for Oil Shock One that began in late 1973 and which completely disrupted our lives in 1974-75. In late 1975, the Congress of the United States passed the Energy Policy and Conservation Act, a complex package of laws designed to bring American appetites into line with the new national political and economic realities.

The act was not greeted with universal enthusiasm; some economists were particularly concerned about the implications of imposing fixed fuel-economy standards on free market demand.

Let me expand on that for a moment.

The Energy Policy and Conservation Act established minimum fleet averages for each manufacturer, domestic and foreign, for each year beginning with 1978. The mileage standard stair-stepped up until it reached 27.5 miles per US gallon, or 8.6 litres per 100 Kilometres, in 1985. I won't go into the methodology for fuel-economy testing, but to put 27.5 mpg into perspective, it was more than double the average being attained by the domestic manufacturers when the legislation was passed by Congress. At about the same time in Europe, manufacturers simply pledged to increase fuel economy by 10 to 20 percent, but here you were starting from a much higher average, of course.

There are a couple of other points worth noting about the Corporate Average Fuel Economy, or CAFE, law. The year-by-year targets are consumption-based rather than car-line-model-based, so harmonic averaging is required in the calculations. This may sound innocuous, but it has quite a tail to it. Let us say you have a three car fleet and each car offers 30 mpg. Your sales people want to offer a roomy station wagon with a rating of 20 mpg in place of one of the 30 mpg cars. Fine – but the fuel economy of the two other 30 mpg cars must be increased by one-third, to 40 mpg, just to maintain your fleet average at 30 mpg.

In short, we must track gallons per mile and

not miles per gallon.

There is one other point in the CAFE law that has a powerful influence on our product planning. All cars included in the fuel-economy calculations for United States manufacturers must have at least 75 percent local content by value. This prohibits us from meeting the standard by importing more fuel-efficient cars that might already be in production at affiliate companies outside the United States.

Needless to say, the CAFE task for American engineers has been formidable and the price of a failure could be hundreds of millions of dollars in fines levied by the government. With that incentive, the industry undertook a massive restructuring of product lines that has no equal in the private sector of the United States economy.

This is what Ford Motor Company alone has done since 1984:

Ford has replaced its entire line of cars - in some instances twice.

We have reduced the weight of the average car by over 1000 pounds.

Engine displacement has been reduced by an average of almost 1.1 litres.

Twelve new engine lines and 14 new transmission lines have been launched.

The technology of our cars has been transformed dramatically.

What are some of the major technological changes at Fords?

All of our new sedans are now front-wheel drive.

84 percent of our cars have electronic fuel injection systems; carburetors will virtually disappear from Ford cars by 1987.

Our latest engines feature state-of-the-art, fuel-efficient combustion and air-fuel management designs.

Over 70 percent of our automatic transmissions have locked-up, or mechanically-linked, torque converters and half of them have overdrive gearing.

Use of lightweight materials, such as plastic, aluminum, high strength steel and composites, has doubled.

The coefficient of drag of the average Ford car has been reduced by 22 percent; we believe Ford's CAFE gain from aerodynamics is almost twice that of the United States industry.

You would expect product investments of that magnitude to generate impressive results and they did - the CAFE of Ford Motor Company's fleet improved almost 70 percent between 1974 and 1983 - from 14.2 mpg to 23.8 mpg.

Despite that, Ford, as well as General Motors, Volvo and BMW, failed to meet the 1983 standard of 26 mpg. Fortunately, each of the companies earned excess credits in earlier years that could be applied to 1983, thus meeting the legal requirement and avoiding the fines I mentioned previously.

Ford and GM did not meet the 1984 model year standard of 27 mpg and, while no official forecast has been made for 1985, it will be even more difficult in the coming year to meet the standard.

You may ask, 'How can this be? How can you possibly be in CAFE trouble after your massive product changes, with down-sizing and new technology?'

The answer, as mentioned earlier, is that fixed fuel-economy standards are inevitably doomed to be out of balance with the changing nature of free market demand.

Free market demand has shifted in parallel with the price of gasoline.

All the computers in the world, programmed with the most sophisticated mathematical models could not have forecast the precipitous decline in gasoline prices. Between 1981 and the second quarter of 1984, unleaded regular gasoline at self-service pumps declined, on an inflation-adjusted basis, from $1.46 per gallon to $1.13. I do not know of another commodity that has dropped 29 percent in three years and it should not surprise you to learn that the car-buying public is turning its collective back on our high-tec small cars and is stampeding toward the ageing V-8 sedans that had been marked for extinction.

Do not mistake my observation for criticism. Quite to the contrary, it is one of the great blessings to businessmen that the free market reacts in a pragmatic way to changing economic conditions. In the auto industry, this means that consumers will not choose small cars over large cars unless economic logic supports the purchase of small cars. When that logic dissolves, you can be sure the majority of American car buyers will opt for the largest cars they can afford to buy and operate.

Currently, three out of 10 purchasers of Ford products select a V-8 engine and the percentage would be even greater if our V-8 engine production capacity would permit. By contrast, sales of diesel engines, for which we once had high hopes, are running at less than one percent.

So that is a snapshot of the fuel-economy scene in the United States today. In summary, we have the most fuel-efficient cars in the 81-year history of our company, yet we do not meet the legally-mandated standards because they are out of balance with free market demand. Parenthetically, if the sales mix by weight class in the United States last year had been the same as that in Britain, our Corporate Average Fuel Economy would have been 32.2 mpg, exceeding the 1985 standard by nearly five miles per gallon.

And keep in mind that our fuel-economy improvement was achieved at very low exhaust emission levels. We probably could have further improved fuel economy by 10 percent if United States emission standards were the same as in Europe.

Let us turn now from socio-economics to engineering - specifically, engineering the vehicle factors that relate to fuel economy improvement.

Vehicle factors consume about one-quarter of the total fuel energy used by a car.

If we focus exclusively on the universe of vehicle factors, we see that energy consumption breaks out like this:

Weight - 32 percent

Wind resistance - 18 percent

Power transmission - 17 percent

Rolling resistance - 14 percent

Electrical load, power steering and exhaust system - 7 percent

Air conditioning - 5 percent

Alternator (less electrical load), wheel bearings and seals - 4 percent

Brake drag, accessory drive belts, constant velocity joints - 3 percent

The energy consumption of vehicle factors has been well understood for a long time, but it has only been in fairly recent years that a concerted, systematic effort has been mounted to attack their power-robbing effects.

Considerable progress has been made and I would like to summarize the more promising developments within each broad area.

Weight, as we have just seen, accounts for about one-third of vehicle energy consumption and therefore it is the factor that received the earliest and most intensive attention from American engineers in the struggle to meet CAFE.

In government fuel-economy tests, all cars are divided into 40 test weight classes. The downwards movement of a car from one weight class to the next lower class can be worth 0.3 to 0.1 mpg in fuel-economy ratings.

Ford has reduced the average car weight from 4260 pounds in 1974 to about 3160 pounds in 1984. One thousand pounds or a half-ton in 10 years - an astonishing accomplishment that the technical community appreciates, but which is only vaguely understood by the general public.

We achieved weight reduction through three strategies: downsizing, lightweight materials and design efficiencies.

Downsizing was the quickest, least costly and most productive means of weight reduction, but we are nearing the limits of what is acceptable to American consumers at today's petrol prices. When Mom and Dad buy a family car, they expect to be able to take the children, the family pets and their luggage on a holiday trip in the space and comfort they are accustomed to in their living rooms.

Lightweight materials have not achieved the use predicted by some advocates, but in percentage terms the increases have been remarkable. Between 1974 and 1984, plastic applications increased 48 percent, aluminum gained 67 percent, and high strength steel literally skyrocketed 151 percent.

The use of lightweight materials will continue to grow and will be a major factor in our plan to reduce average vehicle weight by another 100 to 150 pounds by 1990.

Returning to the energy consumption table, you will recall that wind resistance accounts for 18 percent of the energy consumption attributable to vehicle factors. Learning to deal with wind resistance has catapulted aerodynamic engineering into one of the hottest specialities in the Ford engineering community. What we have learned - and accomplished - is truly astonishing.

As recently as the late 1960s, it was generally held that aerodynamics could not make a significant contribution at speeds below 60 or 70 miles per hour. Consequently, nearly all the work was confined to race cars, such as the Ford Mark II and Mark IV GT prototypes that domnated LeMans during that period.

Work during the past decade has proven how wrong we were. The fact is that at speeds as low as 35 to 40 mph, more than half the engine's output is required just to overcome wind resistance. Simply by bending the sheet metal in a precise way and by tweaking the body, engineers can manage air flow over, under, around and through the car so that substantial fuel-economy gains are made - at virtually no cost.

Ford begins aerodynamic development four years before Job 1 of a new or totally reskinned car. Typically, about 500 hours of wind tunnel testing is conducted with both 3/8 scale models and full-size prototypes.

These efforts have improved the coefficient of drag, or Cd, of the average Ford car from .51 in 1975 to below .40 in 1984. This corporate average drag reduction of 22 percent has resulted in an actual CAFE gain of more than one mpg. Our objective is to achieve an overall drag rating of .33 by 1990 for a total CAFE increase since 1975 of 2.5 mpg.

Power transmission, you will recall from the vehicle factors slide, accounts for 17 percent of energy consumption. It would not be that significant in Europe, where the manual gearbox predominates, but keep in mind that automatic transmissions go into 8 out of 10 cars in the United States.

The obvious technical challenge in automatic transmissions, or transaxles in the case of front-wheel-drive cars, is to get rid of the fluid, which is guilty of pumping losses and friction losses in the torque converter.

Two possible solutions are lock-up or by-passed torque converters and wet start-up clutches. Ford already locks up the torque in top gear and in some of our automatic transmissions we split torque in lower gears. The challenge, which we have not worked-out to our satisfaction, is to develop lower gear lock-up with acceptable smoothness during engine lugging, ratio changes, and power-on/power-off transitions.

Wet start-up clutches have the virtue of eliminating the torque converter/fluid coupling altogether, but they lack the smoothness and freedom from vibration that have made today's automatic transmissions so popular. This is another on-going development project.

The most talked-about concept in power transmission today is the continuously variable transmission - an old idea that is under intensive development here in Europe. In our company, Ford of Europe has the design and engineering responsibility and our responsibility in the United States is limited to engineering the application to our vehicles and engines.

I suspect there are many in this audience who have an intimate knowledge of CVT engineering and manufacturing development work, so I will not cover that area. However, you may not be aware of the extent to which we have consumer tested various CVT designs in the United States. Consumer testing is vitally important to us because people who strongly prefer automatic transmissions are reluctant to give up the silky shifts.

Generally speaking, the CVT has been acceptable to most owners of C-class cars such as the Escort. They are pleased with the improvement in fuel economy and performance, but some of them are bothered by, or at least uncertain about, the different feel of the CVT, compared to the conventional automatic transmission. We have tested both fluid coupling and wet clutch designs and the smoothness of the fluid coupling is clearly preferred, even though there is fuel-economy penalty.

Among all vehicle factors, tire rolling resistance ranks number four in energy consumption, accounting for 14 percent. Tire development is a 'black art' if there ever was one because the trade-offs are so complex and the technical alternatives are so few.

Nonetheless, tire manufacturers are making important progress in reducing rolling resistance and the future designs we have seen are promising. Between 1978 and 1982, tire rolling resistance on Ford cars was reduced by 40 percent and by 1986 we hope to achieve further reductions in the range of 10 to 15 percent.

Incremental improvements in reduced rolling resistance will come hard because there is a strong and contradictory market drive in the United States for cars with much higher standards of handling, lateral stability and braking, all of which require upsized tires that historically have had greater rolling resistance. No one said it would be easy.

As we have just seen, 81 percent of the energy consumption due to vehicle factors can be assigned to four broad areas: weight, wind resistance, power transmission and rolling resistance. The remaining 19 percent is spread over a variety of areas and I will tick through them quickly.

The electrical load in North American cars has risen sharply since the mid-1960s as our customers have ordered very high installation rates of features like air conditioning, premium entertainment systems, power-assisted brakes, steering, door locks, window regulators and so forth. Thrifting the load without compromising function and disappointing customers has been a major challenge until fairly recently when low-power-consumption components became available. The most promising ones include electroluminescent lighting for instruments and interior, all-electronic entertainment systems and advanced computer controls for powertrains.

Power steering components have been improved substantially during the past few years; efficiency has increased 36 percent over 1978 and we believe the pump we adopted for the 1984 model year is best-in-class. Vane pumps, variable displacement pumps and electrically-powered steering have potential for further energy reductions.

Exhaust systems are power robbers, but they can be made considerably more efficient. Ford is applying fluid dynamics technology to improving flow through pipes, calalytic converters and mufflers.

Air conditioning is a way of life for most Americans, not just in booming Sun Belt states like Texas, but even in northern tier states like Michigan and New York. Nine out of 10 private customers - excluding fleet sales to customers like telephone companies and governmental bodies - order air conditioning and they gladly trade the fuel economy for comfort.

We are getting some fuel economy back with refinements like compressor cycling control systems, but we really need new technology for major advances. Among the developments we are investigating are variable-setting, part-load capacity controls, dual displacement compressors and new evaporator blower wheel and housing designs.

New alternator designs could yield energy gains; the most interesting concept is an 'intelligent' alternator that charges as much as possible during deceleration, rather than continuously.

Wheel bearings, seals, brake drag, accessory belts and CV joints round out the picture and their total energy consumption is relatively small within the overall universe. Still, there is room for improvement and we are looking at things like stiffer disc brake calipers for reduced brake drag, serpentine belts or variable speed accessory drives, optimized CV joint angles and new designs and materials for wheel bearings and seals.

I hope these remarks have adequately covered the market and energy scene from the United States perspective. In conclusion, I would like to leave you with two concluding thoughts:

First, the auto market is booming in the United States, for which we are all grateful, but we are having a very difficult time meeting the government's fuel-economy standards because market demand has shifted to larger cars.

Second, Ford Motor Company is determined to meet both CAFE standards and free market demand. Every product decision we make is analyzed in terms of CAFE impact down to one one-hundredth of a mile per gallon. We are searching out every opportunity to improve fuel economy and we have set an internal goal of reducing vehicle-related energy losses by one-third through use of the strategies I have just summarized.

Total vehicle economy — the challenge (Japan)

K MATSUMOTO
Toyota Motor Corporation, Aicho-Ken, Japan

Although the general title is "Total Vehicle Economy -- The Challenge", I would like to speak as "The Engineering Challenge of the Japanese Automotive Industry."

To introduce my speech, I want to touch upon Japan's energy consumption in the transportation field, as well as the special characteristics of transportation in Japan. I believe that an appreciation of these two points will prove effective in understanding the situation of automobiles in my country.

A comparison of Japan's primary energy consumption with that in the major western nations reveals that the percentage of oil consumed in Japan is high, and over 99 percent of the oil is imported. Accordingly, securing energy resources and conserving energy are extremely vital concerns for Japan.

In terms of oil consumption by sector, Japanese consumption for transportation is around 32 percent, very low in comparison to the U.K. and U.S.A., and industrial and residential use is high. It is also true, however, that 80 percent of consumption for transportation is accounted for by automobiles.

Japan ranks below the major nations in terms of annual automobile fuel consumption calculated in per-vehicle and per-head figures. The differences in per-vehicle energy consumption is due to the short trip as well as the better fuel economy of Japanese vehicles.

For per-head consumption, differences depend on the vehicle holding number per-head and the types of vehicles used.

Next, let me mention the characteristics of Japanese automobiles. To reflect upon history for a moment, the automobile was first developed about a century ago in Europe, and the modern automotive industry utilizing mass production came into being approximately 80 years ago in the United States. The Japanese automotive industry patterned itself after that of Europe and the U.S.A., finally was established in the mid-50s. Japan's motorization was definitely late in comparison to Europe and the U.S.A., but the popularization was extremely rapid, with Japan now ranking second behind the U.S.A. in number of vehicles on the road.

A comparison with major nations in ownership on land-area bases shows that Japanese automobile ownership is extremely high. This is a reflection of the fact that 120 million people live within a small land area, which has a major impact on the traffic environment and automotive industry.

A breakdown of Japanese automobile types also reflects the conditions in Japan -- the large volumes of trucks and mini-vehicles in comparison to the West, for example.

To the next, I want to explain the current status and future of fuel consumption improvement technology in Japan. Approximately 80 percent of the total energy consumed by a motor vehicle during its life span is comprised of fuel, and the remaining 20 percent is consumed through materials and vehicle manufacturing. This means that the fuel economy improvement is very important to reduce the amount of energy consumed by a vehicle. Therefore, I would like to focus upon improving fuel economy.

First of all, I will explain the Japanese fuel economy target. The Japanese government has determined passenger car fuel economy guidelines according to the vehicle weight. For passenger cars overall, it was decided to achieve an average of 12.8 km/l by 1985 -- a 40 percent improvement over the 1975 level. Fuel consumption measurements in Japan are conducted with the 10-mode driving cycle.

When a vehicle is running, about 80 percent of a fuel energy is consumed through engine heat and friction loss, several percent through drive train friction loss, and the remainder through tire rolling resistance, aerodynamic drag and acceleration. Therefore, above items are the keys to improve overall vehicle fuel economy. To lower acceleration resistance, vehicle weight must be reduced. The automotive engineers and industry must make efforts to achieve these goals in the most efficient and rapid manner possible.

The Japanese challenge to make these improvements is comprised of the following items.

For the engine, efforts to improve vehicle fuel economy have been carried on for years under the strict emission regulations. These efforts have been targeted engine modifications -- i.e., optimum compression ratio, air/fuel ratio, ignition timing, valve timing and so forth. Also friction and engine weight reduction have been carried out.

More recently, a higher compression ratio and lean combution system are highlighted in spark ignited engines.

The electronic technology has been applied to fuel control systems, ignition timing control and more. As a result, fuel efficiency has been greatly improved.

Recent high power demands in Japan have

resulted to increase the adoption of turbochargers, DOHC engines and four valves per cylinder, aiming for a harmony between fuel economy and power. Efforts have also been turned to the improvements in intake systems, such as adoption of variable induction systems to improve the combustion of low and medium-speed ranges, maintaining the power at high speed range.

Now let's examine the Toyota Lean Conbustion System, a major breakthrough in the move to improve fuel economy.

The current mainstream in Japanese emission control is three-way catalyst systems, using oxygen sensors to achieve combustion with stoichiometric mixture. Such systems made it possible to harmonize the emission requirements and the fuel economy.

The Toyota Lean Conbustion System uses the world first lean-mixture sensor developed by Toyota, to control the air/fuel ratio around 22 for much better fuel economy.

The lean-mixture sensor measures the oxygen concentration in exhaust gas to detect air/fuel ratio continuously, and enables the engine to operate in lean mixture range.

To achieve the better combustion in lean mixture range and the better driveability, the swirl control valve and the electronic spark advance system were employed besides the optimization of fuel injection quantity and timing.

The swirl control valve is located in the one side of helical type swirl intake port, which is divided into two sections.

The valve is operated by the intake mainfold vacuum. In the low and medium speed ranges, the valve is closed to create strong swirl flow for better combustion of lean mixture. In the high speed range, the valve is openned to straighten the flow for higher output.

What about the future of engine technology? In the electronics field, the adoption of LSI and the progress of sensors and actuators will open the door to precise control of engine parameters, with increasingly broad use for engine and emissions control. This should lead eventually to total control of engine and power train.

We may also expect introduction of a wide range of variable mechanism -- including variable displacement volume, variable valve timing and variable induction system. They will be combined with lean combustion and high compression ratio for further improvement of thermal efficiency.

We also assume that direct injection (DI) may be adopted on small diesel engines as well.

In the near future, the new materials like engineering ceramics and composites are anticipated to be a effective means for weight reduction and thermal efficiency improvement.

For the drive trains, the improvement of transmission efficiency and the optimization of gear ratio are performed to achieve better fuel economy and acceleration.

As the percentage of automatic transmission vehicles in Japan is on the rise, the improvement of automatic transmission is another important concern.

The automatic transmissions equipped with lock-up clutch and overdrive are already on the market. Toyota has introduced the Toyota ECT (Electronic Controlled Transmission), offering world first electronic control of lock-up, shift point and other functions. Even more precise control of transmissions will be pursued in the future, and before long we can expect to see continuously variable transmission on the market that will be combined with the control of engine operation for substantial improvements in fuel economy.

The efforts were also continued to reduce tire rolling resistance. The tire structure, air pressure and new materials are the major elements in reducing rolling resistance, although the first two have tradeoff problems in terms of riding comfort. This means that for further reductions the role to be played by new materials.

A great deal of knowledges now exist on the challenge of reducing aerodynamic drag, with increasing improvements integrated into each model changes. At Toyota, for example, the aerodynamic coefficient has been cut by nearly 20 percent over the past decade. The factors to reduce aerodynamic drag include vehicle height, width and length, inclination angle of engine hood and windshield, flush surfacing of side windows and floor, cooling air passage layout and so forth. Each one of these factors, however, also has an influence to vehicle styling and roominess. Electronic technology targeting lower aerodynamic drag includes vehicle hieght, air spoiler controls and others.

Reducing accessory energy consumption is also important to improve fuel economy.

The largest energy-consuming accessory is the air conditioner. Because of the recent rise in demand for higher grade cars, almost all vehicles are now equipped with air conditioner in Japan, and the efforts are continuing to minimize the fuel economy penalty of air conditioning. Although the efficiency improvement for each component is being pursued of course, variable displacement compressors have also been introduced to improve air conditioning system efficiency. As a result, the air conditioner can be operated in proportion to heat loads. With the use of heat reflection glasses, heat insultating interior fittings and the others, the heat load itself will be lowered in the future.

The energy losses with power steering, which is increasingly fitted, also cannot be ignored. The studies are underway on higher pump efficiency and variable displacement pumps. Another interest is to replace the existing hydraulic power system with the direct operation by the electric motor, the idea being to activate power assist only when necessary.

Reducing alternator power consumption is also a key concern. The efforts in this area include attempts to raise the alternator efficiency and to lower the electric loads. We can also expect the use of new magnetic materials in starter, wiper and fan motors to improve efficiency, as well as use of solar batteries as an assistance.

Many lubricants are being used extensively to reduce friction loss, such as engine and axle oil containing friction modifiers. Further moves toward lower viscosity are aoso forecasted, taking care to maintain the other performances.

Vehicle weight has a great influence on acceleration, rolling and friction resistance.

That is, vehicle weight reduction makes a major contribution to fuel economy improvement.

Since Japanese cars have been traditionally small in size, their weight has also been low. The recent demands for more roominess and higher grade cars have led to the efforts to maintain or even reduce the vehicle weight while expanding the cabin space and the equipments.

One approach to weight reduction is the adoption of front wheel drive system. In Japan, FWD cars have increased rapidly over the past few years.

Other efforts to reduce weight include the computer aided structural analysis for optimized design and the extended use of aluminum, plastics, high strength steel and the other lightweight materials.

For further reduction of vehicle weight, it must be followed to establish the noise and vibration reduction technology. Also the cost reduction is a high-impact factor for weight reduction.

Thus far, I have addressed the developments for fuel economy improvements in Japanese automotive industry. With a great efforts of the Japanese automotive industry, the guidelines for overall passenger car fuel economy originally targeted for 1985 were achieved three years ahead of time in 1982. And the more efforts are continuing actively today.

In conclusion, I would like to express my opinion for what the future automobile manufacturing should be in the point of "Total Economy".

After the oil crisis, all the automotive manufacturers in the world have moved actively to improve fuel economy, resulting the vehicles become to have similar technology.

Under these circumstances, the future directions of vehicle manufacturing to satisfy user's needs should be as follows.

(1) High efficiency, especially better fuel economy.
(2) High quality, outstanding durability and reliability.
(3) Reasonable price.
(4) Attractiveness in styling, roomines and comfortability.

In order to meet these demands we must strive for further improvements. The following areas are crucial:

(1) Engineering Capability to balance high fuel economy, performance, comfort, safety and low emissions.
(2) Technology innovations for above factors.
(3) Production flexibility of the vehicles to meet each regulation, in-use environments and market requirements of various nations.
(4) Creative Products to pick out potential user's demands and to pioneer new markets.
(5) Manufacturing innovations to make the most effective use of personnel, facilities and investments.

Japan has made significant progress in design technology and production flexibility.

In the manufacturing field, Japan's production efficiency is truly the world leader -- the Toyota Kanban System is an excellent example. Further advances are forecasted in these areas, including extended use of CAD/CAM systems, FMS to respond quickly to product diversification and volume variation, new production systems to fit in with international automotive manufacturing.

This means that Japanese automotive industry must concentrate their energies on technology and product innovations. The more efforts are necessary, because Japan is certainly not leading Europe or the U.S.A. in these areas to date, and the intensified competition in the industries will certainly have a major impact on innovations.

Not only Japan, but all nations, to greater or lesser degrees, are facing the same challenge. This conference should be an excellent opportunity to discuss and advance the technology and product innovations.

The automobile should be a product which attracts the users forever. And it is our never-ending satisfaction to offer the cars that are fun to drive.

C454/84

SAE 814315

Energy and the vehicle

D A G SIMON, MA, MSA
BP Oil International, London

The petroleum and motor industries have long been associated with one another. Trends and developments in one industry can have a significant impact on the other. We ought to take note, therefore, when one of us points to changes that are likely to take place in the demand or the quality of the products he produces. Today, I would like to give the conference an idea of what types of road transport fuel it can expect from the oil industry over the remainder of this century, and some way into the next.

I assume that both our industries will continue to promote the efficient use of energy in transportation. We should seek to maintain the quality of life by preserving the freedom of movement of the individual at a reasonable cost, without creating unacceptable environmental hazards. There is sometimes a conflict of interests when we try to achieve both objectives simultaneously.

But first, what of the availability of petroleum products in the future? Oil reserves are finite: in spite of present-day gluts, supplies must decline in the longer-term. Of course, new, economically-recoverable reserves are still being found. But generally, these are smaller in size and are more costly to develop than the earlier discoveries. The pace at which exploration and production of new reserves can be sustained will depend in large measure on the prevailing price of oil and, of course, on the level of government tax take.

Even if no more reserves are found, world oil production could continue in today's volumes for about 35 years. Moreover, the availability of conventional road transport fuels - gasoline and diesel fuel - could be extended beyond this timescale. This is possible because oil refiners are able to manufacture more of these products out of each barrel of crude oil, at the expense of producing lower volumes of the heavier fuel oils. The market for the latter has, in any case, been declining in recent years as a result of greater energy conservation, substitution by other fuels and the relative decline of energy-intensive industries. Furthermore, the steady improvements being made in the fuel economy of vehicles will also help to extend the availability of conventional fuels.

The yield of transport fuels, as a proportion of crude oil processed, can range from 40% in a simple refinery, to 85% by employing conversion processes. Thus, if a larger portion of each barrel of crude were to be converted into transport fuels, the period of secure availability of these fuels at a given level of demand could be extended - though not without some penalty, as I shall explain later.

However, there is likely in future to be a shift in the sources and quality of crude oil. The productive life of most non-OPEC oilfields is much shorter than that of the very large oilfields in the Middle East, where over half the world's oil reserves are located. Towards the end of the century there may be a gradual return to dependence on OPEC sources. This could have three consequences: first, a large proportion of our oil would be imported once again from a politically sensitive area of the world; second, producers might once again start to raise the price of oil, both in nominal and real terms; and third, the oil itself would, in contrast to North Sea oil, be heavier and generally contain a higher amount of sulphur - a factor that has environmental implications.

What predictions can be made concerning the future demand for gasoline and diesel fuel? Clearly, the numbers of vehicles coming onto the roads, the fuel efficiency of those vehicles and the number of miles travelled are important factors in any estimate. The general economic environment and the price of fuel must be considered too. I will assume that recovery from the current recession will continue at a steady, though moderate pace, and that major political crises will be avoided. However, I am only too aware that this pattern could be easily overturned by events, in which case a somewhat different outcome could be imagined.

Present forecasts for crude oil prices cover a wide range - depending on the pessimism or optimism of the planning scenarios. But several authorities believe that - in contrast to the shocks of the seventies - the price will remain at its current, nominal level of around $29 a barrel for the next year or so. Then, between 1986-90, increases could take place in line with US inflation; and for the 1990s, the price could start rising in real terms to reach $30-45 a barrel by the year 2000.

With these factors in mind, I expect to see the following broad trends in the demand for transport fuels over the remainder of this century. In the non-communist world, growth in the consumption of diesel fuel will continue, at an average rate of some 3% a year. Modest growth of around 1.5% a year is also expected for jet fuel. But the consumption of gasoline is expected to remain static.

Growth rates will vary between different regions of the world. For example, robust increases can be expected in developing countries, whereas increases may be modest in some mature markets of industrialised nations. Gasoline demand may even decline slightly in the latter.

The higher growth rate for diesel will result from the continuing increase in commercial vehicle sales, and from the gradual conversion of gasoline-powered truck fleets (mainly in the USA) to diesel. The static (and even negative) growth in gasoline will primarily result from the achievement of greater fuel efficiency (and hence more miles per gallon) with gasoline-powered vehicles. For example, the European car stock in the year 2000 is likely to be, on average, about 30% more efficient than the current stock - notwithstanding the effects of unleaded gasoline.

In spite of rising diesel sales, gasoline will remain the dominant road transport fuel by the year 2000. However, the proportion of gasoline to diesel will probably have shifted from the present ratio of about 4 to 1 to a new ratio of 3 to 1.

Within this trend, it is expected that the growth in diesel-engined cars will continue. Their numbers in Europe could double between 1980 and 1990, and again between 1990 and the end of the century. But these large increases will take place on a low base. Even by the year 2000, diesel cars are unlikely to comprise more than 15% of Europe's total car stock. Most customers will still prefer gasoline-powered models on grounds of performance and lower cost. And there is the additional problem of diesel exhaust emissions - primarily oxides of nitrogen and particulates - which the motor industry is currently seeking ways of reducing. Another factor favouring gasoline models is that governments will be reluctant to allow too much loss of revenue through increased use of generally lower-taxed diesel fuel.

In the developed world, the increase in demand for diesel fuel is likely to outweigh the declining demand for middle distillate used for heating purposes. In order to produce more transport fuel from each barrel of crude, the oil industry - particularly in Europe - is embarking on a costly programme installing additional cracking facilities at its refineries. By contrast, there will be less need for this type of investment in America, where the oil industry already produces a higher proportion of lighter products.

In addition to meeting the expected increase in demand for diesel and jet fuel, the oil industry will also be seeking economic ways of maintaining fuel quality. There are, however, two broad areas which present difficulties. These concern the elimination of lead from gasoline and the introduction of stringent regulations on other exhaust emissions. It is not my wish here to engage in an environmental controversy, but in applying stricter regulations, there is a price to be paid - a fact that society must be prepared to accept.

Take the case of unleaded gasoline which, according to EEC draft proposals, is to be introduced generally in Common Market countries at the beginning of 1989. Tighter controls are also anticipated elsewhere in the world. As you know, lead compounds have for several years provided the cheapest means of raising the octane number of gasoline. These grades have made possible the adoption of today's high compression, energy-efficient engines. Simply stopping the use of lead compounds will result in a loss of octane number - approximately 3 RON units for 0.15g of lead per litre, and 5.5 RON units for 0.4g per litre (CONCAWE figures).

Unfortunately, it is not possible for the oil industry to produce unleaded gasoline to the current high octane levels in the large volumes demanded by the present-day car market, without incurring a severe cost and energy penalty. The amount of extra refining required would result in a significant increase in the consumption of crude oil; and a considerable investment in new equipment would also be needed at refineries. For the European scene, these additional costs have been well documented by the ERGA (Evolution of Regulations - Global Approach) study into optimum energy usage. The general conclusion is that the optimum octane number for producing unleaded gasoline is 94.5 RON/84.5 MON. At this level, the increase in total energy consumption is minimised - both in terms of additional crude oil processed at refineries, and increased fuel consumption by vehicles.

The EEC Commission have recently modified their earlier proposal on unleaded gasoline, and the consequences are still being examined. The latest proposal is that there should be a single European grade of 'premium' unleaded gasoline, although there is also a requirement that member countries 'allow' the marketing of a second regular grade - presumably if there is a demand for it. However, we remain concerned that the recommended quality of this European grade is 95 RON/85 MON minimum at the pump, since this implies an overall energy penalty compared with the optimum single grade of 94.5 RON.

It is said that the availability of two unleaded grades would give more flexibility to the motorist and car manufacturer; but I see problems in implementing the policy. In effect, the oil industry would be handling 4 grades, since there would need to be a long transition period when two leaded grades continued to be made available. The larger number of grades is bound to increase the price of fuel through higher distribution, storage and retailing costs. Moreover, the motorists' purchasing choice will be complicated, and the risk of error increased.

The proponents of the two grade approach believe

that market forces will maintain the present balance of premium and regular grades. But this is wishful thinking. The motor industry is already designing cars predominantly for premium grade. If two unleaded grades were to be manufactured at 96 and 92 RON (which was the Commission's earlier proposal, and one still favoured by West Germany), it is hard to believe that the motor industry would not design exclusively for the higher quality in the interests of fuel economy. It is impractical to suggest that any sort of grade split could be maintained in order to keep the pool octane number at the optimum level. Who would want a regular grade car, unless the fuel price differential could be artificially widened to make such a vehicle economically attractive?

The two-grade solution would, I fear, be an expensive, indirect step towards eventually marketing one, unleaded grade of 96 RON premium quality. If Europe is genuinely to control its energy consumption, then production of a single grade at the optimum octane number (94.5 RON) is surely the only effective solution.

On CONCAWE's present estimates, extra fuel will be required at refineries to produce unleaded gasoline, even at the optimum octane level of 94.5 RON. The additional processing for manufacturing unleaded gasoline would add over $5 a ton to the cost of producing low lead gasoline containing 0.15 grammes per litre. This represents an extra $425 million a year. The capital investment in refinery equipment would be even greater - some $1000 million. If, in spite of these penalties, EEC countries do edge their way towards a single, higher octane, unleaded grade, then the oil industry's costs will rise steeply. These figures, incidentially represent the net increase in energy costs. The offsetting effect of the extra miles per gallon enjoyed by the motorist - roughly 1% improvement for each increase in octane number - is already taken into account.

It may seem surprising - given the excess of refining capacity in Europe - that considerable capital investment in refinery equipment is necessary to meet the requirement for unleaded gasoline. The problem is that the existing refinery configurations were established to produce a range of fuels with a lower proportion of gasoline than is consumed today. Moreover, these plants rely on lead compounds to add extra octane numbers. Without the use of lead, more octane upgrading processes are needed; and these are not sufficiently available within our existing types of refinery. The capital investments are therefore needed to correct this deficiency in octane upgrading capacity.

In some quarters I am asked why no acceptable octane boosters are to be introduced as substitutes for lead. We hear a good deal today about the use of oxygenates - ethanol, methanol and the higher alcohols and ethers - in this role. Most of them have good octane ratings and can be blended with gasoline at concentrations ranging from 3% volume for methanol to 10% volume for the higher alcohols and ethers, without requiring adjustment to the carburation of most engines. We know that blends containing 20% or more of ethanol are used successfully in Brazil. However, engines do need adjusting if high percentage blends are used; and neat methanol and ethanol can only fuel specially-modified engines.

What is more, both ethanol and methanol have well-known inherent weaknesses as automotive fuels. Their energy content is only half to two-thirds that of gasoline, so a given quantity of alcohol cannot power a vehicle as far as the same quantity of gasoline. Other disadvantages include: their poor compatibility with some materials used in today's engines and fuel systems; their different volatility characteristics; and, when blended with gasoline, the risk of phase separation of the components, and consequent engine problems, when even small amounts of water are present in the fuel system.

Methyl tertiary butylether (MTBE) is perhaps technically the most promising oxygenate. When blended with gasoline, it does not have the high octane sensitivity of the others; that is to say, it is better able to meet both RON and MON requirements. Nor is it as water-sensitive as other oxygenates. But to produce it requires both methanol and iso-butylene, two products which are in high demand for petrochemical applications. Moreover, manufacturing MTBE is energy-intensive, and the cost is higher than that of the other oxygenates. A recent US estimate (Lundberg) claims that MTBE costs 9 times as much as lead to raise gasoline octane by one number.

But in spite of its relatively high cost, MTBE is already beginning to make a useful contribution to gasoline blends.
A number of oil companies are starting to manufacture the product; and they are also conducting R&D into other oxygenates. With continuing developments, these compounds could account for up to 5% of the gasoline blend by 1990. Thereafter, if their costs come down they will become more significant - although considerable variations are likely to exist between countries.

At this juncture it is perhaps appropriate to say a few words on the opportunities for synthetic fuels. Liquid hydrocarbon fuels obtained from coal liquefaction or natural gas, and alcohols from fermentation of biomass, are currently limited by their high cost. In most parts of the world, the net energy balance involved in producing them is negative, i.e. more energy is needed to manufacture the fuel than is contained in the end product. In general, synthetic fuels are not likely to replace conventional fuels until they can be similarly priced.

However, individual countries will continue to promote these fuels for strategic reasons or from a desire to reduce their balance of payments deficits. South Africa and Brazil may be cited as cases in point. But the technical problems I referred to earlier in the use of methanol and alcohols have to be taken into account.

Another fuel which could be used as a substitute for gasoline is liquefied petroleum gas (LPG).

This is an efficient lead-free fuel, although it does not eliminate the emission of oxides of nitrogen. As you know, there are relatively few technical problems in switching to LPG from gasoline. As a result, sales of automotive LPG - mainly propane - grew substantially during the seventies, particularly in Holland and Australia.

However, the penetration of LPG into the automotive market is closely linked to the taxation policies of governments. Making the fuel available in sufficient quantity presents another problem. The incremental cost of using LPG in terms of additional capital cost of equipment, and concerns over safety, are expected to hold back its rate of growth. For these reasons, LPG cannot be expected to make substantial inroads into gasoline sales - although individual countries could prove to be exceptions.

The use of LNG or CNG (principally methane) in automotive applications presents similar problems to LPG - though in a more extreme form. This is principally because of distribution and storage requirements involving pressurisation and/or refrigeration. Quite apart from these difficulties, natural gas is now in great demand as a chemical feedstock.

At present, synthetic fuels and fuel supplements are at best only partial and costly means of compensating for the elimination of lead from gasoline. For the immediate future, other solutions will have to be sought to meet the generally conflicting requirements of good fuel economy, acceptable performance and low exhaust emissions within the context of unleaded, but otherwise conventional fuels. I am sure that this conference will provide the opportunity to explore some of the feasible solutions.

Having talked about the problems and developments associated with gasoline, I would now like to touch on diesel fuel. As I remarked earlier, exhaust emission problems of a different kind arise from the combustion of this fuel. There are signs, emanating from the US, that regulations to curb diesel exhaust emissions will become more stringent.

I believe the motor industry faces another significant challenge in this area. The search for solutions will not be made easier by the oil industry's increasing difficulty in maintaining present fuel quality specifications. We are led to believe that the forthcoming generation of small direct- injection diesel engines will be even more critical than the present, larger designs as far as ignition quality is concerned. Yet, the increasing levels of crude oil conversion that will take place, as oil refiners crack deeper into the barrel, are going to mean that the current high ignition quality of European diesel fuel cannot practicably be maintained - reflecting the situation in North America and elsewhere.

The modest growth in jet fuel consumption that I referred to earlier will also affect diesel quality. Because of the need to balance the refinery barrel, the resulting diesel fuel will be more aromatic and have a higher gravity and viscosity than today's fuel.

In some countries, vegetable oils may increasingly be used as a means of 'stretching' the available volumes of diesel fuel. Locally-obtained vegetable oils are today used in low proportions in countries as different in character as New Zealand and some Central African states. Ignition quality is not usually impaired by these substances, although unwanted gum-like deposits may quickly be formed. This problem can be overcome by esterification, but the process adds to the price of the fuel blend.

What are the main messages? There will be a role for unconventional fuels, synthetic fuels and fuel supplements; but I see their growth as one of evolution rather than one of sudden development. The day of conventional fuels is far from over although, as I explained earlier, maintaining their high quality in sufficient quantity is becoming increasingly costly and difficult. For the future, automobile engineers would do well to think in terms of designing for maximum efficiency from lower quality fuel.

Regarding the investments likely to be made by the oil industry, it is important to bear in mind that regional differences exist in terms of product demand and refinery technology. The US pattern, for example, should not be seen as a model for the rest of the world. Although US refiners produce a high proportion of transport fuels - mainly gasoline - from the crude oil they process, the quality of these fuels is generally not as high as in Europe. Most US gasoline is manufactured at a lower octane level, and diesel fuel has a lower cetane number. Moreover, the proportion of diesel fuel produced is lower than in Europe.

Although we believe in the principle of free market forces, there are occasions when government action may be necessary to smooth a change in policy. Take the case of unleaded gasoline, for example. If the price of this fuel is allowed to reflect its higher cost differential then there will be no incentive for the motorist to purchase it. In America, where both leaded and unleaded grades are available, many motorists are still purchasing the cheaper leaded grades, and are using them in engines designed for unleaded fuel. As you know, the Environmental Protection Agency is proposing a 90% reduction in the lead content of leaded gasoline to overcome this problem. In order to prevent this situation developing in other countries, some government intervention maybe necessary during the change-over period to avoid an inbalance between the cost of unleaded and leaded grades to the customer.

The oil and motor industries are involved in significant investment programmes. But we must ensure that these investments are not rendered economically inefficient by short-sighted actions taken unilaterally by one of us, or for that matter by governments. It is important, therefore, that our two industries continue with regular discussions, and co-operate closely in efforts to achieve optimum energy efficiencies. We have worked together in the past; and I am sure we will continue to do so in future.

C422/84

SAE 841281

The interaction of fuel economy and emission control in Europe — a literature study

C D DE BOER, BTech
Ricardo Consulting Engineers plc, Shoreham-by-Sea, Sussex
J A JEYES, BEng, CEng, MIMechE
Department of Transport, London

SYNOPSIS Various proposals for more stringent emission legislation have been made for the medium term future in Europe. Set against these proposals is the intensified search for improved vehicle fuel economy and the probable result that overly stringent emission control would decrease fuel economy.

The paper will discuss interaction of fuel economy and emission control in Europe based on a survey of the literature. The performance and operating cost of a number of alternative control strategies and engines will be compared using a current conventional gasoline engine as a basis.

1 INTRODUCTION

In Europe, gaseous emissions from passenger cars are controlled by national regulations which in general are aligned to Council Directive 70/220/EEC, for European Community member states, or to the technically equivalent United Nations (Economic Commission for Europe) Regulation 15 for countries both within and outside the Community. Other countries such as Sweden and Switzerland have adopted different standards.

At the time of the study, (1981/82) the UK and most of the other Community member states were implementing Council Directive 70/220/EEC as amended by 78/665/EEC which is equivalent to the third series of amendments to Regulation 15, generally known as ECE 15-03. Since then a further series of amendments to Regulation 15 has been agreed by the UN, (ECE 15-04), the technical provisions of which have been adopted by the European Community into a revised Council Directive 83/351/EEC. This will become effective for new vehicle model approvals in October 1984 and for existing vehicle models in October 1986. Most member states are expected to apply it.

Various proposals for further tightening of vehicle emission standards in Europe have been made, one of these which was a proposal by the Federal Republic of Germany, first made in 1979, for further amendments to Regulation 15. This 'proposed limit' for ECE 15-05 was specifically considered in this study, as it was estimated to require substantial changes to European vehicles.

Set against these emissions controls is the intensified search for fuel economy and the possible result that overly stringent emission control could increase fuel consumption.

This trend in emission control led the Department of Transport to press for a study to be carried out on a Community basis, to assess all aspects of gaseous emission control, including the technical feasibility, costs, energy implications and environmental benefits. As a result of this proposal, which was supported by other member states, the European Commission decided to set up an ad-hoc working group 'Evolution of Regulations Global Approach - Air Pollution' (ERGA-AP) to study these issues. Ricardo were approached by DTp to contribute towards the study by examining the specific effects of tightening emission regulations on vehicle fuel consumption, cost and performance. The study (1) based on a literature survey was completed in February 1982 and presents and discusses data pertaining to the fuel economy/emissions relationship that has been extracted from published and Ricardo in-house literature to date, and had the objective of assisting with the understanding and potential impact of future emission legislation on vehicle fuel economy and cost. The range of data such as the number of engine types and fuel types was restricted to that regarded as pertinent to the period covering the next five to ten years.

2 OBJECTIVE

The overall objective of the study was to present data relevant to a typical gasoline engined vehicle as a baseline and establish a number of alternative control strategies and engines suitable for reduced exhaust emission levels. The possible options at each emission level were compared on a fuel economy and operating cost basis.

3 SCOPE OF LITERATURE SEARCH

The literature search was based on an examination of the following sources.

(a) Ricardo library catalogues; and indexes including in-house databases. Ricardo reports where confidentiality permits.

(b) Published abstracts and indexes such as Society of Automotive Engineers Index.

(c) External on-line computer databases such as Compendum.

The literature search concentrated mainly on the period from 1970 onwards and a total of over 200 references were used in the main report.

4 BACKGROUND

Control of gaseous exhaust emissions from motor vehicles has existed in the USA since 1966. California was the first state to make vehicle manufacturers meet mandatory limits on the exhaust hydrocarbon and carbon monoxide emissions. Other forms of vehicle pollution, such as HC emissions from the crankcase breather and from fuel evaporation were also controlled. The other 49 States of America introduced similar legislation and test procedures in 1968. Since these early controls the limits on exhaust pollutants have become progressively more stringent and on average limits were lowered every two years or so. Control of oxides of nitrogen was introduced in 1973 and often changes in the test procedures accompanied lowering of limits; these changes were centred on improving sampling and measurement technology and the need to increase measurement accuracy as the level of pollutants being measured gradually became lower.

The United States Clean Air Act of 1970 required the automobile industry to produce vehicles which, in 1975, emitted essentially 90% less pollutants than those of pre-1970 years. In the event, these targets were shown not to be technically feasible for volume production and somewhat less stringent standards were implemented. The Clean Air Act resulted in some far-reaching consequences:

(a) The limits on emissions were so low that automobile manufacturers had to use noble metal catalysts for 1975 model year vehicles.
(b) Unleaded gasoline was made available.
(c) The lead time given to the car companies to try to meet the legislation was very short in terms of vehicle development, which meant that little attention was paid to the effect that emission control technology being applied to 1972-74 vehicles had on driveability and fuel consumption.

This latter point has rapidly gained importance for reasons that centre around the rapid depletion of the world's oil reserves. Indeed, legislation has now been introduced in the USA which requires manufacturers to make rapid improvements in the fuel economy of their vehicles.

The European nations, under the auspices of the Economic Commission for Europe - a United Nations body - developed its own test cycle, sampling and measurement procedures. These were substantially different from those evolved in the USA. Initially concern centred on CO and HC levels and a 20% reduction in total vehicle HC was achieved by the mandatory fitting of positive crankcase ventilation devices.

As in the USA, the limits were progressively lowered and refinements made to the test procedure. At present, after eight years or so with gradual reductions in emission limits, increasingly severe limits are under review. A summary of the ECE 15 exhaust emissions legislation is given in Table 1 below for a typical 1132 kg vehicle.

Table 1 Summary of exhaust emissions regulations in Europe

Legislation		Year of Introduction in Directive for New Approvals	Type Approval Emissions Standard (g/test)			% Reduction from Uncontrolled Vehicle		
			CO	HC	NOx	CO	HC	NOx
Uncontrolled pre-1970 vehicles (average of various surveys)			225	14.5	6.8	0	0	0
UN Regulation	EEC Council Directive							
ECE 15(00)	70/220/EEC	1971	134	9.4	-	40	35	-
15-01	74/290/EEC	1975	107	8.0	-	52	45	-
15-02	77/102/EEC	1977	107	8.0	12.0	52	45	-
15-03	78/665/EEC	1979	87	7.1	10.2	61	51	15 (relative to 02)
15-04	83/351/EEC	1984	67	20.5 combined standard		70	48 (approx)	
West German Proposal For ECE 15-05 (1979)		-	30	10 combined standard		87	75 (approx)	

The above data are presented in figures 1 and 2 in comparison with the emission control trend in the USA. It should be noted here that with the ECE 15-04 Amendment the sampling and analysis methods were changed considerably. The trends show that emission legislation in the USA is some five to eight years in advance of that in Europe and so it was of value to consider the effects of emission control on fuel economy in the USA.

4.1 Historic Fuel Economy and Emission Interaction in USA

The Environmental Protection Agency annually publishes a fuel economy survey of all new cars available in the USA, the data for which come from essentially the same chassis dynamometer test as do the exhaust emission data. If the data for the whole fleet are pooled, then the trend of the average fuel economy can be plotted against year. Figure 3 shows the sales-weighted fuel economy trends for the period 1967 to 1980. The loss in fuel economy from 1967 to 1973 was of the order of 6% and was largely attributed to emission legislation and its side effects such as the lowering of compression ratios to be compatible with the 91 RON gasoline. The data presented in the figures do not, however, give any direct indication of what fraction of the fuel economy penalty may be attributed to various parameters, i.e. vehicle weight, compression ratio, emission legislation, etc. Detailed statistical analyses have been completed in order to identify these variables. Figure 4 shows the relative city fuel economy for 1974, '75 and '76, distinguishing between pairs of US Federal (49 State) and Californian vehicles. Several observations may be made from these data.

(a) Fuel economy is steadily improving and since the same model year cars were selected, these gains may be attributable to technical improvements rather than model mix effects.
(b) With technology fixed at a point in time, fuel economy penalties may be associated with tighter emission levels.
(c) Fuel economy may be improved as advances in design and technology are incorporated.

The improvement in fuel economy after 1974 is shown to be relatively rapid and much of this is only achieved in association with the introduction of catalyst technology; the latter forces a relatively efficient engine tune. This rate of improvement is unlikely to be repeated in the near future since step changes in automotive technology are infrequent. The data presented in Figure 5 are an attempt to isolate the effects of emission legislation from all other vehicle variables. Data on weight, engine displacement, compression ratio, carburation, transmission type, axle ratio, air conditioning and miles driven per month were obtained for each car. Linear regression analyses of these data were made in an attempt to eliminate the variables. It is noteworthy that the dominant factor affecting fuel economy is vehicle weight. Unidentified factors other than emission controls cannot however be ruled out, for example, the fuel economy increase found between 1973 and 1974 models corresponds in time with the imposition of the 55 mile/h speed limit, and this is expected to have a measurable effect on the high mileage service cars used for the survey. The conclusions to be drawn from these data may be similar to those made above in that for a given 'state of the art' emission control fuel economy penalties are associated with tighter emission levels but that these deficits may be minimised as design and technology advances are made.
Although the technology advances may be coincident with emission control changes they can usually be achieved independently.

4.2 Historic Fuel Economy and Emission Interaction in Europe

The in-depth monitoring of fuel economy by EPA in the USA has not been mirrored in Europe, as far as is known. The ECE 15 emission test under stabilized (hot) conditions has more recently been used to give an 'urban cycle' fuel economy and the results are published annually along with the steady state fuel consumption at 90 and 120 km/h. These surveys, published in the United Kingdom first by the Department of Energy and now by the Department of Transport, have been available since 1978. The last change in emission standards occurred in 1979 and therefore little is known about its impact on fuel economy.

A detailed survey of the urban fuel economy data available on the 'typical' 1600 cc engined 1132kg vehicle showed that between 1978 and 1981 the average gain in fuel economy was 5.3% for a constant model mix, a maximum gain in fuel economy for a particular vehicle was 11%. The target improvement of 15% on a constant model mix between 1978-1985 is now reported to have been achieved at the end of 1983. These data do not however identify the loss of fuel economy potential imposed by emission constraints, or the effects of other variables. It is generally accepted that the ECE emission legislation up to 1977 which controlled only CO and HC emissions could as a direct result lead to improved fuel economy since reducing these emissions is best achieved by ensuring that the engine is running at or close to its optimum operating condition.

The ECE 15 Urban Cycle was specifically conceived to simulate congested inner-city conditions where HC and CO emissions are most significant. The mean engine load factor during the test is therefore low compared to average urban driving. It is well recognised that both fuel consumption and emissions are sensitive to drive cycle and the ECE cycle cannot therefore be representative for average driving conditions.

5 FUEL ECONOMY AND EMISSIONS LEVELS

To assess the interaction of fuel economy and emissions a number of engine types and control strategies considered feasible and able to comply with emission legislation in the foreseeable future was examined. The basis of the examination was formed by considering a typical European passenger vehicle, this is represented by a vehicle in the inertia class of 1130 kg having a 60 kW engine, and ECE 15 test cycle. For the purposes of the study the baseline level was formed by a vehicle with a 1.6 litre gasoline engine. This 'typical' vehicle and test cycle were chosen to allow a quantification of the impact of various strategies to be made. The literature survey and consideration of smaller and larger vehicles indicated that although the general trends brought out by this approach are valid, the quantified effects vary slightly as vehicle size or performance is changed.

The results are presented in Table 2. Both the fuel economy and emissions results presented are based on the ECE urban cycle from a cold start. The fuel consumption figures have been compared for the cold start test rather than at the hot, stabilized, condition since the engine control strategy during engine warm-up is a critical factor governing engine emissions and fuel consumption. It should be noted that many of the control devices in operation during the warm-up period are not necessarily employed under stabilized conditions. An example of such a device is the choke which if not used during warm-up would result in increased HC emissions but reduced fuel consumption, not to mention associated driveability problems.

Three data columns are tabled for each engine type. The ECE 15 level is intended to indicate for which amendment the particular engine and control strategy would be suitable, it is also meant to imply that more stringent emission legislation would not be expected to be attainable at that particular level of technology. A current conventional engine (No. 1) is tabled for the 03 amendment and improvements in both the fuelling and ignition strategy would be required to enable 04 regulations to be met, i.e. Engine Type 3. The second column is headed "Emission Build" and indicates the emission levels likely to be attained by a particular engine technology. The ECE 15-03, 04 and 05 type approval limits are listed at the bottom of the table with the 03 levels corrected for HC detection with FID and CVS exhaust sampling. Although 04 and 05 amendments regulate a combined HC + NOx limit, both individual HC and NOx emissions are presented enabling the balance of the individual species to be analysed. The third column "Best Economy Build" is intended to demonstrate the fuel economy which could be attained by a particular technology and also indicates the likely effect on exhaust emissions. The comparison of ´Emission´ and ´Economy´ fuel economy gives an indication of their inter-relationship for a given technology level. A number of engine types is listed without an economy result and this absence is intended to indicate that these engines are inherently not suitable for economy tuning, e.g. Engine Type 8 having EGR at levels which result in a fuel economy penalty.

A limited number of engine options are given and the list is by no means exhaustive. This is intentional as many of the possible combinations of engine, control technology, and aftertreatment are either not complementary or have conflicting requirements. An example of this might be the use of a thermal reactor with a high compression ratio lean burn engine; which would seem to be an attractive choice since the emissions control capability of the engine is limited by excessive HC emissions. However, high compression ratio engines have inherently low exhaust temperatures and this combined with a lean burn strategy makes the application of a thermal reactor impracticable. The application of three-way catalysts is not thought to be a practical route for emissions control in Europe since control of NOx to the levels envisaged can be achieved by running with a lean mixture strategy. In addition, the requirement of stoichiometric mixture control would result in significant fuel economy penalties.

The emissions results are corrected to comply with the ECE 15-04 regulations and the current conventional engine complies with ECE 15-03 and is assessed to have a fuel consumption of 13.5 litre/100km over the cold-start ECE cycle. The potential of the current conventional engine is both for improved fuel economy and reduced emissions by development of the control technology. In this way fuel economy could be improved by around 7% and exhaust emissions reduced to conform with ECE 15-04.

Gains in fuel economy at this level of technology were possible by relaxing the emission constraint; a further gain in fuel economy of 8% could result. Further development of the control technology for the conventional engine by addition of a thermal reactor, adopting a lean mixture strategy, use of exhaust gas recirculation or oxidation catalysts was examined. All resulted in reduced exhaust emissions and varying degrees of fuel consumption penalty compared to the well developed conventional engine. The thermal reactor, which is used in conjunction with secondary air injection and exhaust port liners, was found to be an ineffective strategy for lowering HC and CO emissions over the ECE 15 cycle since an excessive fuel economy penalty has to be made to promote oxidation of the exhaust gases. Adoption of a lean mixture strategy can reduce NOx and CO emissions without a fuel economy penalty, HC emissions are however likely to increase; the survey showed this strategy unsuitable for stringent emission control with a conventional engine. The application of exhaust gas recirculation results in similar emissions to those obtained with a lean mixture strategy but fuel consumption is increased. The ability of noble metal exhaust catalysts to oxidise HC and CO emissions enables conventional engines to comply with ECE 15-05 with a small fuel consumption increase. The application of this technology does however depend on the development of lead tolerant catalysts or the availability of lead free premium fuel if large fuel consumption increases or increased crude oil consumption are to be avoided.

The study indicated that significant savings of fuel can be achieved with high compression ratio lean burn engines and when tuned for best economy, savings of the order of 20% compared to the baseline engine can be made over the ECE 15 cycle. Some of the potential fuel economy is eroded by complying with ECE 15-04 because of the need for hydrocarbon emission control, a fuel saving of around 15% is retained. It was established that compliance with ECE 15-05 required the use of an oxidation catalyst and a revised engine control strategy. This strategy enabled a fuel saving of 9% compared to a conventional engine and catalyst to be made; some of the fuel economy advantage of the high compression ratio lean burn engine was eroded. The use of exhaust gas recirculation with these engines was not found to be an effective emission control strategy since NOx emissions are inherently low and increased hydrocarbon emission and fuel consumption often result.

Prechamber stratified charge engines in combination with a thermal reactor offer a route for very low exhaust emissions and moderate fuel economy. The survey indicated that these engines are able to comply with ECE 15-05 emission levels and reduce fuel consumption by around 6 to 10% compared to the current conventional engine. Open chamber or direct injection stratified charge engines were not considered as feasible low emission engine types within the timescale tend to be high and further work on high efficiency catalysts or special fuel injection equipment is required.

Indirect injection diesel engines inherently have low exhaust emissions and good fuel economy. Specific power output is however low and so swept volume increases and/or boosting are required to maintain vehicle performance. The fuel economy is about 30% better than the baseline gasoline engine over the ECE 15 cycle and exhaust emissions comply with ECE 15-05.

Direct injection diesel engines offer an economy gain of around 10% over the indirect injection diesel but noise remains a problem for all but some single spray versions.

6 COST ANALYSIS

The total cost involved in owning and operating a passenger car is made up of several different elements which can be broadly divided into two categories; standing charges and running costs. The study showed that the factors having the greatest influence on total operating cost are the first cost of the vehicle, maintenance cost and fuel consumption. The effect of emission constraints on total vehicle operating cost was examined by estimating first cost, maintenance cost and fuel cost for the same engine types and control strategies considered feasible for complying with future emission and legislation in Table 2. Total operating costs were examined over two time periods, two years and eight years, to assess the impact to the first owner and to the national economy over the vehicle life.

The cost figures are approximate in absolute terms and significant variations are likely in practice; they were calculated for a special case, a single weight class at a fixed performance level and fuel consumption over a cold start ECE 15 cycle. Nevertheless it is considered that these estimates provide a rational basis for comparison. Uncertainties in the future may alter the findings, e.g. should lead-free gasoline be introduced then increased fuel consumption may result because of reduced octane numbers. The price of diesel and gasoline fuel was considered to be the same whereas in other European countries diesel fuel enjoys a large price advantage. Other uncertain factors concern the likely production costs for technologies now undergoing research and development such as the high compression ratio lean burn engine.

Detailed results of this analysis are presented in Tables 3 and 4 for the two year and eight year vehicle operating periods respectively. The engine of a typical passenger car was found to account for around 20% of the total first cost and alternative vehicle costs were subsequently calculated from the marginal engine cost of each particular engine technology. Analysis showed that significant reductions in exhaust emissions are usually accompanied by substantial first cost penalties. The first cost of a current conventional engine was increased by 50% when reducing exhaust emissions from the current ECE 15-03 to the projected 05 levels. A similar trend was in evidence with high compression ratio lean burn engines. In each case the cost increase was attributed to the more advanced engine control system and/or exhaust aftertreatment.

Stratified charge engines are significantly more complex than conventional engines and the need for sophisticated fuel injection equipment on an unscavenged prechamber engine incurs a further cost penalty.

Diesel engines have a high first cost due to the need for specialized fuel injection equipment, increased engine size and/or boosting to maintain vehicle performance. The cost of a naturally aspirated indirect injection diesel engine was found to be at least 50% higher than a current conventional gasoline engine of the same power. The additional cost of a boosting device, e.g. turbocharger, is partly compensated by the cost saving resulting from the requirement of a smaller basic engine structure. Maintenance costs for the low emission engines considered were found to increase with respect to the current conventional engine. This increase was not necessarily directly related to the relative complexity of emission control systems since many of the concepts result in improved reliability.

Total vehicle operating costs were found to be significantly affected by fuel consumption. The first cost penalty incurred by incorporating more advanced technology tended to be partially or fully compensated for by reduced fuel consumption. The analysis over a two year period, Table 3, showed one high compression ratio lean burn and two types of diesel engine to have reduced total operating costs. On the other hand conventional gasoline engines with add-on emission control devices resulting in poor fuel economy were found to give significant increases in operating cost. Over an assumed vehicle life of eight years, Table 4, two high compression ratio gasoline engine options and all of the diesel engines resulted in cost savings compared to a current conventional gasoline engine.

7 DISCUSSION

The data in Table 2 are further analysed in Figures 6 and 7 where the emissions and fuel economy trends of HC + NOx controls are plotted respectively. Also depicted are lines of constant technology which represent the loss of fuel economy likely to be incurred by a reduction of emissions.

The HC + NOx limit for ECE 15-04 can be met with conventional engines. Fuelling and ignition optimisation minimise fuel economy penalties and in fact a slight improvement of fuel economy is expected as engine controls are refined. Further gains in fuel economy may be made with high compression ratio engines and around 20% fuel saving could be made within the ECE 15-04 CO limit. HC + NOx legislation limits these fuel economy gains to around 15%. As seen in Figure 7

CO control does not constrain fuel economy; in general a fuel economy gain is accompanied by a reduction of CO emissions.

The proposed '05' limits require significantly more advanced technologies. Four approaches seem possible:

(a) conventional engines with high technology engine controls and exhaust oxidation devices.
(b) stratified charge engines with exhaust oxidation for control of HC emissions.
(c) application of oxidation catalysts to conventional or high compression ratio engines.
(d) diesel engines.

Conventional engines are likely to suffer a fuel economy penalty; exhaust aftertreatment, in the form of a thermal reactor or oxidation catalyst, is thought necessary for HC and CO control. The latter option carries a minimum fuel economy penalty.

Stratified charge engines and diesel engines are both likely candidates for the '05' proposal. Present high speed DI diesel development indicates that despite potentially higher HC + NOx emissions (compared to IDI diesels) these engines are able to meet these levels.

The analysis shows that low emission levels may be attained without the need for fuel economy penalties provided that the correct technological routes are chosen. However the cost to the consumer and consumer attributes should also be considered as these may become significant. For example, from a technical standpoint, the diesel engine can result in considerable fuel savings and low exhaust emissions but first cost and diesel acceptability, e.g. noise, must enter into the argument.

The engine options discussed have been considered as having the same power rating so that vehicle performance should not be seriously affected by adopting any of the alternatives. Some reduction of driveability will occur under cold start conditions where exhaust aftertreatment is concerned since the requirements for fast warm-up could lead to reduced performance.

Should vehicle performance be at a different level than that considered above then the emissions and fuel economy relationship may be altered. Vehicle performance is essentially governed by the power to weight ratio and the baseline vehicle has a 53 kW/ton power to weight ratio. Changes to the latter need to be considered, i.e. vehicle weight at a constant performance level and vehicle performance for a constant weight. It is convenient to consider each of the alternatives:

(a) Light vehicle, 750 kg weight at 53 kW/ton - a light vehicle with a smaller engine would tend to show increased specific HC emissions and reduced NOx emissions. The specific duty cycle over the ECE 15 test would be similar with similar gearing. The predominant factor would be reduced engine throughput leading to reduced fuel consumption and emissions.

(b) Heavy vehicle, 1500 kg weight at kW/ton - heavier vehicle has an increased swept volume per cylinder (retaining the same number of cylinders). For a gasoline engine a reduced combustion chamber surface area to volume ratio tends to reduce specific HC emissions. The requirement for a greater amount of work done during the ECE 15 cycle implies that engine throughput is increased and so emissions would increase primarily as a function of fuel flow. This may become a significant factor if the proposed '05' HC + NOx limit is adopted since no allowance for vehicle test weight is made.

If alternative performance levels of 70 kW/ton and 35 kW/ton, achieved by changes in engine capacity, are considered, the effects may be as follows:

(c) High performance vehicle, 70 kW/ton at 1132 kg - The specific duty over the ECE 15 cycle would be reduced with the results that fuel consumption will be likely to increase. The primary effect on emissions would be to increase HC, NOx emissions are likely to remain similar for gasoline engines and increase in the case of diesel engines.

(d) Low performance vehicle, 35 kW/ton at 1132 kg - a small capacity engine implies an increased duty cycle and a reduced fuel flow consumption trend. Higher engine loading and reduced fuel flow would tend to reduce HC and NOx emissions for both gasoline and diesel engines. In the extreme case where fuel enrichment would be required to enable high engine loads to be attained, increased CO emissions would result.

The urban cycle fuel economy test differs from that for the emission test in that it features stabilized engine conditions (fully warmed up) and no specific regard need be made with respect of exhaust emissions. The latter may alter the fuel economy and exhaust emission relationship, because of the effect of engine warm-up and the requirements of exhaust aftertreatment devices. Conventional gasoline engines and high compression ratio lean burn engines show similar differences between 'emissions' and 'fuel economy' test cycle fuel consumptions and a warm-up fuel penalty of 15% is typical. This factor is likely to increase where exhaust aftertreatment devices requiring rapid warm-up are incorporated. Diesel engined vehicles show improved warm-up characteristics since the combustion is not so affected by quench conditions and a fuel penalty of around 10% is typical. Stratified charge engines in this respect are likely to fall between gasoline and diesel engined vehicles.

A comparison between the current conventional engine and the most cost effective engine options for ECE 15-04 and the proposed '05' regulation is made in Table 5, extracted from data in Tables 2,3,4. The conventional gasoline engine was found quite able to meet ECE 15-04 emissions limits provided that the control strategy is developed. Higher technology engine controls lead to increased vehicle first cost which is not entirely offset by reduced fuel consumption. The result is an increased cost to the consumer. The potential of high compression ratio engines is to reduce fuel consumption and this is able to offset the higher first cost; reduced operating costs are evident in under two years. In order to comply with the proposed ´05´ emissions limits, gasoline engines offering cost effective solutions require exhaust oxidation catalysts. Fuel consumption is only slightly higher than at the 04 level but the catalyst equipment results in a first cost premium and increased maintenance cost so that the cost to the consumer is increased. Diesel engines can offer a long term saving in total operating cost to the consumer because of their excellent fuel economy. After a two year period and 20,000 miles the consumer cost is similar to that of a current conventional engine. However after longer periods reduced fuel costs outweigh the increased first cost.

8 CONCLUSIONS

ECE 15-04 EMISSION LIMIT

(1) The current conventional engine can be developed to meet ECE 15-04 emission standards by optimizing fuelling and ignition schedules. A fuel economy gain of around 6% is possible with these engine developments.

(2) Similar development of the conventional engine purely for fuel economy would result in up to a 12% gain in fuel economy.

(3) The application of a thermal reactor, with secondary air and exhaust port liners is not an effective option for emission control with homogeneous charge engines since it incurs increased fuel consumption.

(4) The application of EGR is not a fuel efficient emission control strategy for European emissions levels since NOx control can be achieved by utilizing lean mixtures.

(5) High compression ratio lean burn engines can have a significant fuel economy advantage over conventional engines and are suitable for tuning to ECE 15-04 standards.

PROPOSED ´05´ LIMITS

(1) In order to comply with the ´proposed limits´ a number of choices exist:

(a) A conventional engine with electronic reactor.
(b) Conventional engine with oxidation catalyst.
(c) High compression lean burn engine with oxidation catalyst.
(d) Stratified charge engine with thermal reactor.
(e) Diesel engines.

(2) The conventional engine with electronic controls and thermal reactor is the most expensive solution for these limits.

(3) The high compression ratio lean burn option is the most cost effective solution for gasoline engines, provided that a lead tolerant catalyst or lead free premium fuel can be made available.

(4) The direct injection diesel is the most fuel efficient.

(5) The tightening of emission controls beyond ECE 15-04 need not imply that fuel economy will reduce with respect to the current conventional engine, costs however will increase as the level of technology increases.

GENERAL

(1) Improved engine control strategies can yield improved fuel economy in the absence of emission control.

(2) At the levels considered the control of HC emission is the most significant factor in limiting the fuel economy potential of gasoline engines.

(3) Catalyst technology can allow low emission engines to be developed without large fuel economy penalties provided lead tolerant catalysts or lead free premium fuel can be provided.

(4) Changes in fuel quality affecting octane ratings or sensitivity will modify the findings of this study.

(5) Since the vehicle population consists of a wide spectrum of vehicles and specifications, this study has concentrated on a specific vehicle. The following more general conclusions must also be drawn.

(a) Higher vehicle weight and corresponding increases in engine swept volume are likely to lead to higher exhaust emissions and further fuel economy penalties.
(b) Increased performance vehicles having a higher power to weight ratio will require additional attention to HC emission control with an attendant fuel economy penalty.

(6) Fuel consumption and exhaust emissions are sensitive to drive cycle, and the ECE cycle being designed only to represent congested inner-urban driving, cannot be representative for average driving conditions.

9 ACKNOWLEDGEMENT

The authors acknowledge the support of the Department of Transport, London in sponsoring this project and would like to thank them and the Directors of Ricardo Consulting Engineers plc for permission to publish this paper.

10 REFERENCE

(1) ERGA - AIR POLLUTION (1) DoE 18,
The effects of vehicle emissions controls on fuel usage and cost in Europe.

Table 2 Fuel economy and emissions over ECE 15 cycle

NO.	ENGINE TYPE	AMENDMENT NUMBER	EMISSIONS BUILD				BEST ECONOMY BUILD			
			HC	NOx	CO	F.C. (b)	HC	NOx	CO	F.C. (b)
			g/test			l/100 km	g/test			l/100 km
1	Current conventional engine (1981) (a)	03	13	9	90	13.5				
2	Conventional engine with optimized carburation, mixture preparation and distribution	03	10	11	72	13.1	12	15	60	12.5
3	Conventional engine with optimized carburation, mixture preparation, distribution and optimized ignition	04	8	8	60	12.7	15	12	58	11.9
4	Conventional engine with optimum electronic fuel injection and optimum ignition timing	04	7	6	50	12.5	12	9	45	11.5
5	(3) with SA + PL + TR	04	6	3	46	13.7				
6	(4) with SA + PL + TR	05	5	3	35	13.2				
7	(3) + lean mixture strategy	04	9	4	45	12.7				
8	(3) + EGR	04	9	3.5	55	13.0				
9	(7) + oxidation catalyst	05	3	4	20	12.9				
10	High compression ratio lean burn engine with optimized carburation, mixture preparation, distribution and optimized ignition	04	14	3	35	11.5	18	5	38	10.8
11	High compression ratio lean burn engine with optimum electronic fuel injection and optimum ignition timing	04	11	3	30	11.0	16	4	30	10.5
12	(10) + EGR	04	15	1.5	38	11.8				
13	(10) + oxidation catalyst	05	5	3	14	11.7				
14	Stratified charge engine + TR (scavenged prechamber)	05	6	3	20	12.7				
15	Stratified charge engine + TR (prechamber)	05	5	2.5	20	12.2				
16	Diesel engine IDI (2.5 litre)	05	2	3	6	9.8(e)				
17	Diesel engine IDI boosted (2.0 litre)	05	2	3	6	9.0(e)				
18	Diesel engine DI (2.6 litre)	05	2	4	6	8.9(e)				
ECE 15 Regulation		03(c)	15.6	11.2	96					
" " "		04		20.5	67					
" " " (d)		05		10	36					

Abbreviations

SA = Secondary Air, supplied by pulsed air system
PL = Port liners in exhaust ports
TR = Thermal Reactor
F.C. = Fuel consumption

EGR = Exhaust gas recirculation
IDI = Indirect injection diesel
DI = Direct injection diesel

NOTES

(a) Vehicle weight 1132 kg, 1.6 litre four cylinder gasoline engine.
(b) Fuel consumption over cold start ECE 15 cycle.
(c) Equivalent for 04 measuring techniques.
(d) FRG proposed.
(e) Diesel fuel has a 9% specific gravity advantage over gasoline.

Table 3 Estimated vehicle operating costs over the first two years of ownership and 20,000 miles

	a] Engine First Cost (£)	b] Vehicle First Cost (£)	STANDING CHARGES FOR TWO YEARS (£)				RUNNING COSTS PER MILE (P)						
			c]. Depreciation	d] Interest on Capital	e] Other Standing Charges	Total Standing Charges	f] Fuel	g] Oil & Tyres	h] Maintenance	Total Running Costs	i] Standing Charges	i] Total Operating Costs	j] % Increase
ENGINE TYPES													
(1) Current conventional engine	960	4800	1200	765	948	2913	5.33	0.85	4.00	10.18	14.57	24.75	-
(2) (1) + optimized carburation	1037	4877	1219	777	948	2944	5.18	0.85	4.10	10.13	14.72	24.85	0.4
(3) (2) + optimized Ignition	1085	4925	1231	785	948	2964	5.02	0.85	4.14	10.01	14.82	24.83	0.3
(4) (1) + optimized, electronic fuel Injection & Ignition	1210	5050	1263	805	948	3016	4.94	0.85	4.24	10.03	15.08	25.11	1.5
(5) (3) + secondary air, port liners & thermal reactor	1325	5165	1291	823	948	3062	5.41	0.85	4.30	10.56	15.31	25.87	4.5
(6) (4) + secondary air, port liners & thermal reactor	1440	5280	1320	842	948	3110	5.21	0.85	4.40	10.46	15.55	26.01	5.1
(7) (3) with lean mixture strategy	1114	4954	1239	790	948	2977	5.02	0.85	4.14	10.01	14.89	24.90	0.6
(8) (3) with EGR	1171	5011	1253	799	948	3000	5.14	0.85	4.24	10.23	15.00	25.23	1.9
(9) (7) with oxidation catalyst	1267	5107	1277	814	948	3039	5.10	0.85	4.40	10.35	15.20	25.55	3.2
(10) High compression, lean burn engine with optimized carburation & ignition	1181	5021	1255	800	948	3003	4.54	0.85	4.16	9.55	15.02	24.57	-0.7
(11) (10) + optimized electronic fuel injection & ignition	1306	5146	1287	820	948	3055	4.35	0.85	4.30	9.50	15.28	24.78	0.1
(12) (10) with EGR	1373	5213	1303	831	948	3082	4.66	0.85	4.30	9.81	15.41	25.22	1.9
(13) (10) with oxidation catalyst	1450	5290	1323	843	948	3114	4.62	0.85	4.44	9.91	15.57	25.48	2.9
(14) Stratified charge engine with scavenged prechamber	1421	5261	1315	838	948	3101	5.02	0.85	4.24	10.11	15.51	25.62	3.5
(15) Stratified charge engine with unscavenged prechamber	1517	5357	1339	854	948	3141	4.82	0.85	4.40	10.07	15.71	25.78	4.2
(16) Diesel engine - nat. asp. IDI (2.5 L)	1603	5443	1361	867	948	3176	3.87	0.85	4.20	8.92	15.88	24.80	0.2
(17) Diesel engine - boosted IDI (2.0 L)	1699	5539	1385	883	948	3216	3.56	0.85	4.20	8.61	16.08	24.69	-0.2
(18) Diesel engine - nat. asp. DI (2.6 L)	1728	5568	1392	887	948	3227	3.52	0.85	4.20	8.57	16.14	24.71	-0.2

Table 4 Estimated vehicle operating costs over assumed vehicle life (8 years/80,000 miles)

Engine Types	a] Engine First Cost (£)	b] Vehicle First Cost (£)	STANDING CHARGES FOR 8 YEARS (£)					RUNNING COSTS PER MILE (P)						
			c] Depreciation	d] Interest on Capital	e] Other Standing Charges	Total Standing Charges		f] Fuel	g] Oil & Tyres	h] Maintenance	Total Running Costs	i] Standing Charges	j] Total Operating Costs	j] % Increase
(1) Current conventional engine	960	4800	4800	1836	3792	10,428		5.33	0.85	4.00	10.18	13.04	23.22	-
(2) (1) + optimized carburation	1037	4877	4877	1865	3792	10,534		5.18	0.85	4.10	10.13	13.17	23.30	0.3
(3) (2) + optimized ignition	1085	4925	4925	1884	3792	10,601		5.02	0.85	4.14	10.01	13.25	23.26	0.2
(4) (1) + optimized, electronic fuel injection & ignition	1210	5050	5050	1932	3792	10,774		4.94	0.85	4.24	10.03	13.47	23.50	1.2
(5) (3) + secondary air, port liners & thermal reactor	1325	5165	5165	1976	3792	10,933		5.41	0.85	4.30	10.56	13.67	24.23	4.3
(6) (4) + secondary air, port liners & thermal reactor	1440	5280	5280	2020	3792	11,092		5.21	0.85	4.40	10.46	13.87	24.33	4.8
(7) (3) with lean mixture strategy	1114	4954	4954	1895	3792	10,641		5.02	0.85	4.14	10.01	13.30	23.31	0.4
(8) (3) with EGR	1171	5011	5011	1917	3792	10,720		5.14	0.85	4.24	10.23	13.40	23.63	1.8
(9) (7) with oxidation catalyst	1267	5107	5107	1953	3792	10,852		5.10	0.85	4.40	10.35	13.57	23.92	3.0
(10) High compression, lean burn engine with optimized carburation & ignition	1181	5021	5021	1921	3792	10,734		4.54	0.85	4.16	9.55	13.42	22.97	-1.1
(11) (10) + optimized electronic fuel injection & ignition	1306	5146	5146	1968	3792	10,906		4.35	0.85	4.30	9.50	13.63	23.13	-0.4
(12) (10) with EGR	1373	5213	5213	1994	3792	10,999		4.66	0.85	4.30	9.81	13.75	23.56	1.5
(13) (10) with oxidation catalyst	1450	5290	5290	2023	3792	11,105		4.62	0.85	4.44	9.91	13.88	23.79	2.5
(14) Stratified charge engine with scavenged prechamber	1421	5261	5261	2012	3792	11,065		5.02	0.85	4.24	10.11	13.83	23.94	3.1
(15) Stratified charge engine with unscavenged prechamber	1517	5357	5357	2049	3792	11,198		4.82	0.85	4.20	10.07	14.00	24.07	3.7
(16) Diesel engine - nat. asp. IDI (2.5 L)	1603	5443	5443	2082	3792	11,317		3.87	0.85	4.20	8.92	14.15	23.07	-0.6
(17) Diesel engine - boosted IDI (2.0 L)	1699	5539	5539	2119	3792	11,450		3.56	0.85	4.20	8.61	14.31	22.92	-1.3
(18) Diesel engine - nat. asp. DI (2.6 L)	1728	5568	5568	2130	3792	11,490		3.52	0.85	4.20	8.57	14.36	22.93	-1.2

Table 5 Cost effective engine options for future emission levels

ECE 15 AMENDMENT NUMBER	ENGINE TYPES	FUEL CONSUMPTION ECE 15 (b) (L/100 km)	VEHICLE FIRST COST (c) (£)	TOTAL OPERATING COST OVER TWO YEARS (d) (p/mile)	(% increase)	TOTAL OPERATING COST OVER EIGHT YEARS (d) (p/mile)	(% increase)
03	(1) Current conventional engine (a)	13.5	4800	24.75	-	23.22	-
04	(3) Conventional engine + optimized carburation + optimized ignition	12.7	4925	24.83	0.3	23.26	0.2
	(7) Conventional engine + optimized carburation + optimized ignition + lean mixture strategy	12.7	4954	24.90	0.6	23.21	0.4
	(10) High compression lean burn engine with optimized carburation and ignition	11.5	5021	24.57	-0.7	22.97	-1.1
05	(9) Conventional engine + oxidation catalyst	12.9	5107	25.55	3.2	23.92	3.0
	(13) High compression lean burn engine + oxidation catalyst	11.7	5290	25.48	2.9	23.79	2.5
	(17) Diesel engine - boosted IDI (2.0 L)	9.0	5539	24.69	-0.2	22.92	-1.3
	(18) Diesel engine - naturally aspirated DI (2.6 L)	8.9	5568	24.71	-0.2	22.93	-1.2

'Type Approval' Standards for 1132kg. Vehicles in Europe

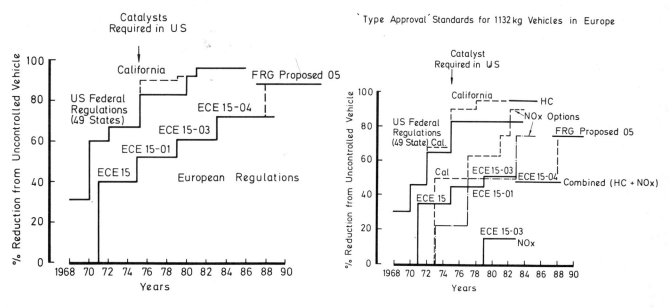

Fig 1 Trends in control of vehicle exhaust emissions: carbon monoxide

'Type Approval' Standards for 1132 kg Vehicles in Europe

Fig 2 Trends in control of vehicle exhaust emissions: hydrocarbons and oxides of nitrogen

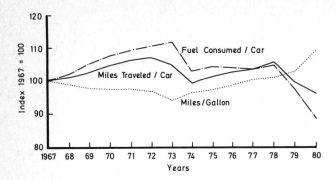

Fig 3   US passenger car efficiency index

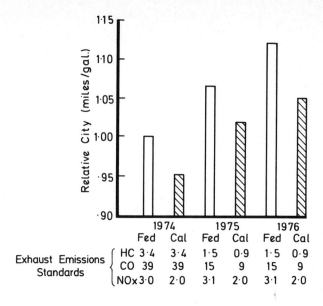

Fig 4   Normalized fuel economies of six groups of certification vehicles (EPA city miles/gal)

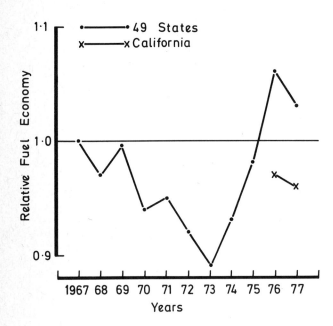

Fig 5   Average US fleet fuel economy with 1967 as base: effect of emission legislation

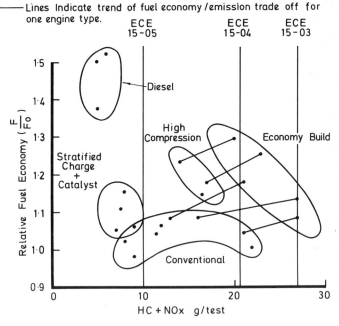

Fig 6   Fuel economy and (HC + NOx) emission for various engine technologies

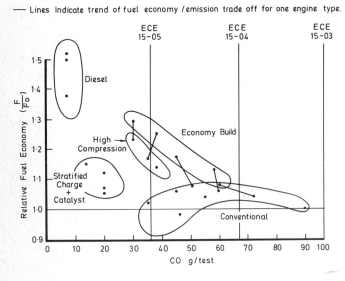

Fig 7   Fuel economy and CO emission for various engine technologies

© IMechE/SAE 1984 C422/84

C441/84

SAE 841282

U S passenger car fuel economy

B H SIMPSON, BSME
Ford Motor Company, Dearborn, Michigan, USA

SYNOPSIS This paper is a study of the changes that have been incorporated in passenger cars sold in the U.S. in recent years to improve fuel economy, and of the significant improvements that have been achieved. Promising areas of new technology are also described that should have the capability of providing further substantial fuel economy improvements. Market acceptance, economic considerations, potential government actions and other factors that may affect fuel economy are also discussed. Although this study is focused on U.S. passenger car developments, the results should also be of relevance in Europe. For consistency with U.S. legislated requirements, fuel economy values in the paper are in miles per U.S. gallon; corresponding fuel consumption figures in litres/100 km are also shown.

1 INTRODUCTION

Mandatory fuel economy legislation was enacted in the USA in 1975 (the Energy Policy and Conservation Act) and was first applicable to 1978 model passenger cars and 1979 model light trucks. A combination of this legislation and consumer demand for vehicles with good fuel economy, spurred by two significant fuel shortage periods and rapid increases in fuel prices, have resulted in dramatic increases in U.S. passenger car fuel economy. While the specific details in this paper apply solely to cars sold in the U.S., those factors that are expected to influence future gains in fuel economy should also have relevance in Europe. This includes items of new technology, consumer demand and acceptance of functional compromise and increased cost, fuel prices, plus possible further governmental goals, incentives or requirements.

All fuel economy figures in this report are in <u>miles per U.S. gallon</u>, with corresponding fuel consumption figures also shown in litres per 100 kilometers. (Miles per Imperial gallon may be calculated by multiplying the miles per U.S. gallon figures shown by 1.2.) Corporate Average Fuel Economy (CAFE) values are based on production volume weighted averages of EPA combined cycle fuel economy values for all vehicles produced for sale in the U.S. by the manufacturer during a given model year. EPA combined cycle fuel economy values are <u>harmonic</u> average values, weighted 55% for the city cycle and 45% for the highway cycle, for a vehicle or class of vehicles. These methods of measurement and calculation are as specified in the U.S. law, and apply to all miles/gallon and CAFE values given in this report, except as specifically stated.

All weighted average fuel economy data in this report were taken from U.S. government reports, since the calculation of such average values requires the use of detailed production volume data or projections which are not publicly available. Sources for these data were the U.S. Environmental Protection Agency (EPA), the U.S. Department of Transportation (DOT), and the U.S. Department of Energy (DOE), as reported in references (1)*, (2) and (3). CAFE values for 1984 and later model years are, of course, projections. Final CAFE values are calculated after the end of the model year, based on actual production volumes.

2 1984 AND PRIOR MODEL FUEL ECONOMY

The increases in U.S. passenger car fuel economy from 1974 through 1984 are shown in Figure 1, along with the applicable CAFE standards. U.S. domestically-produced cars had an average fuel economy of 13.2 miles/gal (17.8 l/100km) in 1974, increasing to 25.4 miles/gal (9.3 l/100km) in 1984, for an improvement of 92%. Average imported car fuel economy improved 39% over this same time period, starting from 22.6 miles/gal (10.4 l/100km) in 1974 and increasing to 31.5 miles/gal (7.5 l/100km) in 1984. The combined fleet average fuel economy of U.S.-produced and imported cars has increased from 14.2 miles/gal (16.6 l/100km) in 1974 to 26.7 miles/gal (8.8 l/100km) in 1984, an increase of 88%.

CAFE standards started at 18 miles/gal (13.1 l/100km) in 1978, increasing to 27.5 miles/gal (8.55 l/100km) in 1985 and beyond. These standards were exceeded by the U.S. domestic fleet from 1978 through 1982, however, this fleet is below the standards in 1983 and 1984. The total domestic and imported fleet average is expected to be somewhat below the standard in 1984.

*Numbers in parentheses refer to references listed at the end of the text.

CAFE compliance applies, of course, to individual manufacturers and a provision in the U.S. law allows manufacturers that are below the standard in a given year to use "credits" they have either earned in prior years for exceeding the standards, or expect to earn for exceeding the standards in future years, to offset the shortage. These credits may be carried forward or carried back for three model years, and are determined on a production-weighted basis. This carryforward credit provision was used by the two largest U.S. manufacturers, General Motors and Ford, to satisfy CAFE compliance requirements in 1983. Credits are expected to be used for compliance in 1984. As discussed in more detail later, U.S. sales of small cars and diesels have declined since 1981, which has lowered CAFE values.

It is significant to note that during the same period that major fuel economy improvements were being made, vehicle emission levels in the U.S. were also being greatly reduced. For example, hydrocarbons and carbon monoxide emissions have each been reduced by 96% and oxides of nitrogen have been reduced by 76%, compared to uncontrolled levels. Stringent emission controls tend to penalize fuel economy, particularly when engine compression ratios must be lowered to accommodate lower octane unleaded gasoline. All cars sold in the U.S. for the past several years have been equipped with catalysts, which require the use of unleaded fuel. The octane of unleaded fuel in the U.S. is typically 91-92 RON, or 87 R+M/2. Leaded regular gasoline in the U.S. is typically 93 RON, or 89 R+M/2.

Fuel economy improvements over the years are more appropriately shown by comparing vehicles within the same weight class. Figure 2 shows such a comparison for U.S. passenger cars made before 1968 (pre-emissions control), and for 1978 and 1984 models. Improvements ranging from about 7 mpg to 17 mpg have been accomplished on vehicles of the same weight from precontrol to the present time. At the same time, vehicles have been getting lighter each year, which has accelerated the rate at which the fleet average has improved.

Consumers are probably more interested in the size and utility of a vehicle rather than its weight, so a comparison by vehicle size is also appropriate. EPA developed a vehicle classification system based on the interior volume of the car, and this classification method can be used to compare vehicles within the same size class. Interior volume under this system is based on the sum of the front and rear seat passenger space plus cargo space. Passenger car classes and corresponding interior volumes are listed below (other classes apply to 2-seaters, station wagons and light trucks).

Table 1 EPA passenger car classes

Name	Interior Volume ft^3	m^3
Minicompact	<85	<2.41
Subcompact	85-100	2.41-2.83
Compact	100-110	2.83-3.11
Mid-Size	110-120	3.11-3.40
Large	>120	>3.40

A comparison of 1978 to 1984 cars, by the above vehicle classes, is shown on Table 2. As shown, the average fuel economy of 1984 cars ranges from 5% to 51% better than 1978 cars of the same size class. Excluding minicompacts, which represent less than 0.5% of the 1984 U.S. market, 1984 class fuel economy is 23% to 51% better than 1978. Over the same time period, engine displacement has decreased by 18% to 42% for the same size cars, while weight has come down by 6% to 22%, excluding minicompacts.

Significant changes in the sales fractions of different vehicle classes have also occurred. For example, sales of mid-size and large cars decreased 27% and 29% respectively from 1978 to 1984, while subcompact and compact car sales increased by 47% and 90% respectively. Through 1980, consumers generally accepted smaller cars with less performance, which helped increase fleet average fuel economy. (The change in this market acceptance trend after 1980 is discussed later.) A general indication of vehicle acceleration performance capability is provided by vehicle weight per cubic inch of engine displacement, and this ratio has increased significantly for all 1984 classes except minicompacts.

Another useful measure of vehicle fuel efficiency, although not commonly used, is a vehicle size fuel consumption ratio—a measure of the interior volume efficiency of a vehicle as a function of its fuel economy. This ratio is related to the number of miles a unit volume of interior space is transported per unit of fuel. It is calculated by dividing the average interior volume for each class by the average fuel consumption (gal/mile) for that class times 1000. By this measure, 1984 vehicle classes range from 25% to 47% more efficient than the same classes in 1978, excluding minicompacts.

All comparisons thus far have been on a sales-weighted fleet or class average basis; however, since most people buy and compare vehicles one-at-a-time, individual vehicle comparisons are also appropriate. Table 3 compares several individual models produced by Ford in the U.S. in 1975 and in 1984. Comparisons are made of full size, 6-passenger base sedans, and of the lowest and highest fuel economy models offered by Ford in those years. For full size sedans with standard engines, the 1975 entry had an EPA label (city) rating of 11 miles/gal (21.4 l/100km) compared to 17 miles/gal (13.8 l/100km) for the comparable 1984 model, an improvement of 55%. At the same time, the 1984 model has 5% greater interior volume and a 15% better 0-60 mile/h acceleration time. The lowest fuel economy model Ford offered in 1975 had a 9 miles/gal (26.1 l/100km) label (city) rating compared to 17 miles/gal (13.8 l/100 km) for the lowest 1984 fuel economy model, an increase of 89%.

Looking at Ford's highest fuel economy models, the Pinto was the fuel economy leader in 1975 with a city rating of 19 miles/gal (12.4 l/100 km), which compares to 37 miles/gal (6.4 l/100 km) on the 1984 gasoline-powered Escort/Lynx, an improvement of nearly 100%. If the fuel economy of the 1984 diesel-powered Escort/Lynx is compared to the best 1975 entry, the improvement in 1984 is over 140%.

3 MAJOR CHANGES AFFECTING FUEL ECONOMY

What changes have been responsible for the fuel economy improvements just described? Two major changes are the reduction in average vehicle weight and use of smaller displacement engines. Figure 3 compares these design factors from 1974 to 1984, showing that average vehicle weight has been reduced nearly 1000 pounds (454 kg) and average engine displacement has been reduced by approximately 110 cubic inches (1.8 1) over this period of time. These changes represent reductions of 24% and 38%, respectively, for vehicle weight and engine displacement. Although engine displacement has been reduced by a greater percentage than has vehicle weight, vehicle acceleration performance has not suffered proportionately because of improved powertrain technology.

Along with the reduction in engine displacement has come more four and six cylinder engines. Eight cylinder engines, which were the most popular engines in the U.S. in prior years, represent less than one-quarter of the 1984 fleet, while four cylinder engines represent over half of the 1984 fleet. Note the close parallelism between the vehicle weight and engine displacement curves. These are the predominant factors, along with improved technology, responsible for the fuel economy improvements that have been achieved.

Figure 4 compares average weights of vehicles produced by U.S. domestic manufacturers and those of U.S.-sold vehicles produced by European and Japanese manufacturers. As seen, the weight of U.S. domestic vehicles has dropped very significantly, whereas the weight of European vehicles over the past several years has increased and Japanese vehicle weight has remained relatively constant. Average engine displacement by location of manufacturer is shown in Figure 5. The displacement of Japanese engines, on average, has remained relatively constant from 1975 through 1984, whereas European engine size has increased by 20% and U.S. domestic engine size has decreased 38%.

An interesting weight comparison is seen from the distribution of sales in the different weight classes for several model years in the U.S. This is shown in Figure 6 for precontrol models, 1975 models and 1984 models. For precontrol and 1975 model cars, weight classes of 4000 pounds (1814 kg) and above represented substantial sales fractions, whereas with 1984 models most vehicles were produced in the range of 2750 to 4000 pounds (1247 to 1814 kg), with nothing heavier than 4500 pounds (2041 kg).

Weight reductions have been the result of both vehicle downsizing and the use of weight-saving materials. The average weights of different materials used on Ford cars in the U.S. for 1975 and 1980, and those projected for use on 1985 models, are shown on Table 4. The average weight of lightweight materials (plastics, aluminum, and high strength steel) has increased 84% in the 1975-1985 decade on Ford vehicles. The use of more lightweight materials reduces the amount of vehicle downsizing required to meet given weight reduction targets. The average weight of other materials used in Ford cars has decreased 39%, and total vehicle dry materials weight has been reduced 28%, over this same time period.

Another significant trend is the increasing use of front wheel drive (FWD) and this is shown on Figure 7. From a fuel economy standpoint, FWD is important in that it generally results in a weight reduction for a given interior size of vehicle. On a total fleet basis, FWD has increased from 9.6% in 1978 to 52% in 1984. Until 1982, European makes sold in the U.S. had the highest percentage use of FWD; however, this percentage has been decreasing since 1980. Japanese manufacturers have increased FWD usage significantly to over 75% in 1984, and U.S. manufacturers are now approaching 50% FWD.

The popularity of diesel engines has changed considerably over the past few years in the U.S., as shown on Figure 8. Contrary to the situation in Europe, the market share for diesel engine passenger cars in the U.S. has declined significantly since the 1981 model year. This diesel cutback has been attributed largely to the stabilization of fuel prices in the U.S., plus consumer dissatisfaction with certain diesel characteristics, reported problems, fewer service stations and increased initial cost. The main reason for purchasing a car with a diesel engine is for improved fuel economy, and the relative advantage of diesels over gasoline engines has been going down over the past five years. For example, the fuel economy advantage of best-in-class 3000 lb (1361 kg) diesel passenger cars relative to best-in-class gasoline cars has declined from approximately 29% in 1980 to 13% in 1984 (6). Although diesel fuel economy has improved over time, gasoline engines have improved faster with the introduction of fast burn, low friction designs with electronic controls.

On a total fleet basis, diesel engines peaked at 6% of the passenger car fleet in 1981, dropping to 4.7% of the 1982 fleet and 2.4% of the 1983 fleet. For 1984, diesels are projected at 3% of the fleet. By country of origin, both European and domestic diesels peaked in 1981, representing 36.4% and 5.0% of their respective fleets as sold in the U.S. European diesels are expected to drop to 13.8% of European imports in 1984, domestic diesels to 2.2%. Japanese diesels entered the U.S. market in 1981 and are expected to represent 2.3% of the Japanese import fleet in 1984.

Significant changes have taken place in transmissions in the U.S. market since 1978 to improve fuel economy. As shown on Figure 9, in 1978 nearly three-fourths of all transmissions were automatics that incorporated neither lockup or overdrive. In 1984, 73% of all automatics have lockup, and about half of them have overdrive. Automatic transmissions of all types have dropped somewhat in installation rate since 1978, from about 80% to 72% in 1984. Manual transmissions have increased the number of forward gears and most now incorporate overdrive in top gear. Considering both automatic and manual transmissions, overdrive usage has averaged about 60% since 1982.

Another significant trend is the increased use of fuel injection, as shown on Figure 10 for domestic, European and Japanese cars sold in the

U.S. On a total fleet basis, less than 5% of the 1975 fleet was equipped with fuel injection, compared to nearly 40% on 1984 cars.

Under steady speed driving conditions, engine power is consumed primarily in overcoming chassis resistance, which is mainly tire rolling resistance and aerodynamic drag. These two items have received a great deal of attention so as to improve fuel economy. The rolling resistance of tires used on Ford cars has been reduced an average of 27% from 1978 through the use of increased inflation pressure (11%) and improved construction materials (16%). An additional 12% reduction in rolling resistance accrued from improvements in construction and materials on 1983 tires. As the result of reduced vehicle weight, tire rolling resistance has also been reduced. These improvements in tire rolling resistance have all been accomplished without compromising vehicle ride and handling characteristics or tire wear and traction. The fuel economy improvements which Ford has realized from tire rolling resistance reductions are summarized on Figure 11. In 1983, Ford's total CAFE benefit from tires amounted to nearly 2 miles/gal, made up of 0.58 miles/gal from reductions in vehicle weight and another 1.4 miles/gal from tire pressure, construction and material changes.

Aerodynamic drag is of major importance when traveling at highway speeds or when encountering headwinds, since aero horsepower increases as the cube of vehicle air speed. The term C_D, or drag coefficient, is a measure of vehicle efficiency as an aerodynamic shape and is useful for comparing different vehicle designs. C_D is a dimensionless expression obtained by measuring the drag force of the vehicle and dividing it by the product of dynamic pressure and vehicle frontal area. C_D values of current production cars sold in the U.S. range from approximately .52 to .30.

The average drag coefficient of Ford U.S. cars from 1975 projected through 1990 is shown on Figure 12. The reduced aerodynamic drag from 1977 through 1984 has resulted in a CAFE improvement of over 1 mile/gal, and a further CAFE improvement of approximately 1.5 miles/gal is expected to accrue from aerodynamic improvements by 1990. Aerodynamic fuel economy improvements are actually the result of reductions in aero horsepower, which is a function of both the drag coefficient and the frontal area of the vehicle. The aero horsepower at 50 miles/h (80 km/h) of typical 1984 vehicles ranges from approximately 10.4 hp (7.8 kW) to 5.3 hp (4.0 kW). By contrast, a heavy duty truck could have an aero horsepower as high as 100 hp (75 kW) at 50 miles/h (80 km/h). It is estimated that the reduction of one aero horsepower is the fuel efficiency equivalent of reducing vehicle weight by 300 pounds (68 kg), based on combined city and highway driving.

Effects of Changing Vehicle Mix

The factors affecting fuel economy discussed so far have all been related to the design of the product--vehicle size, weight, powertrain configuration, etc. However, another major factor over which the manufacturer has much less control also importantly affects CAFE. This factor is market mix, which of course affects CAFE, since CAFE is a production-weighted average. A definite increase in consumer preference for larger cars has occurred in the U.S. over the past several years. Figure 13 shows the month-to-month changes and trends in consumer demand for small cars and large cars in the U.S. from late 1979 to the end of last year. As seen, the demand for small cars increased from approximately one half of the market to a peak of over 60% in early 1981, coincident with the peaking in fuel prices. Since early 1981 there has been a steady decline in the trend line demand for small cars, with the small car segment dropping to just over 50% by the end of 1983. The picture on large cars is the mirror image of small cars. Also plotted on this figure is fuel price, which correlates directly with the small car segment share trend and inversely with the large car segment share trend.

A significant change in market mix, such as that which has been experienced since 1981, has a major effect on CAFE for any manufacturer that produces a full range of products. For example, in December 1980 Ford projected that U.S. market demand for 1983 models would be such that 49% of its 1983 models sold would be those with fuel economy ratings above the 1983 CAFE standard of 26 miles/gal (9.05 1/100km). Two years later, in December 1982, fuel prices had come down and actual sales of Ford cars above the standard were only 35%. This lowered Ford's CAFE by 1.2 miles/gal compared to the earlier projection.

U.S. Domestic vs. Foreign Vehicle Fuel Economy

Before leaving current CAFE it would be well to displace a commonly held but incorrect belief in the U.S. that imported cars typically have better fuel economy than U.S. domestic cars. While it is true that the fleet average of foreign cars is higher than that of domestic cars, because foreign cars are smaller and lighter on average, it is not true that foreign car fuel economy is better than domestics when cars are compared on an equal basis. Figure 14 compares the fuel economy of foreign and domestic cars on an equal weight basis.

There are 17 EPA equivalent test weight classes in which both foreign and domestic cars are produced in 1984, and in 14 out of these 17 classes domestic cars on average have higher fuel economy than foreign cars of the same weight, with an average advantage of 1.8 miles/gal (9.4%) per class. This comparison is based on a nonweighted harmonic average of the fuel economy of all cars tested in each weight class for both foreign and domestic makes. Thus, it is an average domestic car compared to an average foreign car in each weight class, without regard for sales volume. The same type of comparison has also been made based on equal vehicle size, using the previously described EPA vehicle classifications. There are six vehicle classes in which both foreign and domestic cars are produced in 1984, and domestic cars show fuel economy superiority in each of these classes, with an average advantage of 2.3 miles/gal (10.2%) per class.

4 WHAT LIES AHEAD FOR CAFE?

The improvements that have been made in U.S. passenger car fuel economy from 1974 through 1984 have now been defined, identifying the main factors that have changed and the new technology that was introduced during this period, and how mix changes have affected CAFE. Next, what is CAFE likely to be in the future, and what are the main factors that will determine how much further improvement may be accomplished? The answers to these questions depend on three main factors: technology; economic and marketing factors; and other political, legislative, and competitive factors. Each of these factors will be discussed in turn.

New Technology

The availability of technology, per se, should not be a limitation on making further substantial improvements in fuel economy. However, practical limitations on function, cost, production feasibility, plus market demand and product acceptance, will certainly apply in determining future CAFE levels. The following examples are cited to dramatize this point.

First, two extreme examples. Two groups of Ford engineers, one in Britain and the other in Australia, each recently designed and tested what might be described as all-out, no-holds-barred, super fuel economy vehicles. In other words, vehicles that define the technological limits of fuel economy at this time. The fuel economy achieved by these vehicles was phenomenal. The British Ford research vehicle (Figure 15) achieved 2100 miles/gal (0.112 1/100km) on its first run at the Silverstone Grand Prix circuit, and they expect to do better after further refinement. This vehicle weighs just 65 lb (29 kg) and was driven by a 95 lb (43 kg) young lady, at an average speed of 15 miles/h (24 km/h). It is powered by a 15 cm^3 engine with electronic fuel injection. The vehicle has a C_D of 0.13.

The Australian Ford team of engineers did even better. They set a world's mileage marathon record of 2457 miles/gal (0.096 1/100km) in an event held at Amaroo Park near Sidney, Australia. The functional compromises and limitations required to achieve these remarkable fuel economy levels are severe, and of course completely unacceptable for any commercial vehicle. These vehicles only prove what is possible with an all-out technological approach.

Another demonstration of technology capability, focused just on reduced weight, was a research vehicle built by Ford in 1978 in which the entire body, frame and certain other chassis components were fabricated from graphite fiber material. A basic design (size and shape) of the vehicle was unchanged from that of a standard 1978 Ford LTD, so function was not compromised. A total of 1200 lb (544 kg), or one-third of total vehicle weight, was reduced through the use of graphite fiber and other alternate material substitutions and powertrain downsizing to match the reduced weight. The net result was a fuel economy improvement of 7 miles/gal, or 40% over the base LTD vehicle. The limitations on this vehicle were not function (except possibly for crashworthiness, damageability, and durability which were not confirmed), but rather production feasibility and cost. Many of the graphite components were hand fabricated and the cost of this single research demonstration vehicle was several million dollars.

Areas of new technology which appear to offer definite potential for fuel economy improvements in the future, on a practical basis, include the following:

- Spark ignition engines with improved thermodynamics and reduced friction
- Improved and new concept diesel engines
- Continuously variable ratio transmissions
- Increased use of electronic controls
- Further reductions in aerodynamic drag
- Further weight reductions
- Improved fuels and lubricants
- Other (tires, electrical systems, hydraulic systems, air conditioning)

Gasoline Engines

Significant changes and improvements are being incorporated in spark ignition gasoline engines at this time, and further application of these changes can be expected in the future. These changes include improved combustion efficiency, improved volumetric efficiency, reduced mechanical friction, possible application of variable valve timing, increased charge density, application of electronic controls, and reduced engine weight. Increased combustion efficiency is being accomplished through the use of lean burn/fast burn combustion designs, which provide improved fuel economy while still meeting emissions control requirements. This feature depends on achieving a high turbulence in the cylinder charge, plus modification of the combustion chamber for compact shape and minimum surface to volume ratio.

Reduced mechanical friction is accomplished through many modifications in engine design. These include reduced friction rings, narrower rings, use of roller tappets, reduced crank journal diameter, lubricant and coolant pumps with reduced parasitic loss, and general attention to all areas involving mechanical friction within the engine. Increased charge density is achieved by either turbocharging, possibly coupled with ram manifolding, or supercharging. Further application of these techniques can be expected in order to either provide increased engine output without increasing basic engine size, or providing the same output with a smaller engine to achieve better fuel economy.

Diesel Technology

Despite the downturn in U.S. passenger car diesel sales in recent years (at the same time that European diesel sales have been increasing), diesels were projected to attain a 10% U.S. installation rate by 1990 in the University of Michigan 1983 Delphi Survey (7). It is noted however, that prior Delphi Surveys projected substantially higher diesel rates (e.g., 25% in the 1979 survey, and 20% in the 1981 survey), and the 1983 forecast of 10% diesels by 1990 may now be optimistic. Near-term improvements in diesel engines will be the result of further combustion chamber design innovation, friction reductions, adaptation of direct fuel injection, use of electronic controls, and further application of turbocharging.

A significant added challenge that diesel engines have to face in the future is that of very stringent particulate emission standards, applied first in California and then possibly in all other states. The California particulate standard for 1989 model diesel cars is 0.08 g/mile. This level of control will mean that very efficient particulate traps must be used, and these traps will cause some fuel economy penalty because of increased exhaust back pressure. Some trap designs also use extra fuel to "light off" the trap regeneration process, further penalizing fuel economy.

An important new concept, which still requires considerable development at this time, could further enhance the rate of diesel engine application. This is the so-called adiabatic diesel, in which the conventional cooling system is virtually eliminated and the engine runs much hotter. This minimizes heat losses from the combustion chamber and increases overall engine efficiency and fuel economy. Adiabatic diesel engines are projected to provide a 10-15% improvement in fuel economy over liquid cooled diesel engines. Approximately 5-10% gain is attributable to improvement in thermal efficiency while the remaining 5% is due to elimination of the fan and water pump cooling system components.

To allow the engine to operate without a conventional liquid cooling system, alternate materials and lubricants must be developed that will tolerate higher temperatures. A number of design concepts to accomplish this are being studied, as shown on Figure 16. These include the application of ceramic thermal barrier coatings such as zirconia oxide to relatively standard metal engine designs, and the use of low thermal conductivity solid ceramic materials such as partially stabilized zirconia to form high temperature surfaces of the pistons, cylinder liners and cylinder head which are exposed to combustion gases. High strength ceramic materials are also being investigated for the fabrication of the entire engine structure, thereby eliminating conventional cast iron configurations. Each of these approaches is under study at this time, as described by Wade (8) and Kamo (9).

Other potential advantages associated with the adiabatic diesel include reduced noise levels, because of elimination of the cooling fan and because of reduced ignition delay, and the increased tolerance for low quality fuels. This fuel tolerance occurs as a result of the higher end compression temperatures that exist during fully warmed-up engine operation, reducing ignition delay times. The development of an effective cold starting assist mechanism could provide an overall operational tolerance for lower quality level diesel fuels.

Transmissions

Further refinement of present types of automatic transmissions will take place through the application of electronic controls and further incorporation of overdrive and lockup features. The major change coming on transmissions, however, will be the introduction of CVT—continuously variable transmissions that automatically change ratio over a wide range and do so in a continuous manner rather than by discrete steps or gears as at present.

The theoretical benefits to be derived from being able to ideally match the engine speed and load to vehicle operating demands have long been recognized, and numerous patents have been issued over the years on various mechanical and hydraulic CVT concepts. The designs now being developed for future production are based on VanDoorne's adaptation of an old concept in which variable sheave pulleys are connected by a segmented belt. A cross-section drawing of a CVT is shown in Figure 17.

Through use of appropriate sensors and electronic controls in the CVT, it will be possible to do a much better job of providing the optimum transmission ratio for each particular driving condition. This should result in substantial fuel economy benefits with CVTs, probably in the range of approximately 15-20% over present automatic transmissions. The greatest fuel economy benefits will accrue under light load operating conditions, where CVTs will automatically shift to the equivalent of a super overdrive ratio, greatly slowing down the engine speed and reducing engine friction losses. At the same time, CVTs will allow improved acceleration performance because of their wide range of ratios and smooth, stepless acceleration performance capability.

Electronic Controls

Electronic engine controls, already discussed, allow the engine to operate at more nearly optimum conditions under all driving conditions, while maintaining low exhaust emission levels. Current electronic engine control systems instantaneously gage crankshaft position and speed, throttle plate position, engine coolant temperature, exhaust gas oxygen levels, air conditioning operation, intake manifold vacuum, barometric pressure, exhaust gas recirculation valve position and throttle position. This input information is processed by the electronic engine control system, and output signals are sent to the fuel control and emissions control systems. The electronic engine control system on 1984 Ford cars has the capability of performing one million mathematical operations per second and can process the input information so as to provide output signals in less than one crankshaft revolution.

In addition to controlling engine functions and components, electronic controls will also be used increasingly to control other functions. These include transmission shifting and lockup, integrated speed control, anti-lockup brakes and suspension leveling devices, as well as enhanced engine controls, possibly integrated with transmission controls.

Aerodynamics

Further reductions in aerodynamic drag will be incorporated in future vehicles to improve fuel economy. To explore the possibilities of improved vehicle aerodynamics, Ford has constructed a series of concept vehicles, the latest of which is Probe IV, pictured on Figure 18. This is the world's most aerodynamic four-door car, with a C_D of 0.15--which is comparable to that of a jet fighter aircraft. Aero horsepower at 50 miles/h is 2.5 hp (1.9 kW at 80 km/h). Significant design features of this running vehicle that help achieve its outstanding aerodynamics, in addition to its overall "slippery" contours, include fully shrouded wheels, flush glazing and headlights, smooth underbody, a rear-mounted engine cooling system (to eliminate the need for a front grille), a moveable front air dam that automatically reduces nose ground clearance as vehicle speed increases, and a vehicle suspension that automatically adjusts vehicle altitude to suit vehicle speed.

From a ground clearance of 6.5 in (165 mm) at rest, Probe's front clearance reduces to 3.3 in (84 mm) and the rear lowers slightly to 6.0 in (152 mm) at speeds over 40 miles/h (64 km/h). This nose-down altitude reduces the lift and drag caused by undercar airflow and improves road holding. The onboard computer controlling these functions can be manually overridden to suit special conditions. While not all of these special features can be expected to show up on production cars in the near future, some of them will--with vehicles such as Probe IV pointing the way toward further aerodynamic improvements.

Reduced Vehicle Weight

U.S. domestic manufacturers are expected to continue reducing their average car weight in the future, but probably at a slower rate than heretofore. This slower rate is due to the facts that the most cost-effective lightweight material substitutions have already been made, and the increasing market demand for larger cars. On the other hand, the average weight of Japanese and European cars may hold relatively constant or even increase somewhat in the future, if the weight trends of the past several years continue (see Figure 4).

Some of the additional applications of lightweight materials that may be incorporated include all-plastic (polycarbonate) bumpers, plastic fuel tanks and body panels, increased use of high strength steel throughout the vehicle, and wrought aluminum radiators. Within the engine, forged aluminum or composite materials may be used for reciprocating components to reduce friction and vibration. Ford expects that by 1990 its average car use of lightweight materials will be increased over 1985 approximately as follows: plastics +12%; aluminum +5%; and high strength steel +35%. As pointed out by Wheeler (10), the total life cycle energy associated with the use of lightweight should also be considered when making materials trade-off decisions.

Improved Fuels and Lubricants

Regarding gasoline octane, Adams (11) indicates a reduction in vehicle fuel consumption of about 6.5% for a 5 RON increase, assuming engine compression ratio is increased to accept the higher octane and that acceleration capability is held constant. While recognizing that higher octane provides the opportunity for improved fuel economy, it is also recognized that the most appropriate octane level is dependent on many factors in addition to the preferred engine appetite. Also important are such considerations as gasoline yield per barrel of crude, demand for middle distillate fuels and other products, the use of antiknocks/octane improvers, available refinery facilities, crude quality and regulatory requirements. The use of ethanol and methanol as octane (and profit) improvers is a topic of considerable interest in many countries at this time. For example, the Sixth International Alcohol Fuels Technology Symposium, held in Ottawa Canada on May 21-25, 1984, attracted participants from 27 different countries.

The use of near-neat methanol fuel (usually 90% methanol, 10% unleaded gasoline) offers definite potential advantages in the future, based on Ford's experience with two fleets of cars designed specifically for this fuel (12). One fleet of 40 Escort vehicles has been opera-operating in California for three years and in mid-June 1983 had accumulated a total of 1.2 million miles (1.9 million km), with several vehicles over 100000 miles (161000 km). The other fleet of about 580 Escorts was put into service in mid-1983, with 500 in use in California. As of mid-June 1983 these methanol vehicles had accumulated from 3000 to 20000 miles (4800 to 32000 km) each.

This combined fleet experience has identified the hardware needed to run satisfactorily on near-neat methanol, and at the same time has demonstrated a noticeable performance improvement over comparable gasoline vehicles. The fuel economy of these methanol vehicles, on an energy-equivalent basis, is about 8% better than on comparable gasoline vehicles. This improvement is primarily because of increased compression ratio (11.9:1 vs. 8.5:1 with gasoline). With the engine and fuel system optimized for methanol, fuel economy should be at least 15% better than with 91 RON gasoline.

Two significant problems were encountered during the operation of these methanol fleet vehicles. First, problems of clogged or plugged fuel filters and tanks, and corrosion of refuling system components, were encountered due to methanol fuel contamination. Second, hot restart problems (vapor lock) were encountered after operation at ambient temperatures of approximately 95 F (35 C) or higher. Both of these problems are now being corrected. The contamination problem has been corrected by cleaning the fuel storage and transport facilities and eliminating an identified source of fuel contamination. The vapor lock problem is being corrected by adding an electric fuel pump in series with the mechanical pump to ensure fuel flow at high temperatures.

Improved lubricants represent another important means by which further fuel economy gains may be achieved. While most oil suppliers now offer "energy conserving" engine oils, it is believed that significant additional fuel savings should be possible with the development of "second generation," and perhaps "third generation," energy conserving lubricants. In fact, a formal proposal was made to the SAE Fuels and

Lubricants Technical Committee in January 1984 to determine the need for increasing the fuel economy improvement limits on energy conserving oils, or to establish a higher level classification, based on the need for improved or second generation energy conserving oils.

The potential for further improvement was demonstrated in valve train friction studies reported by Staron and Willermet (13). These studies showed a marked reduction in friction when using a friction modified oil, as shown on Figure 19. The results of these studies indicate that the proper lubricant may give levels of friction reduction comparable to those obtainable by costly component redesign (i.e., roller tappets or roller cam followers).

5 MARKET/ECONOMIC FACTORS

Technology makes it possible to provide improved fuel economy, but the resulting products must be accepted in the market in order to improve CAFE. If potential purchasers consider that vehicle size has been reduced too much, or that performance has been compromised too much, or that the vehicle costs too much, they simply will not purchase such products. Also, changing fuel prices have a significant influence on CAFE, as previously discussed. Abrupt changes in price have the most impact, especially if accompanied by changes in fuel availability.

Actual fuel shortages, resulting in long refueling lines and other personal inconveniences, or just the fear of shortages, cause drastic changes in consumer demand. The surge in demand for diesel engines which followed the last gasoline shortage, in spite of substantial price premiums for diesels, is a dramatic example of this consumer reaction phenomenon. Whereas fuel shortages, or the fear of them, and rising fuel prices increase the market demand for smaller cars with high fuel economy, the converse is also true. Stable or declining prices and assured availability stimulate the sale of larger products and diminish consumer emphasis on fuel economy. This is evidenced by the increased demand for larger vehicles and vehicles with higher performance capability over the past two model years in the U.S.

Another important consideration is the declining value of incremental increases in fuel economy. As fuel economy levels continue to increase, the incremental savings in consumer fuel cost and petroleum consumption will diminish. Figure 20 shows plots of annual consumer fuel costs as a function of fuel economy and fuel price. Curves representing three different fuel prices are shown. The $1.30/U.S. gal level represents the mid-1983 average U.S. price for unleaded regular gasoline, $1.60/U.S. gal represents the 1990 U.S. price of unleaded regular predicted by the 1983 Delphi Survey (7), and the $2.25/U.S. gal price represents approximately the mid-1983 price of premium gasoline in Britain. (Corresponding fuel prices in Pounds Sterling per Imperial gal are also shown on Figure 20.)

As seen, the savings from incremental increases in miles/gal drop off rapidly as fuel economy increases. For example, with fuel at $1.30/gal, increasing fuel economy from 10 to 20 miles/gal results in an annual savings of $780.00, whereas a 10 miles/gal increase from 40 to 50 miles/gal saves only one-tenth as much. These fuel costs and incremental savings are also tabulated on Table 4.

The decreasing benefits associated with increasing fuel economy levels are of particular significance when considering national fuel conservation measures. The combination of decreasing per vehicle fuel savings, and the gradual rate at which new vehicles replace older ones in the total fleet of cars, indicate ever-smaller national fuel savings as new car fuel economy levels continue to increase.

6 OTHER FACTORS AFFECTING FUTURE CAFE

In addition to all of the above factors, future fuel economy levels could be affected significantly by such things as additional CAFE legislation or other legislative or taxation requirements intended to stimulate higher fuel economy. Present U.S. legal requirements maintain passenger car CAFE at 27.5 miles/gal (8.55 1/100 km) for 1985 and later model years. The law also requires that future light truck CAFE standards be reviewed by DOT and established for each model year at "maximum feasible" levels. Proposed new fuel economy legislation was introduced in the U.S. Congress in April 1984 that would establish 45 mile/gal (5.23 1/100 km) passenger car and 35 mile/gal (6.72 1/100 km) light truck CAFE standards by 1995, plus other fuel conservation measures. There has been no committee action on this bill as of mid-1984. Other types of fuel economy standards, such as vehicle class standards, are also being discussed in Washington, D.C.

New technology has the capability of providing additional fuel economy benefits in the future; however, such factors as fuel price and assured availability, the overall economic condition of the nation, market acceptance of higher vehicle prices necessitated by the use of higher cost, high technology components, or functional compromises resulting from smaller or lower powered vehicles, and the fuel economy effects of added emissions or safety/damageability regulatory requirements, will all have an important influence on the actual fuel economy levels that will be achieved. Also, continuing consumer demand for improved fuel economy, plus competitive pressure, will provide vehicles manufacturers with strong incentives to increase CAFE levels.

Factors other than technology also appear to be key in deciding whether or not neat methanol fuel will be used on any wide-scale basis. Here, the essential item appears to be the development of a viable economic plan for the production and distribution of methanol and vehicles that are designed for methanol. Recently proposed legislation regarding neat methanol and regulatory relief proposals should, if enacted, be helpful in providing more visibility via demonstration fleets and avoiding some regulatory obstacles; however, any large scale use of methanol will need much greater assistance to get launched.

Whether or not additional, more demanding CAFE legislation is enacted probably depends on the possibility of disastrous world events, such as those which triggered Congress into passing the present fuel economy legislation (Energy Policy and Conservation Act of 1975). While this Act has accelerated the introduction of fuel efficient vehicles, there are those who say that its usefulness has been served and that it now deserves to be retired. This position was stated in a House Republican Research Committee Report issued in September 1983 (14).

While overall fleet average fuel economy is expected to increase in the future as new technology is phased in on a practical basis, motivated strongly by consumer demand and competitive pressure, there are definite risks associated with forcing CAFE too high or too soon. For example:

- Functional compromises and added costs associated with "forced" fuel economy increases would result in consumer resistance and retaining of older cars for longer periods of time. This, in turn, would depress the auto industry and its suppliers and increase unemployment.

- Expected national energy conservation savings would not be realized if consumers refuse to purchase high fuel economy vehicles that did not satisfy their needs, or cost too much. Also, theoretical energy savings may not be achieved since there is evidence that higher fuel economy stimulates more driving.

- Economic, market demand and other factors affecting fuel economy can not be predicted with any degree of accuracy, so any future CAFE standards would be either too high or too low, resulting in continuing industry/government conflict.

- Continually increasing CAFE requirements would eventually result in the availability of only smaller, higher priced new cars that would seriously limit consumers' choice, in addition to putting a severe and undue burden on full-line vehicle manufacturers.

As seen above, future CAFE levels are dependent on a great many factors, most of which cannot be predicted with accuracy and some of which are impossible to predict, e.g., fuel shortages or scares caused by international problems, and/or possible further CAFE legislation. Because of these many uncertainties, predictions of future CAFE levels inherently have very low confidence, which essentially come down to mere guesses.

7 SUMMARY

1. The average fuel economy of passenger cars sold in the U.S. has increased dramatically over the past decade. The U.S. domestic fleet average is up 92%, the imported fleet average is up 39% (from a higher base than domestics), and the total fleet average is up 88%.

2. To accomplish these fuel economy improvements, new vehicles, new powertrains and many other new components have been introduced, resulting in 1984 vehicles that are 24% lighter and with engines that are 38% smaller than in 1974. Since 1978, domestic auto makers have spent $80 billion to make these changes.

3. Half of the 1984 U.S. fleet of passenger cars is projected to be front wheel drive, over half the fleet will have 4-cylinder engines, fuel injection usage will be about 40%, three-quarters of all automatics will have lockup and half of these will have overdrive, over 90% of manual transmissions will have overdrive, and only 3% of the 1984 fleet is expected to have diesel engines.

4. The fuel economy of 1984 U.S. domestic cars is better, on average, than that of foreign cars in 14 out of 17 weight classes, and in 6 out of 6 vehicle size classes, in which both domestic and foreign cars are offered.

5. Increasing market demand for larger cars over the past three years, concurrent with decreasing or stabilized gasoline prices, has significantly lowered CAFE levels.

6. Future significant fuel economy improvements should be possible through the development of new vehicle technology, including the refinement of lean burn, low friction gasoline engines, improved diesels and adiabatic diesels, introduction of continuously variable ratio transmissions, wider use of electronic controls, further reductions in aerodynamic drag and vehicle weight, and other improvements. However, practical limitations on function, cost, production feasibility, etc. will certainly apply in determining those items which can be incorporated in production vehicles, and the timing of their incorporation.

7. Improved fuels and lubricants also offer the possibility for worthwhile fuel economy gains. Near-neat methanol fuel, because of its higher octane, should yield a 15% energy equivalent fuel economy improvement on vehicles designed to take advantage of the added octane. The development of improved energy-conserving engine oils appears both feasible and worthwhile.

8. The actual levels to which CAFE will increase in the future will depend on many factors beyond the availability of new technology for vehicles, or of improved fuels and lubes. Possible new fuel economy legislation, fuel price and availability, consumer product demands and acceptance of new technology, higher prices and reduced vehicle size or performance, general U.S. and world economic conditions, and many other factors will also importantly influence future CAFE levels.

REFERENCES

(1) HEAVENRICH, R. M., MURRELL, J. D., CHENG, J. P., and LOOS, S. L. Passenger Car Fuel Economy...Trends Through 1984. SAE Paper 840499, March, 1984

(2) DENNISON, R. J. (DOT/NHTSA). Summary Fuel Economy Performance. Printouts dated Sept 9, 1983 and Feb 7, 1984

(3) PATTERSON, P. D., WESTBROOK, F. W., GREENE, D. L., and ROBERTS, G. F. Reasons for Changes in MPG Estimates, Model Year 1978 to the Present. March, 1984

(4) EPA 1984 Federal Test Car List Printout, Sept. 26, 1983

(5) EPA 1984 Federal Fuel Economy Gas Mileage Guide Printout, October 3, 1983

(6) WADE, W. R., WHITE, J. E., JONES, C. M., HUNTER, C. E., and HANSEN, S. P. Combustion, Friction and Fuel Tolerance Improvements For The IDI Diesel Engine. SAE Paper 840515, February, 1984

(7) COLE, D. E., and HARBECK, L. T. The UMTRI Research Review; Selected Results From A 1983 Delphi Survey. *Volume 14*, Number 3, November-December, 1983

(8) WADE, W. R. Light Duty, Uncooled, Diesel Engine Research. Presented at Workshop on Combustion and Heat Transfer in Adiabatic Diesel Engines, December 13-14, 1983

(9) KAMO, R., and BRYZIK, W. Cummins/TACOM Advanced Adiabatic Engine. SAE Paper 840428, February, 1984

(10) WHEELER, M. A. Lightweight Materials and Life Cycle Energy Use. SAE Paper 820148, February, 1982

(11) ADAMS, W. E., and FRENCH, B. J. World Trends in Passenger Car Production and Engines - 1983. Ethyl International, August, 1983

(12) NICHOLS, R. J. Further Development of the Methanol-fueled Escorts. *Proceedings of the Second International Pacific Conference of SAE*, Tokyo, Japan, November 6-10, 1983

(13) STARON, J. T., and WILLERMET, P. A. An Analysis of Valve Train Friction in Terms of Lubrication Principles. SAE Paper 830165, March, 1983

(14) LEWIS, J. and OKUN, R. Are the CAFE Standards Out-to-lunch? House Republican Research Committee, September 15, 1983

Table 2 1978 to 1984 comparisons, by vehicle size class

	EPA Vehicle Class				
	Minicompact	Subcompact	Compact	Mid-Size	Large
1978 Sales-Weighted Fleet Averages					
Fuel Economy, combined MPG	27.7	24.7	19.8	18.6	16.8
Engine Displ., in^3	119	163	242	293	358
Weight, lb	2551	2864	3613	3800	4391
Sales Fraction	.056	.170	.123	.327	.194
Lb/in^3 displ.	21.4	17.6	14.9	13.0	12.3
Interior Volume, ft^3	79.4	89.7	105.4	112.9	128.5
Interior ft^3/GPM·10^3	2.20	2.22	2.09	2.10	2.16
1984 Sales-Weighted Fleet Averages					
Fuel Economy, combined MPG	29.1	30.7	29.8	24.3	20.7
Engine Displ., in^3	136	133	140	210	292
Weight, lb	2653	2690	2819	3332	4018
Sales Fraction	.0046	.2500	.2340	.2400	.1370
Lb/in^3 displ.	19.5	91.3	20.1	15.9	18.8
Interior Volume, ft^3	75.9	94.0	103.0	113.4	130.6
Interior ft^3/GPM·10^3	2.21	2.89	3.07	2.76	2.70
1984 Compared to 1978, Per Cent					
Fuel Economy, combined MPG	+5.1%	+24%	+51%	+31%	+23%
Engine Displ., in^3	+14	-18	-42	-28	-18
Weight, lb	+4.0	-6	-22	-12	-8.5
Sales Fraction	-92	+47	+90	-27	-29
Lb/in^3 displ.	-8.9	+15	+35	+22	+12
Interior Volume, ft^3	-4.4	+4.8	-2.4	+4.4	+1.6
Interior ft^3/GPM·10^3	-	+30	+47	+31	+25

Sources: (1), (2) and (3)

Table 3 1984 vs. 1975 individual model comparisons

(Ford U.S. 49 State Sedans)

		1975	1984	% Change, 1984 B/(W) than 1975
I.	Comparable 6-Passenger Models	Ford LTD 5.8L, A3	LTD Crown Vic. 5.0L, A4OD	
	- City MPG	11	17	55%
	- Highway MPG	16	27	69
	- Inertia Weight, lb	5000	4250	15
	- Interior Volume, ft^3	106	111	5
	- 0-60 MPH, sec	16.3	13.8	15
II.	Lowest Fuel Economy Models	Ford LTD 6.6L, A3	LTD Crown Vic. 5.0L, A4OD	
	- City MPG	9	17	89%
	- Highway MPG	14	27	93
	- Inertia Weight, lb	5500	4250	23
	- Interior Volume, ft^3	106	111	5
	- 0-60 MPH, sec	13.6	13.8	(1.5)
III.	Highest Fuel Economy Models	Pinto 2.3L, M4	Escort/Lynx 1.6L, M4OD	
	Gasoline Engine			
	- City MPG	19	37	95%
	- Highway MPG	28	56	100
	- Inertia Weight, lb	3000	2375	21
	- Interior Volume, ft^3	74	85	15
	- 0-60 MPH, sec	14.8	14.3	3.4
	Diesel Engine	None	2.0L, M5OD	
	- City MPG		46	142%
	- Highway MPG		68	143
	- Inertia Weight, lb		2625	12
	- 0-60 MPH, sec		14.5	2.0

Table 4 Average U.S. Ford passenger car dry materials weights, by model year

	Average Weights, Lb			% Change, 1985 O/(U) 1975
	1975	1980	1985	
Plastic	157	227	223	48%
Aluminum	85	140	141	66%
Steel - HSS	105	228	264	151%
Sub-Total Lightweight Materials	347	595	638	84%
Steel-Non HSS	2290	1643	1375	(40%)
Cast Iron	630	472	355	(44%)
Glass	90	79	70	(22%)
Rubber	180	145	127	(29%)
Sound Deadener	85	59	63	(26%)
Copper/Brass	35	35	25	(29%)
Lead	30	25	27	(10%)
Zinc	35	12	11	(69%)
Minor Materials	65	55	53	(18%)
Sub-Total Other Materials	3440	2525	2106	(39%)
Average Car Total Dry Materials Weight	3787	3120	2744	(28%)

Table 5 The diminishing value of incremental gains in fuel economy

		Annual Fuel Costs and Savings, $					
		$1.30/gal		$1.60/gal		$2.25/gal	
MPG/(L/100km)	GAL	Cost	Saving	Cost	Saving	Cost	Saving
10/(23.5)	1200	$1560	–	$1920	–	$2700	–
20/(11.8)	600	780	$780	960	$960	1350	$1350
30/(7.8)	400	520	260	640	320	900	450
40/(5.9)	300	390	130	480	160	675	225
50/(4.7)	240	312	78	384	96	540	135
60/(3.9)	200	260	52	320	64	450	90
70/(3.4)	171	222	38	274	56	385	65
80/(2.9)	150	195	27	240	34	338	47

Assumes 12,000 miles per year

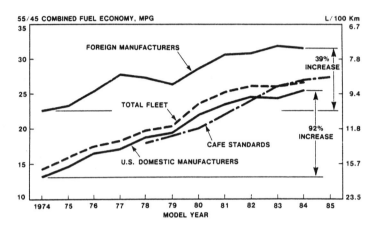

Fig 1 Passenger car fuel economy, by model year

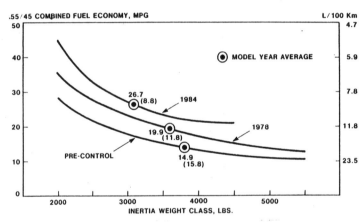

Fig 2 Sales-weighted fuel economy by weight class for pre-control, 1978 and 1984 model passenger cars

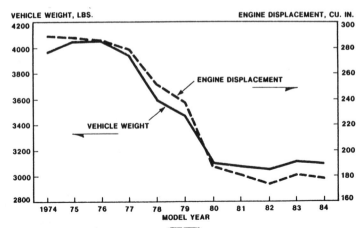

Fig 3 Fleet average vehicle weight and engine displacement by model year

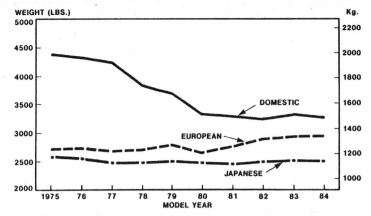

Fig 4   Weight of domestic, European and Japanese cars

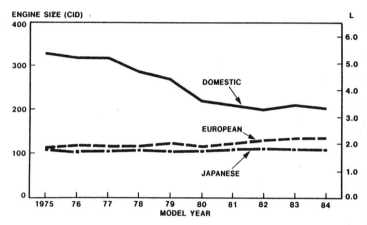

Fig 5   Engine size of domestic, European and Japanese cars

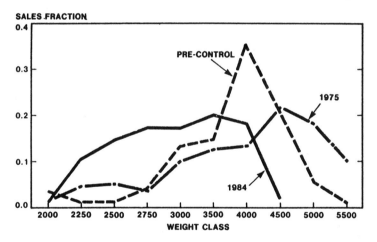

Fig 6   Sales fraction by weight class for three model years

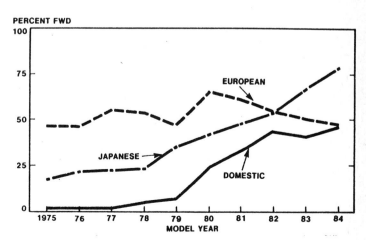

Fig 7   Front wheel drive usage: domestic, European and Japanese cars

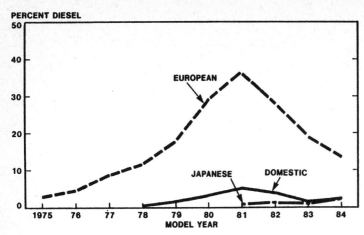

Fig 8 Diesel engine usage: domestic, European and Japanese cars

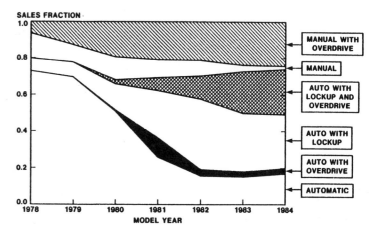

Fig 9 Sales fraction by transmission type

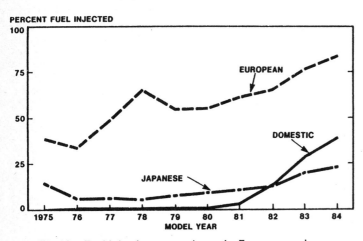

Fig 10 Fuel injection usage: domestic, European and Japanese cars

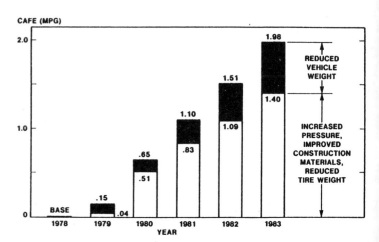

Fig 11 Tyre contribution to Ford CAFE improvement

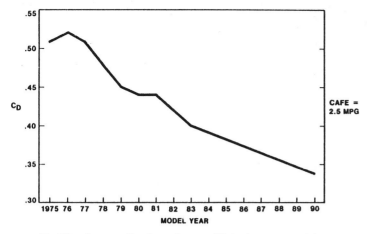

Fig 12 Average Ford car drag coefficient versus model year

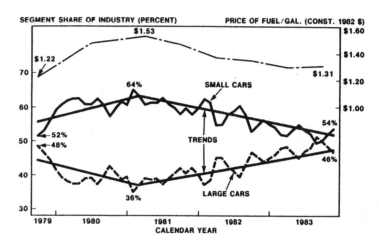

Fig 13 US market segmentation trends

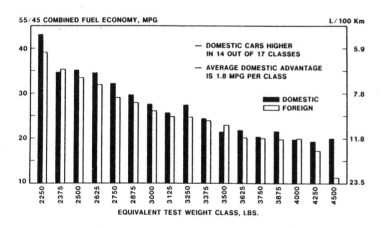

Fig 14 1984 domestic versus foreign average passenger car fuel economy, by weight class

Fig 15 British Ford 'Super MPG' vehicle

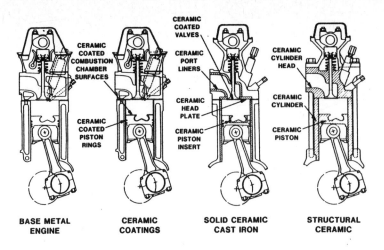

Fig 16   Adiabatic diesel engine design concepts

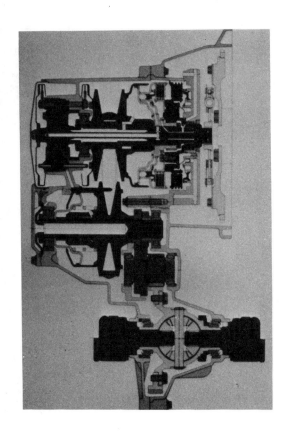

Fig 17   CVT cross-section

Fig 18   Ford 'Probe IV' aerodynamic concept vehicle

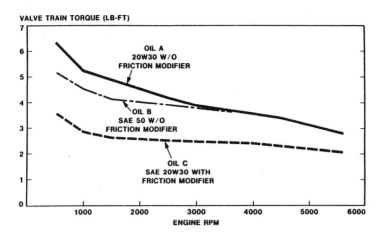

Fig 19 The effect of oil viscosity and friction modifier on friction torque loss

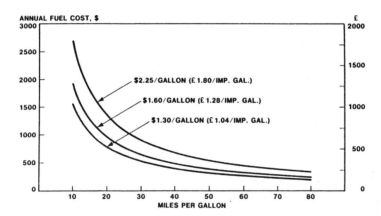

Fig 20 Annual fuel cost as a function of fuel economy and fuel price (12 000 miles/year)

Design of a high-efficiency, high-BMEP, small car powerplant

M A PULICK, BSME, MSME, MBA, G F LEYDORF, Jr, BSE(ME), MSME, R C INNES, BSE(ME), MSE(ME), MBA,
R A STEIN, SB, SM and B A S SHANNON, BSC, MMarE
Powertrain Research Laboratory, Ford Motor Company, Dearborn, Michigan, USA

SYNOPSIS An inline, three-cylinder engine was designed to serve as a basis for comparative studies of alternative engine subsystems and features with ultimate application in a research fuel economy concept car. The objective was to provide a fuel-efficient gasoline powerplant with high BMEP, good packageability and a high degree of manufacturing commonality for DI Diesel convertibility. The selected configuration has a square bore/stroke ratio with vertical inline valves offset 5.5 mm from the cylinder centerline.

Two basic chamber configurations were designed, a bowl-in-piston/near-central plug and an open-chamber/dual-plug geometry. Intake ports having high swirl and no swirl were provided for each configuration. High port flow coefficients were achieved, permitting low overall specific head loss in the induction system.

A multi-piece experimental cylinder head structure was conceived, to permit testing of two alternate valvetrains: direct-acting "bucket" tappets and roller-finger followers. This iron/aluminum concept offers an advantageous balance of cost and weight.

For balancing of the primary rocking couple, a balance shaft was provided. A low-friction oil pump design was evaluated and included in the engine program.

Test results to date are summarized, covering induction system optimization, camshaft selection, valvetrain friction studies, and combustion evaluation of the first (bowl-in-piston/swirl port) combustion system alternative. The BIP/helical intake port geometry with optimized, timed sequential fuel injection provided excellent dilute operating capability and thermal efficiency. Unburned hydrocarbon levels were somewhat high.

1. INTRODUCTION

This paper describes an inline, three-cylinder engine which was designed at the Ford Powertrain Research Laboratory (PRL), to serve as a basis for comparative studies of alternative engine subsystems and features. The primary criterion for comparison was fuel economy, which was to be maximized during a dynamometer development/evaluation program.

Further refinement and verification of U.S. Federal test procedure results were carried out after installation in a research fuel economy concept car. The vehicle program was a substantial one, in which a number of drivetrain and vehicle variables were studied, in addition to the effects of engine brake thermal efficiency gains.

The scope of this paper is limited to a discussion of the research engine, focusing on its design, and including some of the program alternatives which, while not all fully evaluated at this time, offer an insight into PRL's view of concepts which merit exploration.

Initial development data is presented for some of the design variants.

2. PROGRAM OBJECTIVES/ASSUMPTIONS

The program objectives and principal assumptions were set forth as follows:

- Include "diesel convertibility" as a cylinder head design requirement. This involves establishing valve centerline locations, with vertical valve orientation, which would serve for a direct-injection diesel cylinder head. These centerlines are then maintained in all spark-ignition engine cylinder head variants.

- Study fuel economy/performance tradeoffs obtained with alternative fast-burn/lean-burn combustion systems.

- Specifically, determine the feasibility of obtaining adequate high-speed performance (37.5 kW/liter), along with excellent low-to-midrange performance and good idle speed/quality; all of this with the constraints/advantages of:

  - fast-burn/lean-burn systems
  - vertical valves/direct-injection diesel convertibility
  - 1.0 bore/stroke ratio
  - tuned induction system and electronic fuel injection (EFI)
  - overhead camshaft (OHC) valve gear.

- Study performance/friction tradeoffs obtained with alternative OHC valve trains.
- Develop a cylinder head configuration which has a favorable balance between weight and cost reduction.
- Provide an engine for application in a concurrent program on a research fuel economy concept car.

3. ADDITIONAL DESIGN REQUIREMENTS

In addition to the features listed above under "Program Objectives/Assumptions," the following design requirements were laid down for this experimental engine:

- Use I-3 configuration, for vehicle "packageability."
- Provide cylinder head flexibility to facilitate evaluation of:
  - alternative combustion chambers/intake ports
  - alternative OHC valve trains
- Provide a family of camshafts for each valve train system, to facilitate evaluation/selection.
- Design the experimental reciprocating/rotating assembly for flexibility to change pistons during combustion chamber evaluations.

  Expedient methods of producing piston variants were to be acceptable, as development of the engine "bottom-end" was not envisioned. Rather, this system was to serve as a reliable, durable base while engineering efforts were concentrated elsewhere.

- Provide a low-friction oil pump for development in the dynamometer program and application in the research fuel economy concept car.
- Provide a balance shaft assembly to cancel out the primary inertia couple inherent in an I-3 engine. This assembly need not be integrated into the cylinder block assembly.

4. BASIC ENGINE PARAMETERS

Based on a theoretical powertrain/vehicle matching study, a displacement of 1.2 liters was selected to meet the research concept car requirements. The frontal area and tread width constraints of a car this size (680 kg curb, 850 kg inertia test weight) were a strong incentive in the selection of an in-line three cylinder arrangement. The I-3 configuration would remove engine overall length and "packageability" as factors in selecting a bore/stroke ratio.

Bore/stroke ratio was then set at approximately 1, with the selection of 80 mm bore and 79.5 mm stroke. This was judged to be the lowest ratio at which adequate high-speed breathing could be achieved, given the constraints on valve arrangement.

Compression ratio was set at 9.3 to 1, for operation on 91-RON unleaded fuel.

5. CYLINDER HEAD ASSEMBLY

Evaluation of the valvetrain, combustion chamber, and intake port variants was facilitated by constructing the cylinder head assembly in two sections:

1. A cast iron structure that contains the combustion chamber, water jacket, ports, and valve assemblies. (For comparative studies of octane requirement and heat rejection, aluminum castings are readily made with minimal pattern changes.)

2. An aluminum structure that contains the camshaft and the valvetrain components. Depending upon the specific valvetrain variant, this structure consists of one or more aluminum pieces designed for die-casting but plaster-cast for this experimental engine series.

This multi-piece construction was conceived as an advantageous balance between low cost and light weight. The use of cast iron for the lower section eliminates the considerable cost of separate valve guides and seat inserts. The use of die-cast aluminum upper housings is an effective weight-control measure, at a modest cost penalty.

Table 1 indicates the cylinder head variants which were made.

Figure 1 shows the direct-acting bucket tappet (DAB) valvetrain, which is carried in a single aluminum casting. Cast iron tappet-bore liners are cast in place to minimize the oil leakage clearance path between the hydraulic tappet and carrier bore wall. Four bearing caps retain the camshaft, each providing a 25mm bearing diameter. The camshaft carrier is retained by the cylinder head bolts and sealing is accomplished with an interposed embossed steel gasket.

Figure 2 shows the roller finger follower (RFF) valvetrain. The hydraulic lash adjusters are housed in a lower cast-aluminum support retained by the head bolts. The camshaft is housed in a cast-aluminum cover with four individual bearing caps which permit camshaft journals of 25mm diameter. This construction allows the camshaft to be offset substantially from the cylinder center line without interference from the cylinder head bolts located below.

As explained in the valvetrain discussion, freedom to independently select camshaft lateral location allows the finger follower ratio to be optimized, with benefits in camshaft lobe design.

6. COMBUSTION CHAMBER

Two alternative combustion chambers were selected: a compact, high-squish Bowl-in-Piston (BIP) chamber and a two-spark-plug open chamber. The BIP chamber, shown in Figure 1, has one near-central spark plug with its gap located 14.5mm from the bore center. In order to achieve this favorable spark plug location, the valves are placed in-line and offset 5.5mm from the bore center line with uni-sided porting ("U flow"). The valves are recessed in the flat-face cylinder head. The piston bowl is centrally located in the piston, with the bowl sides slightly tapered and having a generous radius at the base. The bowl diameter at the piston crown is 56.5mm, providing 50% squish.

The open chamber, shown in Figure 2, maintains the same valve centerline locations as the BIP chamber. The cylinder head casting has increased height, to accommodate the entire combustion chamber. The piston crown is flat. Two broad, angled bosses depress the roof of the chamber locally, and allow the spark plug shell and electrodes to project into the chamber, the plug gaps reaching a midpoint between chamber roof and piston crown.

The adoption of crossflow porting provides space for two spark plugs at opposite sides of the chamber, but constrains the plug gap radial positions to greater distances from the cylinder bore centerline than for single-ignition/"U-flow": 53 and 62 percent of the 40 mm bore radius.

7. PORTING

Two alternative inlet ports were designed and flow-bench tested to provide high and low in-cylinder swirl levels. Air flow and in-cylinder swirl results for the two alternative (helical and straight) ports appear in Figure 3. Design of the complete induction system is discussed in a later section.

8. VALVETRAIN

As stated earlier in this paper, a decision was made to design this engine for the maximum degree of diesel commonality. This necessitated the selection of vertical valves, in order that diesel compression ratios could be reached. After reviewing the various types of valvetrains, a direct-acting bucket tappet (DAB) was selected as the first choice. The DAB is the most rigid of the common valvetrains, and therefore can provide the most aggressive events in terms of valve event flank accelerations, while being the least susceptible to dynamic problems. The rigidity also assures that lower spring loads can be used for a given event. Valvetrain friction tests had also shown that the DAB was the lowest of the non-roller valvetrains for parasitic losses.

Placement of the valves was selected as the best compromise for the valve head diameters (37mm intake and 32mm exhaust), minimum spacing between the valve heads in the bridge area (6.0mm), valve-to-bore clearance (1.5mm), and the closest position of the spark plug to the center of the bore (14.5mm). The valve-to-valve spacing is 40.5mm. For North American operation, hydraulic lash compensation is required, and a 35mm diameter hydraulic DAB was selected as the prime design component.

Four alternative valve events were seen as adequate for engine tests: 232, 240, 248, and 256 degrees. (Valve event is defined here as beginning and ending at the tops of the ramps.) A polynomial design program was used to define the lift capability of each event in accordance with the velocity limit of the 35mm DAB and stress limitations of the camshaft material. The initial valve events and lifts used for both intake and exhaust valves were 232-8.5mm, 240-9.1mm, 248-9.4mm, and 256-9.7mm.

While the DAB was acceptable for the spark ignition and indirect-injection diesel engines, it was inadequate for the direct-injection diesel. One requirement of the diesel application which was significant for the valvetrain was the need to accommodate unit injectors for direct injection. The unit injector requires another lobe on the camshaft. This lobe must be between the intake and the exhaust lobe, due to the valve placement described above and the need to place the injector as close to the center of the bore as possible. These requirements are difficult to meet using the DAB, because the third cam lobe would activate the bucket tappets of the intake and exhaust valves, unless the injector cam base circle diameter is inconveniently small.

An end-pivot rocker arm (finger-follower) valvetrain was selected as the best candidate to accommodate both the required valve motion and the unit injector. Space was available between the valves to put a third rocker arm.

Another important factor, however, was the ability of the finger follower to accept a roller as the cam follower element. The most significant factor in valvetrain friction is the high contact load and coefficient of friction between the cam and the follower. The simplest solution to reducing this area of high friction is to replace the sliding-contact pad with a needle-bearing roller which is a common, highly-developed machine element.

Thus, the roller-finger follower (RFF) valve train has a potential application where engine friction is to be minimized (with an increase in cost). At this point in the program, it was decided to add the roller-finger follower to the spark ignition engine program.

The challenge was to duplicate the valve events of the DAB valvetrain without creating a concave flank on the cam lobes. The key issue for the RFF was the rate of flank acceleration (0.019685 mm/deg/deg) utilized in the

DAB valvetrain. According to the equation[1]/ which describes the curvature of the cam surface for a cam with a roller follower, the rocker ratio and the sum of the roller radius and the cam base circle radius are the two factors which offset the acceleration rates that would otherwise lead to concave flanks. For this design, a 2.0:1 rocker ratio, a 20mm base circle radius, and a 12.7mm roller radius were selected. The position of the roller and the base circle radius relative to the rocker arm pivot were carefully balanced while moving the camshaft substantially off the bore center, so that the difference of instantaneous rocker ratios of the opening and closing sides of the cam lobe was not a factor in producing concavity. Thus, all the DAB valve events were achieved for the RFF valvetrain, so that a direct comparison between the friction of the two designs could be made. The effective valvetrain weight of the RFF at the valve was less than the DAB. The initial tests of the RFF were nevertheless conducted with the DAB spring.

Other significant features of the RFF valvetrain design include a camshaft contained in the cover, so that the offset camshaft bearing cap bolts do not interfere with the head bolts, and an oil galley cast in the cover to lubricate the cam journals and act as a spray bar to lubricate the rocker arm roller assemblies.

9. VALVE TRAIN FRICTION

Friction tests of the two valvetrains were conducted on a bench test rig, so that the test conditions could be closely controlled. 93° C oil and water were circulated through the cylinder head. Camshaft speed was closely controlled by an electric motor. Torque was carefully measured by a load cell. The test results displayed in figure 4 show that at 12.5 rev/sec cam speed (25 rev/sec engine), the friction of the RFF was about 25% of the DAB. Other tests showed that the DAB was sensitive to spring load, while the RFF, at the low friction measured, was relatively insensitive because of the high mechanical efficiency of the RFF.

---

[1]/The occurrence of concavity is best explained by examining the equation which describes the cam surface for a translating radius follower:

$$\rho = \frac{[(R_F + R_{BC} + Y)^2 + V^2]^{3/2}}{(R_F + R_{BC} + Y)^2 + 2V^2 - (R_F + R_{BC} + Y)a} - R_F$$

where ρ = radius of curvature of the lobe profile
R_F = follower radius, fixed or roller
R_{BC} = cam base circle radius
Y = follower lift = valve lift ÷ rocker ratio
a = follower acceleration = valve acceleration ÷ rocker ratio

Source: ROTHBART, H. A., CAMS, John Wiley and Sons, Inc., New York, pp.80-84(1956).

Test data obtained on engine dynamometer were in agreement with the bench test rig results. At an engine speed of 25 rev/sec engine FMEP was 10 kPa lower with the RFF valvetrain at both part throttle and WOT. This improvement gradually diminished with increasing engine speed, until at 75 rev/sec engine speed, friction results obtained with the two valvetrains were similar.

10. RECIPROCATING/ROTATING ASSEMBLY

As mentioned above, this assembly was to serve as a reliable, durable base, while engineering efforts were concentrated on the engine "top end."

Accordingly, an aluminum forging was identified which could serve as the basis for pistons for both combustion systems described above under "Combustion Chamber." For the BIP system, the bowl is machined in the piston crown. For the open chamber system, the piston crown is flat, and excess material is machined from the underside of the crown.

Compression height is common at 39 mm. A "sled runner" skirt profile is used, with skirt clearance of 0.08mm at the minimum point. Cam drop (at 90 degrees) is 0.40mm. The lands are elliptical and tapered.

Since pistons would be changed at intervals during the experimental program, a "floating" pin was used, retained by wire snap rings. A forged steel connecting rod from the production Ford "CVH" 1.6 liter engine was used, giving a L/R of 3.31. As this rod is normally used with a shrunk-in pin, no bushing is provided. The floating pin is hard-chromed and ground to give 0.12mm chrome thickness, and a lubrication hole is drilled into the rod small end. Pin clearance is 0.02mm.

The crankshaft design was influenced by the decision to provide balancing mass to cancel out the primary inertia couple inherent in an in-line three cylinder engine. This couple is of magnitude

$M = 1.732 \times 10^{-9} W \omega^2 ra = 1073$ Nm at 83.3 rev/sec,

where
W = reciprocating mass weight of one cylinder, 620 grams
ω = crankshaft angular velocity, 523.4 radians/sec
r = one half the engine stroke, 39.75 mm, and
a = cylinder bore center distance, 91.8 mm.

It is a vertical "rocking" couple, and this requires two counter-rotating balance shafts to cancel its effects. Each balance shaft must produce a couple equal to one half the peak magnitude of the inertia couple; and phasing of the shafts must, of course, cause the counterbalancing couples to add vertically and cancel horizontally. The crankshaft can serve as one of the counterbalancing shafts.

The desired crankshaft couple is provided, in this instance, by a system of counterweights which were symmetrical with respect to their crankpins. Part of the counterweights were designed to counteract the rotating mass of the crankpin, arms, and large end of the rod, to reduce high-speed bearing loads. Additional counterweighting is added, equal to one half the reciprocating piston assembly mass, at the crankpin radius, to each pair of counterweights for each cylinder. A couple equal to one half the reciprocating couple is thereby created, but in the opposite direction to the piston couple. A further refinement of the counterweight design would be to add the desired weight only in the direction of the desired couple, to reduce overall crankshaft weight.

The small group of experimental crankshafts for this series of engines was machined from "billets" of steel. Threaded plugs of "Heavy-met" tungsten alloy were used as necessary to obtain the desired unbalance in the space available.

The counter-rotating portion of the counterbalancing system (one half the reciprocating couple) was provided in the form of an external balance shaft assembly. This is bolted to bosses on the left-hand side of the engine, and driven by a 25.4mm width, 9.52mm pitch timing belt.

The balance shaft was machined from steel, with a counterweight radius of 35mm. Antifriction bearings were used, to minimize parasitic losses. After engine installation in the research concept car, a subjective evaluation was carried out as to the value of eliminating the primary couple. The consensus of raters was that the benefits were substantial, and that, for this engine, mounting system and vehicle, the balancing function was essential.

11. OIL PUMP

To improve the friction characteristics of the 1.2L I3 engine, an offset gerotor oil pump was designed to replace the crank-mounted gear segment oil pump. A cross section of the oil pump is shown in Figure 5. Salient design features are:

- 44mm outside diameter of the outer gerotor element, compared with the 98 mm outside diameter of the production 1.6L I-4 engine's internal/external gear pump. This diameter reduction was expected to reduce pump friction, as well as leakage from the pressure side to suction side.

- Internal bypass to improve the suction velocities and to reduce the occurrence of pump cavitation at high-speed conditions.

- Smaller displacement (7.2 cm^3/rev versus 8.9 cm^3/rev for the production 1.6L I-4 pump) to reduce the pump input torque (lower FMEP).

While the oil pump design capacity was reduced 18.5 percent, the pump drive torque (for 275 kPa delivery pressure) was reduced by 33 percent (to 0.71 Nm) at 8.3 rev/sec, improving to 70 percent reduction (to 0.95 Nm) at 83.3 rev/sec.

12. INDUCTION SYSTEM DESIGN

As explained above, the selection of vertical valves and an approximate 1.0 bore/stroke ratio threatened to reduce volumetric efficiency in the upper range of engine speed. The design philosophy for the induction system was then to minimize flow resistance in the system due to all factors other than valve diameter.

Within the cylinder head, this was accomplished by achieving a good port flow coefficient, and hence good air flow, over the range of valve lift. The ports were laid out using empirical methods based on prior experience, and refined using flow models on a steady-state air flow stand. Experimental cylinder heads were verified in this same manner, and those results are shown in Figure 3.

The remainder of the induction system was tailored to produce the desired "ram" effects and volumetric efficiencies during the dynamometer development phase (discussed below). Steady-state flow resistance was held at a low level by providing adequate flow areas through the throttle body and manifold.

The merit of the complete induction system design, judged for its contribution to volumetric efficiency at the higher engine speeds, can be determined from a resistance factor/specific head loss analysis, using parameters proposed by Huber and Brown.[2]

In this analysis, a resistance factor K is determined for each component of the induction system. Then the specific head loss h_s can be determined:

$$h_s = (K_{p,v} + K_m + K_{c,t}) \, Q^2 \; (kPa)$$

where:

$K_{p,v}$ = resistance factor of port and valve at 10.16mm valve lift (kPa/litres^2 per sec^2)

K_m = resistance factor of manifold

$K_{c,t}$ = resistance factor of carburetor or throttle body

Q = displacement of one cylinder at rated engine speed (liters^2/sec^2)

In this analysis,

$$K = \frac{\rho_a}{A^2 \, C^2 \, 2 \, \rho_m}$$

---

2. HUBER, P. and BROWN, J. R., Computation of Instantaneous Air Flow and Volumetric Efficiency, S.A.E. paper no. 812B, January, 1964. See Appendix, particularly equations 4a and 5.

where

A = flow area, based on a selected characteristic diameter

C = measured discharge coefficient, based on the characteristic diameter

ρ_a = density of air

ρ_m = density of mercury

The specific head loss can be used to compare induction systems from engines of differing displacements. Table II shows a comparison of measured values for current-production engines versus the 1.2L engine. (While data for the straight-port variant is listed, this engine has not completed its experimental evaluation and assessment versus the swirl-port variant.)

Data such as these show that the low head losses of the intake manifold and throttle body, combined with the good flow coefficient of the intake ports, have effectively offset the relatively small valve diameters inherent in this cylinder head design. In the case of the swirl port, this was accomplished while achieving a significant level of in-cylinder swirl.

13. PRELIMINARY INDUCTION SYSTEM DEVELOPMENT

An initial dynamometer development task was the optimization of the induction system and camshaft events to obtain a broad, high level BMEP curve. The use of electronic fuel injection allowed the optimization of the intake system within vehicle packaging constraints.

Intake system development was performed using a series of straight pipes and plenums which were easily fabricated and interchanged. The effect of bends on the inlet wave dynamics was investigated subsequently. Pressure versus crank angle profiles measured in an intake runner near the head face showed that bends with radii as small as 50mm for a runner diameter of 35mm had an insignificant effect on the inlet dynamics and on the overall system restriction.

Design parameters investigated included primary runner diameter and length, plenum volume, throttle body flow area, and the diameter and length of the ducting between the throttle body and air cleaner. Vehicle constraints limited the primary runner length to about 45 cm and required the use of an air meter and air cleaner.

14. CAMSHAFT EVALUATION

Following some initial induction system development, the family of camshafts described in the valvetrain section was evaluated. Cylinder pressure versus crank angle data, as well as conventional engine data, were used in the evaluation of the valve events. Analysis of pumping loops is useful in the optimization of the exhaust valve opening point (EVO) and the overlap timing points intake valve opening (IVO) and exhaust valve closing (EVC), and in the determination of the overall PMEP. An example of a pumping loop for the 248° event at 75 rev/sec WOT is shown in Figure 6.

When used in conjunction with tuned inlet manifolding, the 248° event resulted in higher BMEP levels than either the 232° or 256° event camshafts, as shown in Figure 7a. In addition, the EVO timing of the 248° event is a reasonable compromise between high speed exhaust stroke pumping work and low speed expansion work. The 28° overlap is sufficiently low to result in acceptable idle quality and speed, especially when the fast inherent burn rate of the engine is considered. Because of the emphasis on high BMEP and vehicle fuel efficiency, the 248° event was selected for use with the final induction system.

15. FINAL INDUCTION SYSTEM RESULTS

BMEP results at WOT are shown in Figure 7b for three inlet systems for the Bowlin-Piston engine with helical intake porting, the direct-acting bucket valvetrain, and the 248 degree event camshaft. The upper two curves were obtained using the final vehicle intake system, with and without the secondary length which consisted of the air meter, air cleaner, and connecting ducting. Comparison of these curves illustrates the extreme sensitivity of the WOT BMEP of an I-3 engine to the inlet ducting upstream of the throttle body. The lower curve was obtained using only the fuel injector mounting section of the induction system. This was the shortest length that could be used under firing conditions, and represents the case where the effect of wave dynamics was minimized in the engine speed range of interest. Relative to this minimum-tuning case, the full vehicle system resulted in BMEP improvements of 6% to 18%.

For the full vehicle system, the results show 9.46 bar BMEP at 25 rev/sec, maximum BMEP of 10.83 bar at 41.7 rev/sec, and 45.4 kw (approximately 37.8 kw/liter) at 75 rev/sec, with volumetric efficiencies at these speeds of 85%, 93%, and 92%, respectively. The full-load performance of the engine is acceptable, considering the restriction on valve sizes imposed by the design constraints, and the camshaft selection based on the emphasis on low-to-midrange performance and fuel economy.

16. COMBUSTION EVALUATION OF THE BOWL-IN-PISTON SYSTEM, WITH SWIRL PORT

One of the main design objectives was the attainment of a fast burn combustion system which would allow stable dilute engine operation. In order to improve accuracy, part-throttle data for the combustion evaluation were acquired through the use of spark sweeps rather than at subjective MBT settings. Because torque is relatively insensitive to spark timing in the vicinity of MBT, especially at dilute (slow burn rate) conditions, MBT spark timing and therefore the emissions at MBT cannot be exactly determined. Because of this, constant throttle spark sweeps were run at each A/F and EGR rate. Data values were then interpolated from each spark sweep at the spark timing which resulted in a 1% increase in BSFC relative to the minimum BSFC of the sweep.

Part-throttle data obtained in this manner are shown in Figures 8a and 8b for the Bowl-In-Piston engine with helical intake porting, the roller finger follower valvetrain, and optimized timed sequential fuel injection at the speed/load point of 25 rev/sec/2.62 bar BMEP. Two curves are shown in each graph of Figure 8: one curve was acquired at lean air-fuel ratios without EGR, the other was acquired at stoichiometry with EGR. As shown, the engine can be operated at lean air-fuel ratios or high EGR rates to achieve low BSNO levels, without excessively long burn times. Previous test results have indicated that engine stability is related to the initial flame development and approximately correlates with the 0-10% burn time. At a 0-10% burn time of 35 degrees, BSNO levels of about 4 and 2 g/kW hr were achieved by operation at 21:1 A/F without EGR and 14.7:1 A/F with 25% EGR, respectively. As is well known, lower BSNO levels can be achieved at equal stability via stoichiometric-with-EGR operation than by lean-without-EGR operation, but with somewhat higher BSFC.

At 4 g/kW hr BSNO and 21:1 A/F, BSFC was 316 g/kW hr. FMEP was 1.34 bar and PMEP was 0.46 bar; thus mechanical FMEP was 0.88 bar. The dilute capability, efficiency, and friction of the engine compare favorably with current production engines.

BSHC levels are somewhat higher than current production engines. This is probably due to the relatively high top land/ring crevice volume of the BIP geometry. In addition, the nearly-central spark plug results in increased loading of the land/ring crevice with unburned mixture.

17. CONCLUSIONS

Regarding this engine design (and realizing that the test evaluation of combustion systems is incomplete), the following conclusions can be stated:

An S.I. engine designed with vertical, inline valves and DI diesel convertibility can produce high BMEP over an engine speed range to 75 rev/sec, when the following features are present:

- 1.0 bore/stroke ratio (or greater)
- Tuned-length/volume induction system and port fuel injection
- High average port flow coefficients, permitting low overall specific head loss
- DAB or RFF overhead camshaft valve gear with optimized kinematics and valve lift curve design.

The BIP/helical intake port geometry with optimized, timed sequential fuel injection results in excellent dilute operating capability and thermal efficiency. Unburned hydrocarbon levels are somewhat high. Conclusions regarding the remaining combustion system alternatives must await the completion of the dynamometer evaluation program.

A multi-piece cylinder head construction has been conceived, which is an advantageous balance between low cost and light weight. The use of cast iron for the lower section eliminates the considerable cost of valve guides and seat inserts. The use of die-cast aluminum upper housings is an effective weight-control measure, at a modest cost penalty.

Design technology has been developed which makes feasible the optimization of roller-finger follower valve train kinematics, while controlling parameters which affect manufacturing feasibility.

An offset-camshaft arrangement has been devised, which uses bearings in the camshaft cover to provide design freedom for the kinematic optimization mentioned above.

The roller-finger follower valve train offers a significant reduction in friction, at the possible expense of a limitation on maximum acceleration and area under the valve lift curve.

The offset, geared-gerotor oil pump design offers a significant reduction in engine friction over the alternative "crescent" type pump with its internal/external, large-diameter gears.

ACKNOWLEDGEMENTS

The authors wish to acknowledge the substantial contributions of Messrs. T. R. Stockton, R. D. Moan, and F. J. Morris with their geared-gerotor oil pump design. We would also like to acknowledge the contributions of Messrs. A. O. Simko for the evolution of the engine design concept, M. Schrader and M. Mahoney for the design layout, S. R. Whitter and J. W. Cooper for hardware preparation, Mr. D. W. Roper for dynamometer evaluation, and D. Z. Menzik and C. C. Warren for data analysis.

REFERENCES

(1) ROTHBART, H. A., Cams, John Wiley and Sons, Inc., New York, 1956, 80-84.

(2) HUBER, P. and BROWN, J. R., Computation of Instantaneous Air Flow and Volumetric Efficiency, SAE paper no. 812B, January, 1964, 14.

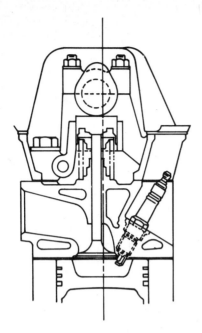

Fig 1   DAB valve train — bowl-in-piston chamber

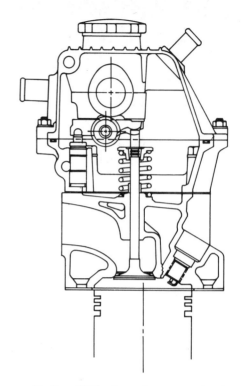

Fig 2   RFF valve train — open chamber

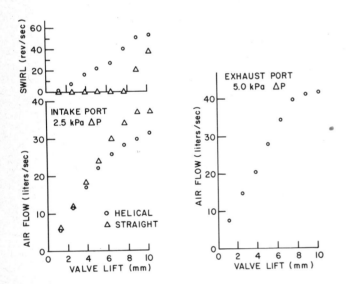

Fig 3   Summary of steady-state air flow and swirl results

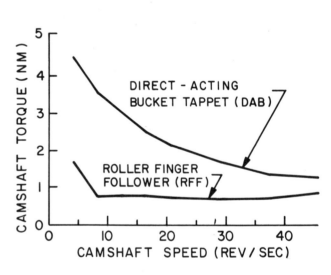

Fig 4   I-3 valvetrain torque

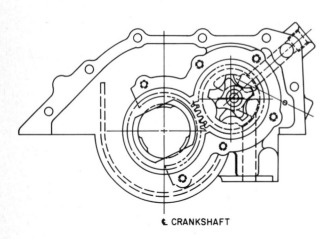

Fig 5   Offset-geared gerotor oil pump

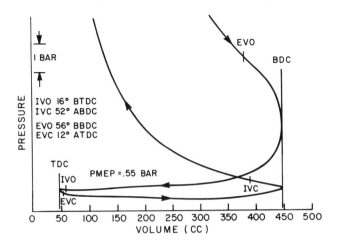

Fig 6   1.2ℓ I-3, BIP, helical intake port: 248° cam, 75 r/s, WOT

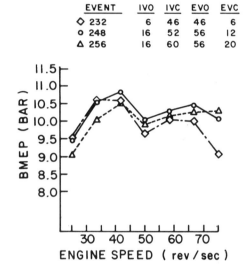

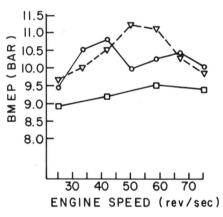

Fig 7   1.2ℓ I-3, BIP, helical intake port. Direct-acting bucket valvetrain — wide open throttle.
13:1 AF
MBT spark timing
Minimum back pressure
Data corrected to 760 mm Hg and 20°C dry air

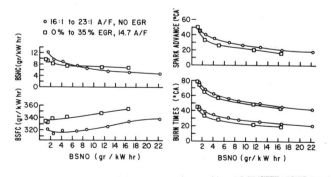

Fig 8   1.2ℓ I-3, BIP, helical intake port. Roller finger follower valvetrain: 25 r/s, 2.62 bar BMEP

C443/84

SAE 841284

Torque characteristics and fuel efficiency of various gasoline engine concepts

D GRUDEN, Dr Techn and H RICHTER, Dr Techn
R & D Propulsion Systems, Dr Ing hcF Porsche AG, Stuttgart, West Germany

1. INTRODUCTION

In recent years development of spark ignition (SI) engines for passenger car application has been focussed on the improvement of fuel economy and the simultaneous reduction of noxious exhaust emissions. The availability of high-octane fuels for many decades has created the necessary basis for increasing the compression ratio of European SI engines by an average of 1,5 to 2 units in the course of about 10 years. Thanks to modern mixture formation and ignition systems the engines could be operated with air-fuel ratios and ignition timings tuned for almost optimum fuel consumption. This development was paralleled by optimisations in all areas of vehicle design, so that at the same time the emission of noxious exhaust components and fuel economy could be clearly improved.

Though it is generally known that better fuel efficiency values and lower exhaust emissions are most easily achieved with small displacement engines installed in small and light-weight vehicles, the share of big powerful engines on the European market has continued to increase. The consumers' desire for more powerful engines is also reflected in the increasing number of exhaust gas turbocharged engines. Between 1973 and 1983, the number of supercharged engines for passenger cars increased from 1 to more than 45.
This trend indicates the main targets of engine development for the future, i.e. the further reduction of fuel consumption and exhaust emissions in engines with high specific power output. It must be kept in mind, however, that as a consequence of the present discussions about European exhaust gas regulations, emphasis in the years to come will have to be placed on the reduction of noxious emissions. In doing so, the potential introduction of unleaded fuels must also be taken into account, which will entail fuel quality changes.

2. DEMANDS ON A PASSENGER CAR ENGINE'S TORQUE RESPONSE

In the last few years experts entered into discussions about which are the optimum torque characteristics for passenger car application.

Would it be better to lay the engine out for maximum torque
- either at relatively low engine speeds
- or at high speeds? (Figure 1).

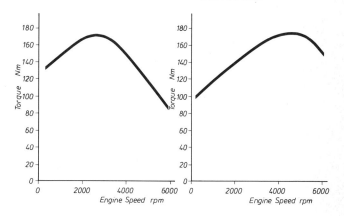

Fig 1   Torque characteristics of the Otto-engine

The criteria used to evaluate the torque characteristic of the engine at wide open throttle are : fuel consumption, exhaust emissions, dynamic vehicle porperties such as top speed and acceleration times, as well as the subjective impression of driveability produced by the car. The present paper reports on investigations performed by Porsche for a couple of years in order to optimize the torque response of SI engines.

3. PARAMETERS INFLUENCING THE TORQUE CHARACTERISTIC

3.1. Compression ratio

The compression ratio is the design parameter which has a decisive influence on all engine data.
Contrary to Diesel engines with their relatively limited compression ratio range there is no generally applicable optimum C.R. value for the SI powerplant. The optimum compression ratio is not only influenced by the combustion chamber shape, cylinder bore, materials used and fuel quality but also by the desired torque characteristic.

As far as fuel economy in the part-load range is concerned, relatively high compression ratios are most favourable (Figure 2).

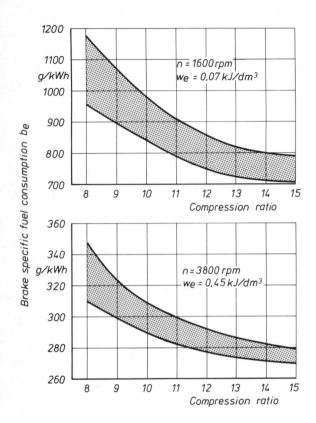

Fig 2   Influence of compression ratio on brake specific fuel consumption

Fuel consumption changes by up to 8 % per compression ratio unit. Under full-load conditions, however, the compression ratio increase is limited by the knocking combustion. Maximum torque and specific work

$$w_e\ [kJ/dm^3] = \frac{\text{mean effective pressure [bar]}}{10}$$

at high engine speeds is obtained with a compression ratio of about 13, provided that premium fuel of RON = 98 be used (Figure 3 right side).
For maximum torque at low engine speed, compression ratios of 9 to 10 should be used (Figure 3, left side).

If the engine is to be operated on lower quality fuel (RON = 91) the compression ratio for maximum specific work decreases by 1 to 2 units. This is true only if the ignition timing is adjusted for optimum power or close to the borderline of detonation respectively.

Sometimes the production tolerances of the engine and its mixture formation and ignition systems make it necessary to adjust the ignition timing distinctly below the knock limit.

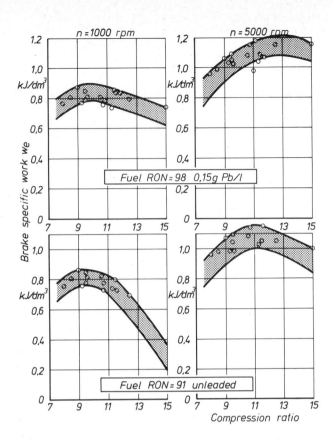

Fig 3   Compression ratio and brake specific work at WOT (ignition timing MBT and BLD respectively)

In such cases, the engine torque and other characteristic data deteriorate clearly, if the compression ratio deviates from its optimum value (Figure 4).

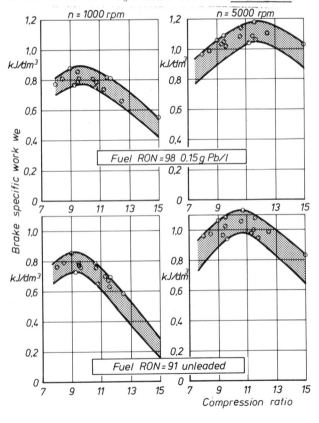

Fig 4   Compression ratio and brake specific work at WOT (ignition timing BLD − 6° CA)

Consequently, engines required to perform in an optimum way when operated with regular fuels, necessitate tightening of the production tolerances, thus causing higher production costs than powerplants designed for operation with premium fuel. The lower the octane number of the fuel, the higher the demands on the production quality of engines and their mixture formation and ignition systems are.

3.2. Combustion chamber shape

The combustion chamber shape is the second design parameter which - next to the compression ratio - has the greatest influence on the brake specific engine data [1]. In order to obtain high compression ratios and an efficient combustion of lean mixtures, a combustion chamber shape with short flame propagation paths and intensive charge turbulences is required. Using a single-cylinder test engine Porsche have examined the influence of different combustion chamber shapes (Figure 5) on engine operation.

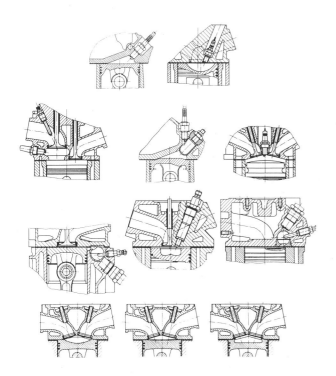

Fig 5 Combustion chamber shapes investigated at Porche's

The investigations cover not only various conventional SI engine combustion chambers but also devided chamber configurations with controlled combustion as well as various Diesel combustion processes. Those variants after having performed well in the single-cylinder test engine, where then installed and tested in multi-cylinder engines. Among the results of this work number the Porsche SKS (stratified charge engine), the 924 TOP engine, as well as combustion chambers for the Porsche 944 and 928 engines [2].

For a long while it has been known from the experience with high-performance sports and racing car engines that 4-valves per cylinder increase the maximum power output, as compared with the 2-valve variants. However, in comparison with normal production engines, these powerplants have been designed for maximum torque and power at extremely high engine speeds. Investigations with 4-valve combustion chambers performed on the single-cylinder research engine were aimed at improving the torque output throughout the speed range of production passenger car engines and at an increase of fuel efficiency in the part-load range. This variant complies best with the various demands on a modern passenger car engine [3,4].
The central spark plug position in the 4-valve engine allows the ideal spherical combustion chamber configuration to be approached. An added advantage, in the light of potential future trends in engine development, is the lower octane requirement of the 4-valve engine, which - as compared with the 2-valve variant with identical compression ratio - is 4 to 6 octane units lower. Figure 6 shows the specific fuel consumption characteristics of the 2- and 4-valve versions of a 4-cylinder SI-engine.

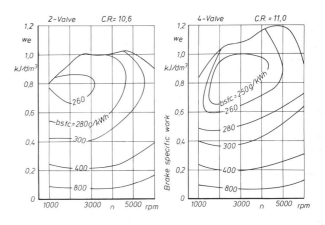

Fig 6 Engine maps of specific fuel consumption

The 4-valve variant returns a better specific fuel consumption throughout the entire operation range with the engine torque and or the specific work being at least 10 % higher at all engine speeds.

The Porsche 944 vehicle with 4-valve engine consumes 8 % less fuel than the 2-valve variant. The increased torque has positive effects on the dynamic vehicle properties. The exhaust gas emissions of the Porsche 944 with 4-valve engine are comparable to those of the 2-valve version.
Both engine variants comply with the ECE 15-04 limits without necessitating any exhaust gas aftertreatment.
For compliance with the US limits a catalyst is also required for the 4-

valve variant.

3.3. Supercharging

The most efficient method to increase the charge density in the cylinder and the torque and power output of combustion engines is to provide for a supercharger. In the area of heavy duty vehicles and their engines with medium and big displacements, the naturally aspirated engine has been almost completely superseded by turbocharged Diesel engines. In 1973, Porsche presented the first passenger car SI engine with exhaust gas turbocharging system, which is still being produced today.

There are several potential supercharging techniques available:

- exhaust gas turbocharging
- mechanical supercharging
- pressure wave supercharging.

Of these three possibilities, turbocharging is used nearly exclusively for passenger car engines. This helps to increase the average specific power output of SI engines by more than 50 % from 38,3 kW/dm³ to 58,9 kW/dm³. Maximum specific work was 1,56 kJ/dm³ for supercharged SI engines as compared with 1,05 kJ/dm³ for the respective naturally aspirated powerplants. Turbocharging of SI- and Diesel-engines does not change the speeds at which the maximum torque and the maximum power output are reached. The drawbacks resulting from the free interaction between the piston engine and the two turbo machines, i.e. the compressor and exhaust turbine, could be improved by thorough turbocharger development and by providing for a boost pressure control at the exhaust end. Nevertheless, today's exhaust-gas turbocharged engines are still characterized by lower torque values at low engine speeds than naturally aspirated engines with the same maximum power output. In addition, the turbocharged engine performs less well under transient operating conditions when acceleration [5,6].

As far as supercharging systems are concerned, these drawbacks can be avoided since the mechanical coupling between the engine and the supercharger also allows a corresponding boost pressure at low engine speeds. This boost pressure is available for acceleration almost immediately [7,8,9,10,11].

Figure 7 shows to which degree specific work or torque can be increased by exhaust gas turbocharging or mechanical supercharging systems respectively. As the compressor efficiency of mechanical superchargers available today is clearly inferior to that of exhaust gas turbochargers, a charge air cooling system is required for mechanical supercharging, in order to realize a clear power increase as compared with the naturally aspirated variant.

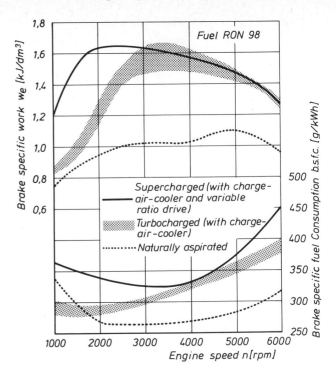

Fig 7 WOT-characteristics of naturally aspirated and pressure-charged SI-engines

With supercharging a distinct torque increase at low engine speeds can be obtained by providing the supercharger drive with a variable transmission ratio. It allows to compensate the boost pressure drop caused by the decrease of the volumetric efficiency in this speed range, by increasing the supercharger speed. At high engine speeds, the boost pressure and the required supercharger drive power can be limited by reducing the transmission ratio.

All hitherto introduced supercharged engines have been derived from corresponding naturally aspirated engines, by making some minor modifications. The question is, whether the optimum combustion chamber shape for naturally aspirated operation is also the optimum solution for a supercharged engine. Therefore, Porsche are performing an investigation on a single-cylinder engine in order to optimize the combustion chamber shape and compression ratio as a function of the type of supercharging system used, the boost ratio, the charge air cooling system, and the fuel quality.

Figure 8 shows the first results of this investigations. It compares the specific work or torque and fuel consumption values of two different combustion chamber shapes at various boost pressures. The bowl in piston combustion chamber returns higher torque values under naturally aspirated and supercharged conditions compared to a chamber in head version, while fuel consumptions are almost identically.

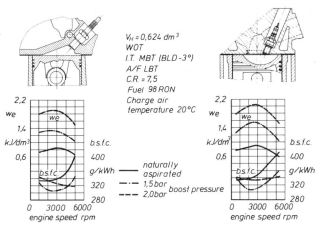

Fig 8   Influence of combustion chamber shape on WOT-characteristics

3.3. Engine cooling

As already mentioned, the attainable torque at wide open throttle is often limited by the occurrence of knocking combustion. The borderline of detonation can be influenced by corresponding combustion chamber design and by combustion chamber surface temperatures. A general reduction of coolant temperature increases the friction losses in the engine. Therefore a separate cooling circuit for the cylinder head has been introduced to maintain high coolant temperatures in the area of the engine block to keep friction losses low. The second cooling circuit is operated at lower temperatures in order to reduce combustion chamber surface temperatures.
Figure 9 shows that the knock limit can be shifted by separate cooling of the cylinder-head and of the intake air.

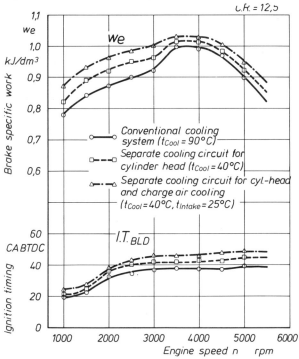

Fig 9   Influence of cooling conditions on borderline of detonation and brake specific work

This improves maximum specific work or torque mainly in the lower speed range.

3.4. Emission limits, fuel quality and torque

The air fuel ratio strongly influences the maximum engine torque, fuel economy, and all exhaust gas emission components (Figure 10).

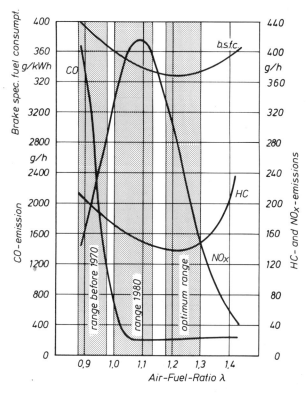

Fig 10   Influence of air-fuel ratio on exhaust emissions and fuel consumption

Until about 15 years ago, passenger car SI engines were operated on rich air-fuel mixtures, resulting in good full-load performances and excellent driveability. Fuel economy and exhaust emissions were of secondary importance at that time. European developments carried out since 1970 led to the step-by-step introduction of leaner air-fuel mixtures, reducing at the same time the fuel consumption and the amounts of carbon monoxide and unburnt hydrocarbons. If the passenger car emission regulations had continued to develop in the same direction, even leaner air-fuel mixtures would have been required in the near future, with corresponding reduction of NO_x-emissions by 40 - 50 %. Based on latest discussions about exhaust emission limits, extremely low values might soon be introduced in Europe. According to the actual state of the art, exhaust gas aftertreatment by means of a three-way catalyst and an oxygen sensor would be absolutely necessary to comply with such limits. This, however, is possible only by using unleaded fuel. Ideally the quality of the unleaded fuel should be the same as today's premium gasoline.

High conversion rates for all of the emission components (CO, HC and NO_x) in the catalyst are obtained only by operating the engine with a stoichiometric air-fuel mixture. Under such conditions the fuel consumption is about 6 % higher than that of a powerplant tuned for optimum fuel efficiency.

The majority of European SI engines having been designed for operation with premium fuel, it will be necessary to use reduced compression ratios if the unleaded fuel has lower quality with the aforementioned effects on fuel economy and torque.

Figure 11 shows the specific fuel consumption of the European and US versions of a modern SI engine.

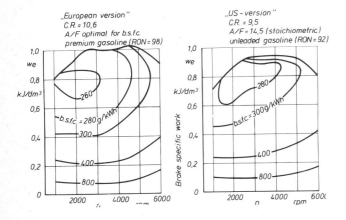

Fig 11 Brake specific fuels consumption of nowaday's Otto engines

Due to the aforementioned measures, the specific fuel consumption of the US variant was 10 % higher throughout the entire operation range. In city traffic, the European version consumes by up to 15 % less. The torque of the European engine is at least 10 % higher in the entire speed range.

4. SUMMARY

In recent years the development of passenger car SI engines has been forcussed on improving fuel economy while reducing the amount of noxious exhaust emissions. These improvements came along with an continuous increase of the mean engine displacement and the mean maximum engine power output.

Among the parameters to be taken into consideration in engine design, the compression ratio has the most decisive influence on the effective engine data. Relatively high compression ratios (C.R. = 13) favour fuel economy in the part-load range and help to realize high engine torques at high speeds. If development is focussed on obtaining maximum torque at low engine speeds, compression ratios of C.R. = 9 to 10 should be used for SI engine applications. By laying the engine out for operation with regular gasoline, the optimum compression ratio decreases by 1 to 2 units. In order to obtain optimum results also with low-octane fuels the production tolerances must be reduced.

Porsche's extensive investigations to optimise the combustion chamber shape have led to the conclusion, that the various requirements on a modern passenger car engine, such as torque characteristic, maximum power output, fuel consumption, exhaust emission and octane requirement are most efficiently met by an engine variant with 4 valves per cylinder.

The specific power output of combustion engines can be distinctly improved by supercharging systems. The method used almost exclusively today on passenger car SI engines is the exhaust gas turbocharging. Mechanical supercharging allows to improve the engine torque at low engine speeds and the vehicle response during acceleration.

Among the operating parameters, the air-fuel ratio has the most decisive influence on the SI engine characteristics. Development during the past 15 years has led to the stepwise introduction of leaner air-fuel mixtures which has had positive influence on exhaust emission and fuel consumption, too. Latest developments might entail the introduction of extremely low emission limits in Europe. According to the actual state of the art, such limits require an exhaust gas aftertreatment by means of a 3-way catalyst and an oxygen sensor. This, however, must come along with the introduction of unleaded fuel.
As the required conversion rates for CO, HC and NO_x can be obtained only with stoichiometric air-fuel mixtures, fuel consumption will increase as compared with the air-fuel ratios tuned for optimum fuel efficiency.

5. REFERENCES

[1] Gruden, D.: Combustion Chamber Layout for Modern Otto Engines. SAE-Paper 811231.

[2] Gruden, D., Richter, H., Wurster, W.: Combustion Chamber Investigations at Porsche - Paving the Way for the 4-Valve Engine; XX FISITA-Congress Vienna 1984.

[3] Downs, D.: The four-valve engine therefore respresents a potentially rewarding area for future development; XIX FISITA Congress, Melbourne 1982.

[4] Gruden, D., Wurster, W.: Entwicklungstendenzen bei Porsche : 4-Ventilmotor als PKW-Antriebsaggregat; 4. Wiener Motoren-Symposium Mai 1982, VDI-Fortschritt-Bericht Reihe 6 Nr. 103

[5] Widhalm, G., Hiereth H.: Besonderheiten und Probleme des Ottomotors mit Abgasturboaufladung. Aufladung von Verbrennungskraftmaschinen, Technische Akademie Esslingen, November 1982.

[6] Mc. Elroy J.: Turbochargers versus Superchargers; Automotive Industries, Nov. 1981.

[7] Walzer, P., Rottenkolber, P.: Supercharging of Passenger Car Diesel Engines; The Institution of Mechanical Engineers C 117/82.

[8] Walzer, P., Geiger, I.: Ottomotor-Antriebskonzept für hohe Leistungen und niedrigen Verbrauch; Automobil-Revue Nr. 50, 9. Dez. 1982.

[9] Collins, D.: Petrol Engine Development Trends; VII European Automotive Symposium (AGELFI) Brussels, October 1983.

[10] Mayr, Y.: Quelques aspects de la suralimentation volumétrique; Ingénieurs de l'automobile 82/4

[11] Lampredi, A.: Suralimentation Volumétrique; Ingénieurs de l'automobile 82/4

C425/84

SAE 841285

Comparison of a turbocharger to a supercharger on a spark ignited engine

T G ADAMS, BSME, MSE, PhD, PE
Ford Motor Company, Dearborn, Michigan, USA

SYNOPSIS — Experiments were conducted to determine the effect of turbine inlet temperature and intake manifold volume on turbocharger/engine response. Then with what is believed to be an optimized turbocharger installation, the performance characteristics of an exhaust turbocharged engine are compared to a mechanically supercharged engine during both steady state and transient operation. Also, an analytical procedure is presented that predicts the steady state, wide open throttle performance improvement of a turbocharged or supercharged engine, with or without an intercooler, over a naturally aspirated engine.

1 INTRODUCTION

The continuing trend toward small displacement passenger car engines has led to an increased use of turbochargers as a means of improving engine performance. Currently available small turbocharged engines have demonstrated equivalent steady speed performance to much larger naturally aspirated engines but have reduced performance during transient operation. This phenomenon, usually referred to as turbo lag, is partially dependent on the manner in which the turbocharger is matched to and installed on the engine. In addition, the turbocharger's rotational speed, and hence its boost, depends on the flow and the temperature of the exhaust gas passing through its turbine. These, in turn, depend on engine speed and load, but not in a linear way. Thus, boost is generated only after the exhaust gas accelerates the turbine and the compressor generates its own column of compressed air.

In contrast to the turbocharger's non-linear spin rate, a supercharger's speed is directly proportional to engine speed at whatever ratio is dictated by the drive mechanism. Although some power is lost turning the gears, belt or chain which form the drive mechanism, the direct drive allows the supercharger to respond immediately to the driver's command. The only delay in getting boost is the time required for compressing the air column between the supercharger's discharge port and the engine's intake port.

If both a turbocharger and a supercharger were to compress an equal column of air, the difference in response time, or turbo lag, would be solely due to the time required for the exhaust gas to accelerate the turbine. Therefore, if the turbocharger installation were optimized the boost characteristics of the turbocharger and hence, the transient performance characteristics of the engine should closely approximate the performance obtained with a supercharger.

The engine tests described in this paper examine the effect of turbine inlet temperature and intake manifold volume on turbocharger/engine response. Then with what is believed to be an optimized turbocharger installation, the performance characteristics of an exhaust turbocharged engine are experimentally compared to a mechanically supercharged engine during both steady state and transient operation. Finally, an analytical procedure is presented that accurately predicts the steady state, wide open throttle performance improvement of a turbocharged or supercharged engine, with or without intercooler, over a naturally aspirated engine. This procedure, when used in a reiterative manner, allows the supercharger drive ratio to be optimized. In addition, the wide open throttle performance characteristics of a turbocharged engine may be compared to a supercharged engine without costly, time-consuming engine tests.

2 EFFECT OF TURBINE INLET TEMPERATURE ON ENGINE/TURBOCHARGER PERFORMANCE

The power of an exhaust turbine is directly related to the turbine inlet temperature; the higher the temperature the more power available from the turbine to drive the turbo compressor. This leads to higher boost pressures at low engine speeds and/or better turbocharger response.

The 1979/80 Ford turbocharged 2.3ℓ engine uses a long pipe to route the exhaust gases from the exhaust manifold to the turbine inlet. This results in a relatively large thermal energy loss. Preliminary calculations indicated that rerouting the cross-over pipe could reduce the length by 60 percent, increase the turbine inlet temperature by 56°C (100°F) and increase the power by about 4 percent. The tests described herein were designed to compare both the steady state and transient engine/turbocharger performance with the production cross-over pipe to the performance with a short cross-over pipe.

The turbocharger selected for these tests was an IHI-RH06A, selected because of the availability of different housings which were to be used in subsequent turbocharged engine performance studies. The wastegate actuator on the turbocharger had been pre-set by IHI and was calibrated to give a maximum boost pressure of approximately 30 kPa (9 in. Hg). The actuator was not readjusted during these tests. The remainder of the engine was production which included a carburetor upstream of the turbocharger and an electronic pressure spark retard system. Both the transient and steady state performance tests were run using the production fuel and spark calibrations.

The transient tests were conducted at 2500 rev/min with the dynamometer in the speed control mode. This speed was selected since it is the lowest engine speed at which the turbocharger was certain to develop full boost. The engine throttle was positioned by an electronically controlled actuator with the throttle "open" and "closed" set points adjusted to the appropriate values. The full throttle set point was programmed to coincide with maximum engine throttle opening while the initial or "closed throttle" position was arbitrarily chosen to give 30 kPa (9 in. Hg) MAP.

The throttle opening rate was controlled via a selector switch on the actuator. For these tests a ramp speed of 1/4 second was used which is the time required for the throttle to move, at a linear rate, from the fully closed to the fully open position. This rate was selected as being typical of the rate at which a driver might actuate the throttle if sudden power and/or acceleration was desired.

The steady state performance tests were full throttle engine as installed tests. Both the steady state and transient engine tests were performed with an exhaust backpressure representative of the 1979/80 turbocharged 2.3ℓ engine.

The results of the steady state and transient response tests are shown in Figures 1 and 2, respectively. Of significance in the steady state tests are the turbine inlet temperatures which were increased about 56°C (100°F), at engine speeds above 2500 rev/min, by the use of the shorter cross-over pipe. The turbocharger speed was marginally higher with the short pipe which accounts for approximately one-half of the manifold absolute pressure (MAP) difference. The rest of the difference was due to a 1.4 kPa (0.4 in. Hg) change in barometric pressure between the time the long and short cross-over pipe were tested. The production (long) pipe was evaluated on the day having the lower barometric pressure. The MAP increase, created by the improved turbocharger performance, resulted in a slight torque improvement.

The effect of improved exhaust energy conservation with the short cross-over pipe is especially noticeable on the transient test results shown in Figure 2. The time required for the turbocharger speed to increase to 60,000 rev/min was 2 seconds with the long pipe and 1-3/8 seconds with the short pipe. The boost, which is directly related to turbo speed, was approximately 17 kPa (5 in. Hg) at this time. Correcting for the differences in atmospheric pressure we find, at this state point, the turbocharger/engine transient response is improved by 31 percent. At higher boost levels the comparative response is improved still further. The steady speed turbocharger/engine performance improvement is about 4 percent.

3 EFFECT OF INTAKE MANIFOLD VOLUME ON ENGINE/TURBOCHARGER RESPONSE

As was previously noted, supercharger lag and some of the turbocharger lag is due to the time required to compress the column of air between the compressor discharge and the engine intake port. The volume occupied by this air column is composed of the tube (or pipe) volume connecting the compressor discharge to the intake manifold, the intake manifold plenum and the intake runners. Obviously, the smaller the total volume the better the response. The question is, how much better? To determine the effect of volume on response it was decided to vary the volume over a wide range and compare the results.

The effect of intake and exhaust manifold volume on the response of a turbocharged 2.3ℓ engine was theoretically calculated. With assumed manifold volumes of both 0.7 and 16.4ℓ (40 and 1000 cu. in.) each and constant engine speeds of 2000 and 3000 rev/min, the transient engine/turbocharger response was studied from an initial condition of 48 kPa (7 psia) manifold absolute pressure (MAP) to a wide open throttle condition by simulating an instantaneous opening of the throttle. The analytical results were as expected in that the response in power, manifold pressure and turbocharger speed is faster with the smaller volume. The computed differences, however, were not as large as initially expected since a 25 fold increase in volume changed the response time by less than 0.1 seconds. Because of these surprising results it was felt that experimental work was required to verify the theoretical study.

Dynamometer engine tests were conducted with an electronic fuel injected (EFI) turbocharged 2.3ℓ engine to experimentally compare the turbocharger/engine response with small and large volume intake manifolds. The turbocharger was an IHI-RH06A. The base or small volume intake manifold had a volume of 2.5ℓ (154 cu. in.) which included the four runners, 1.7ℓ (103.4 cu. in.) and the plenum, 0.8ℓ (50.6 cu. in.). Provisions were made to the intake system to attach an auxiliary plenum having a volume of 16.7ℓ (1020 cu. in.) thereby increasing the total volume to 18.2ℓ (1174 cu. in.). A 1979/80 production ignition module and ignition timing switch assembly were used to provide the auto spark calibration. A simulated vehicle exhaust system produced an exhaust backpressure similar to the 1979/80 turbocharged Ford Mustang.

To reproduce the theoretical conditions the engine tests were conducted with the dynamometer in the speed control mode. Tests began at 1000 rev/min and were increased to 3000 rev/min in 500 rev/min increments. The engine throttle located upstream of the manifold was positioned by an electronically controlled actuator; the throttle ramp was 1/4 second. The first series of tests were run using the base intake manifold. The auxiliary plenum was then added to the intake system and the tests were repeated. The transient data were recorded with a Honeywell 16 channel fibre-optic recorder. Steady state performance tests were also conducted to verify that the additional manifold volume would not affect the steady state results.

The transient test results at 2000 and 3000 rev/min are shown in Figures 3 and 4, respectively. Noteworthy are the engine torque, turbocharger speed, MAP and spark during the first 1/3 second. The rapid throttle opening results in a sudden change in the signal to the EFI system which relies on engine speed and manifold pressure for its control. As a result, the air-fuel ratio goes momentarily lean causing the engine torque to drop and affecting, to a lesser degree, the turbocharger speed. The spark decrease is due to the change in distributor vacuum.

The best indicator of the effect of manifold volume on engine/turbocharger performance is given by the spark signal which is sensitive to boost pressure. At 2000 rev/min (Figure 3) the spark is retarded by the pressure switch at 1-1/8 seconds with the base manifold and at 1-1/4 seconds with the larger manifold. The torque curves are coincident through the first 1-1/8 seconds and become coincident again, following the pressure activated spark retard, at 2 seconds. The turbocharger speed is slightly affected by the large plenum for the first few seconds but reaches a steady speed of approximately 59,000 rev/min, for both cases, at the end of six seconds. The MAP, which lags by approximately 3.4 kPa (1 in. Hg) with the larger manifold volume, reaches an equilibrium value of approximately 115 kPa (34 in. Hg) in 6 seconds.

Intake manifold volume has almost no effect on transient response at 3000 rev/min. The differences that do exist show there is a 0.1 to 0.125 second delay with the large manifold volume. As expected, there were no observable differences in the steady state results.

These results confirm the theoretical results predicted by the transient turbocharger program and show that transient engine/turbocharger performance is virtually unaffected by the plenum volume of an intake manifold. Increasing the volume over 16.4ℓ (1000 cu. in.) resulted in a transient response delay of 0.1 to 0.125 seconds compared to the base manifold. Thus, it is concluded that supercharger/turbocharger lag is only slightly affected by the volume between the compressor discharge and the engine's intake port.

4 COMPARISON OF THE PERFORMANCE CHARACTERISTICS OF A SUPERCHARGED ENGINE TO A TURBOCHARGED ENGINE

Most turbocharger engineers would agree that turbo lag has been a problem because many OEM's have chosen less than the optimum turbo installation. One consideration is the use of either an upstream or downstream carburetor/throttle. Placement is dictated somewhat by the choice of compressor, charger speed at part load and regard for the response behavior of the entire charge assembly. If throttling is done after the compressor, air at approximately atmospheric pressure enters the compressor, is compressed and is then throttled by the throttle valve to the desired pressure before the intake valve. In some cases the desired pressure may be considerably less than atmospheric pressure. At moderate intake manifold pressure this would result in small volume flows, poor turbocharger efficiency and under certain conditions, requirements for compressor operation beyond the surge limit.

If throttling is done before the compressor, a much lower pressure exists at the compressor inlet. Air at low pressure (high specific volume) is compressed to the desired pressure before the intake valve. This takes place with larger volume flows and thus with better compressor efficiency and higher compressor speeds. This improves the response time of the turbocharger when increasing the load and, since the part load supply does not take place at the surge limit, slightly moderates the requirements of compressor flow range. As a pre-condition for an upstream throttle, however, is a turbocharger design which remains oil tight to the compressor housing even with a high vacuum in the compressor housing.

Engine design may also influence the turbocharger installation. For example, with a carbureted engine having a cross-flow head many automakers have used what is called a draw-through installation. Here the carburetor is placed upstream of the turbocharger and the fuel-air mixture is drawn through the turbocharger. As we have just noted an upstream throttle improves performance but positioning the turbocharger on the intake side of the engine increases the length of plumbing required to connect the turbocharger to the exhaust manifold. The overall effect is to increase the underhood temperature and degrade engine/turbocharger performance.

The best installation is to connect the turbocharger directly to the exhaust manifold, put a throttle on the upstream side of the compressor and use a fuel injection system. This type of installation will eliminate much of the plumbing and reduce turbo lag.

To accommodate this type of installation a new exhaust manifold was fabricated for the Ford 2.3ℓ engine. The manifold, of tubular design, carries the exhaust gas from each cylinder directly into the turbine housing. The turbocharger is placed directly on the manifold and the compressor discharge is directed across the top of the valve cover into the intake manifold designed to accommodate port injection. This configuration results in slightly less plumbing than is required for the supercharger installation but the difference in volume, less than 0.16ℓ (10 in^3), is considered insignificant based on the results of the previous tests. Both the turbocharger and supercharger use the same intake manifold; only the tube from the compressor discharge to the engine throttle, located at the manifold inlet, was different.

Dynamometer engine tests were conducted with an electronic fuel injected (EFI) 2.3ℓ engine using a supercharger and a turbocharger. The supercharger was a Bendix 1.8ℓ (110 cu. in.) clutched, positive displacement engine driven device installed so that the electric clutch was engaged whenever the intake manifold vacuum was 10 kPa (3 in. Hg) or less. The turbocharger was an IHI-RH06A, specification number 600 IIP 15NRR39. Both devices were adjusted to provide a maximum boost pressure of approximately 44 kPa (13 in. Hg). The 1979/80 Ford ignition module and ignition timing switch assembly produced for the turbocharged Mustang, were used to provide the auto spark calibration for the transient tests. The steady state tests were conducted using minimum for best torque (MBT) spark. The fuel system was calibrated over the entire operating range to be slightly rich of lean for best torque (LBT). A simulated vehicle exhaust system produced an exhaust backpressure similar to that of the 1979/80 turbocharged Ford Mustang.

Transient tests were conducted with the dynamometer in the speed control mode. Testing began at 1000 rev/min and the engine speed was increased to 3000 rev/min in 500 rev/min increments. The engine throttle was positioned by an electronically controlled actuator. At a given engine speed the throttle closed setting was selected to give an engine output torque of approximately 6.8N-m (5 lb-ft). The throttle was then opened at a pre-set rate by the actuator. The throttle rate used for these tests was 1/4 second which was the time required for the throttle to move from a full closed to the full open position. The transient data were recorded with a Honeywell fibre-optic CRT recorder. "Engine as Installed" steady state performance tests were also conducted using both the supercharger and the turbocharger.

Transient test results at 2000 and 3000 rev/min are shown in Figures 5 and 6, respectively. Noteworthy is the response of the turbocharger and supercharger, as indicated by the MAP, to a rapid throttle opening. The microswitch closure, which energizes the electric clutch of the supercharger was set at 88 kPa, MAP (3 in. Hg vacuum or 26 in. Hg, MAP). From the plot of MAP versus elapsed time it can be seen there is a 0.1-0.2 second delay after the supercharger clutch is energized before a positive boost pressure begins to develop. The delay results from the need to accelerate the supercharger rotor and the filling of the intake manifold volume and connecting pipes. The total time for the supercharger to reach full boost is generally about 0.75 seconds.

With the turbocharger, the MAP shows a quick rise to near ambient pressure as the throttle is opened but a comparatively slow rise in boost pressure thereafter. Full boost at both 2000 and 3000 rev/min is reached in 1.9 seconds; about 1 second longer than is required for the supercharger. At 2000 rev/min the maximum boost pressure is 13 kPa (4 in. Hg) less with the turbocharger than with the supercharger. Equivalent boost pressures were achieved at engine speeds of 3000 rev/min and above.

The comparative torque levels were not as different as might be expected. With an engine speed of 3000 rev/min (Figure 6), the "turbo lag" was 1/2 second or less to reach the same torque level. At 2000 rev/min (Figure 5), the response of the turbocharged engine, as measured by the output torque, was equal to that of the supercharged engine for the first 0.6 seconds. Thereafter, because of the lack of equivalent boost pressure, there was a definite torque lag. This was highlighted by the slow rising boost pressure triggering the ignition timing switch at 1.9 seconds which resulted in 6 degrees of spark retard accompanied by a loss of 13.6 N-m(10 lb. ft.) of torque. The turbocharged engine did not reach the same torque level as the supercharged engine due to the 13 kPa (4 in. Hg) difference in boost pressure and a 2 degree difference in spark timing.

The "Engine as Installed" steady state performance test results are shown in Figure 7. Significant are (1) relative boost levels as a function of engine speed, (2) the increased temperature rise across the supercharger due to compressive pumping and internal friction and (3) the increased exhaust backpressure, measured at the exhaust manifold flange, created by the exhaust turbine. The net result is that the high speed performance of the turbocharged engine with its lower charge air temperature but higher exhaust backpressure is slightly better than the supercharged engine with a higher charge air temperature, lower exhaust backpressure and increased parasitic losses. See Figure 8. The supercharged engine was slightly better at engine speeds below 1500 rev/min due to the higher boost and lower exhaust pressure.

5 ANALYTICAL PROCEDURE

The foregoing results gave rise to speculation that it may be possible to predict the WOT steady state performance of turbocharged and supercharged engines. To test this premise, a format was established listing the knowns and the unknowns. Basic calculations were performed to "fill-in-the-blanks" and the results were compared with experimentally determined values. In spite of, or maybe as a result of, some rather simplistic assumptions the calculated values compare favorably with the experimental values.

The calculations are based on the steady state, WOT performance of the naturally aspirated engine and, where applicable, the power required to drive a supercharger as a function of boost pressure and supercharger speed. It should be noted that the predicted results will be at the same equivalent spark setting as the base engine, i.e., if the naturally aspirated (base) engine data is at KL spark then the predicted results for the turbocharged or supercharged engine will also be at KL spark. The step-by-step calculation procedure is as follows:

1. Determine the brake mean effective pressure (BMEP) and indicated mean effective pressure (IMEP) for the naturally aspirated engine. A reasonable cross section can be obtained by choosing several engine speeds. For the examples shown later three engine speeds were selected, 1500, 3000 and 4500 rev/min.

2. Specify realistic boost pressures and exhaust backpressures for the boosted engine. Generally, both the turbocharged engine and the supercharged engine will reach full boost at 3000 rev/min. If a supercharger is used and driven at a speed significantly faster than engine speed then the engine may reach full boost at a lower engine speed. Turbochargers usually do not produce boost at engine speeds below 1500 rev/min. The exhaust pressures of a turbocharged or supercharged engine may be approximated as follows:

   - Supercharged engine: at 1500 rev/min the exhaust backpressure is approximately equal to that of the naturally aspirated engine; at engine speeds of 3000 rev/min and above, the exhaust pressure is approximately 1.3 times the naturally aspirated engine. Between 1500 and 3000 rev/min the backpressure may be assumed to increase linearly.
   - Turbocharged engine: at 1500 engine rev/min the restriction created by the exhaust turbine creates a backpressure about twice that of the naturally aspirated engine. At 3000 rev/min, the exhaust backpressure is equal to the boost pressure (lbf/in²) +0.5. For engine speeds equal to or greater than 3500 rev/min the exhaust backpressure (EBP), in lbf/in², gauge, may be approximated from the formula

   $$EBP = \frac{(MBP + 3) N}{3000}$$

   where MBP is the maximum boost pressure in lbf/in² and N is the engine speed in rev/min.

3. Calculate the intake manifold density ratio (ρ_2/ρ_1). The initial step in this calculation is to determine the isentropic absolute temperature, T_{2s}, due to the compression by either the turbocharger or supercharger.

   $$T_{2s} = T_1 \left(\frac{\rho_2}{\rho_1}\right)^{\frac{\gamma-1}{\gamma}}$$

   where T_1 = initial temperature, °R = 540°R
   P_1 = initial pressure, in Hg absolute = 29
   P_2 = boost pressure, in Hg absolute
   γ = ratio of specific heats = 1.4

   The intake manifold air temperature (T_2) in °R, may then be calculated from the expression

   $$T_2 = T_1 + \frac{T_{2s} - T_1}{n_c}$$

   where n_c = compressor efficiency = 0.6
   If an intercooler is used the heat exchanger effectiveness (E) must be known, or a reasonable value assumed, in order to calculate the intake manifold air temperature (T_{2ic}).

   $$E = \frac{T_2 - T_{2ic}}{T_2 - T_1}$$

   The manifold density ratio (ρ_2/ρ_1) is then calculated from the ideal gas equation of state, where

   $$\frac{\rho_2}{\rho_1} = \frac{P_2 T_1}{P_1 T_2} \quad \text{without intercooler}$$

   or

   $$\frac{\rho_2}{\rho_1} = \frac{P_2 T_1}{P_1 T_{2ic}} \quad \text{with intercooler}$$

4. Calculate the change in indicated mean effective pressure ΔIMEP (in lbf/in²), due to turbocharging or supercharging.

   $$\Delta IMEP = IMEP_{NA} \left(\frac{\rho_2}{\rho_1}\right) - IMEP_{NA}$$

5. Determine the pumping mean effective pressure, PMEP, (in lbf/in²) of the boosted engine.

   $$PMEP = P_i - P_e$$

   where P_i = boost pressure in lbf/in² and P_e is the exhaust backpressure in lbf/in².

6. If the engine is supercharged, determine the equivalent friction mean effective pressure, $FMEP_{S/C}$ (in lbf/in²), due to the supercharger.

   $$FMEP_{S/C} = \frac{79200 \, (BHP)}{N \, V_d}$$

   where N is the engine speed, rev/min
   V_d is the engine displacement, cu. in.
   BHP is the brake horsepower required to drive the supercharger at the specified speed and boost conditions.

7. The change in friction mean effective pressure, FMEP (in lbf/in²) for the boosted engine is

   $$FMEP = PMEP - FMEP_{S/C}$$

8. The change in brake mean effective pressure, BMEP, (in lbf/in²) of the boosted engine from the base or naturally aspirated engine is

   $$BMEP = IMEP + FMEP.$$

   Note: The FMEP will always be equal to or less than zero.

9. The calculated brake mean effective pressure, $BMEP_{calc}$, of the supercharged or turbocharged engine is

   $$BMEP_{calc} = BMEP_{NA} + BMEP$$

10. The percent increase in BMEP of the boosted engine compared to the base engine is

    $$\text{Percent Gain} = \left(\frac{BMEP_{calc}}{BMEP_{NA}} - 1\right) 100.$$

6 ANALYTICAL RESULTS

This method of predicting steady-state WOT engine performance of a turbocharged or supercharged engine based on naturally aspirated performance data has been verified on several engine/boost system combinations. These are the 1.6ℓ supercharged engine, 2.0ℓ turbocharged engine (with and without intercooler), the 2.3ℓ supercharged engine and the 2.3ℓ turbocharged engine. All of the calculated results have shown excellent agreement with experimental results.

The table of Figure 8 compares the predicted BMEP of the 1.6ℓ supercharged engine with the experimental results at three engine speeds. The percent difference between the experimental and predicted values range from less than 1 percent to approximately 3.5 percent. Figure 8 is also used as an illustrative example to predict the performance of a 1.6ℓ turbocharged engine and to compare these results with the predicted performance of the supercharged engine. At the low engine speeds the higher boost pressure obtained with the supercharger results in better engine performance. As the engine speed increases, the power required to drive the supercharger begins to become significant and the turbocharged engine shows slightly better performance. Above 4500 rev/min, the exhaust backpressure created by the exhaust turbine of the turbocharger becomes dominant and the performance of the turbocharged engine drops below the supercharged engine.

The table of Figure 9 compares the predicted and experimental results of the 2.0ℓ turbocharged engine, with intercooler, at three engine speeds. The difference between the predicted and experimental values range from 0.5 percent to 2.1 percent. Figure 9 also compares the performance of a turbocharged 2.0ℓ engine to a supercharged 2.0ℓ engine, both equipped with an intercooler. The boost pressures were assumed and the exhaust backpressure determined as detailed in the Procedure. The intercooler effectiveness and the power requirements for the supercharger were determined from the manufacturer's specifications. As before, the supercharged engine performs better at the low engine speeds due to the higher boost. At the high engine speeds, the power required to drive the supercharger can be equated to the pumping losses, due to the high exhaust backpressure created by the exhaust turbine, of the turbocharger. These parasitic losses result in equal engine performance for the two systems.

7 CONCLUSIONS

1. Turbocharger/engine response can be improved by exhaust heat conservation. Peak performance will be obtained by close coupling the turbocharger to the exhaust manifold (Figures 1 and 2).
2. Experimental studies with an EFI (port injected) turbocharged engine show that the transient engine/turbocharger response is virtually unaffected by intake manifold plenum volume. Steady state tests are not sensitive to manifold volume (Figures 3 and 4).
3. The turbocharger is not able to produce as much boost as a supercharger at low engine speeds. For the combination of engine speeds and boost levels used in these tests it was found that the "turbo lag" was 1 second or less relative to the supercharged engine. The power required to drive the supercharger is comparable to the "power loss" of the turbocharged engine due to the backpressure increase created by the exhaust turbine. The internal friction and pumping effort of the supercharger results in sufficient energy transfer to the charge air to raise its temperature an average of 10°C (18°F) above that supplied by the turbocharger (Figures 5, 6 and 7).
4. The WOT performance of both turbocharged and supercharged engines, as predicted by the analytical procedure, show excellent agreement with experimental test data. Based on the accuracy of the calculations it is concluded that the analytical procedure may be used to compare the performance of supercharged and turbocharged engines and/or optimize the supercharger drive ratio (Figures 8 and 9).

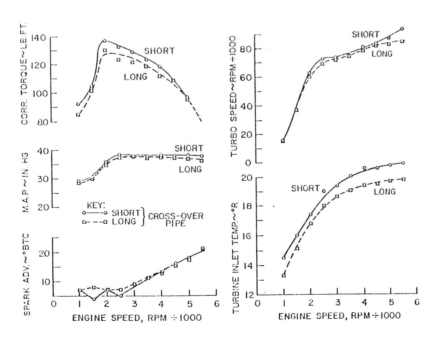

Fig 1   Effect of exhaust cross-over pipe length on steady state engine turbocharger performance

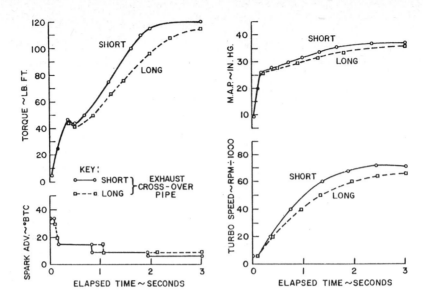

Fig 2 Effect of exhaust cross-over pipe length on transient engine turbocharger performance (2500 rev/min; 1/4 second throttle ramp)

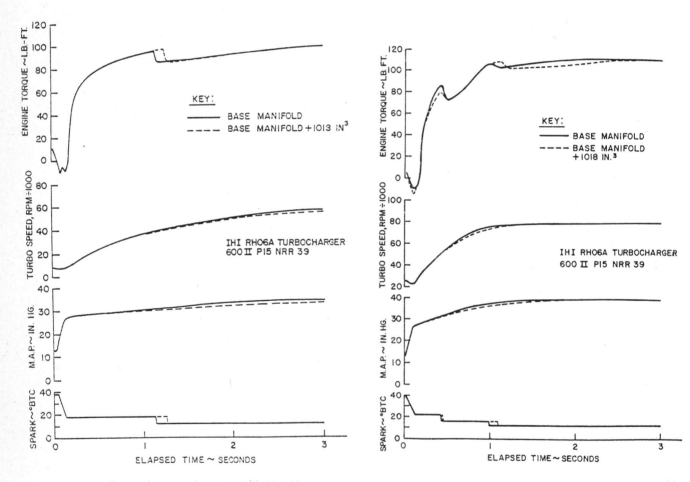

Fig 3 2.3ℓ turbocharger/engine response, EFI, auto fuel/auto spark (2000 rev/min; 1/4 second throttle ramp)

Fig 4 2.3ℓ turbocharger/engine response, EFI, auto fuel/auto spark (3000 rev/min; 1/4 second throttle ramp)

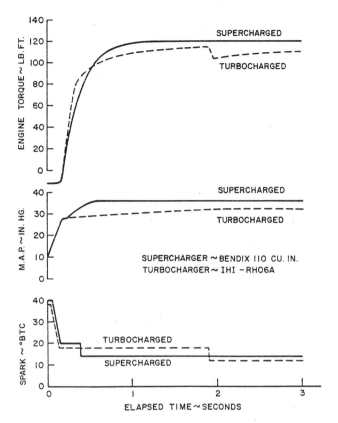

Fig 5 Comparative turbocharged/supercharged 2.3ℓ EFI spark-ignited engine performance (2000 rev/min; WOT transient; 1/4 second throttle ramp)

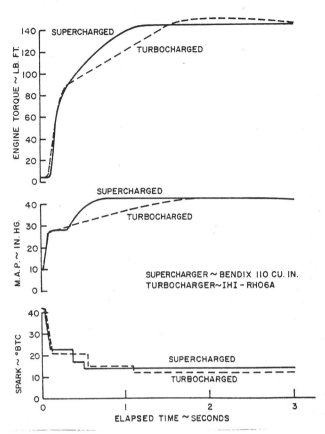

Fig 6 Comparative turbocharged/supercharged 2.3ℓ EFI spark-ignited engine performance (3000 rev/min; WOT transient; 1/4 second throttle ramp)

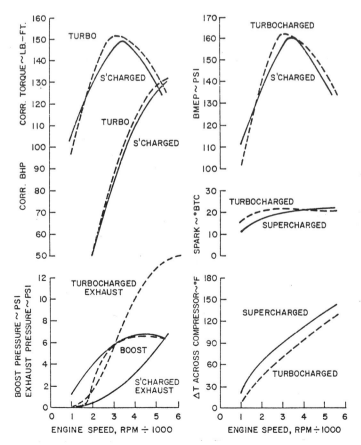

Fig 7 WOT performance; simulated vehicle exhaust, Bendix 1.8ℓ (110 cu. in.) supercharger; IHI-RH06 turbo

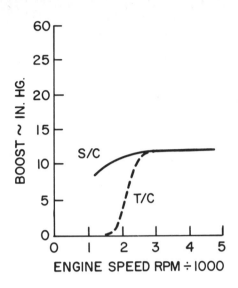
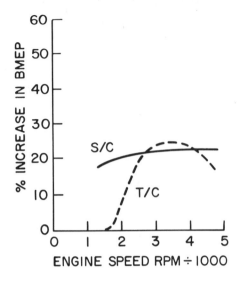

SUPERCHARGED*			
RPM	1500	3000	4500
BMEP N.A.	125	141	139
IMEP N.A.	142	164	172
MANIFOLD DENSITY RATIO	1.169	1.206	1.206
BOOST, PSI	4.9	6.0	6.0
EXHAUST, PSI	0	0	0
PMEP	4.9	6.0	6.0
S/C HP	1.0	3.2	5.7
FMEP S/C	(5.6)	(8.7)	(10.4)
ΔFMEP	(0.7)	(2.7)	(4.4)
ΔIMEP	24.0	33.8	35.3
ΔBMEP	23.3	31.1	30.9
BMEP, CALC.	148.3	172.1	169.9
BMEP, EXP.	146.9	174.7	176.0
% GAIN BMEP CALC.	18.6	22.1	22.2

TURBOCHARGED			
RPM	1500	3000	4500
BMEP N.A.	125	141	139
IMEP N.A.	142	164	172
MANIFOLD DENSITY RATIO	1.0	1.206	1.206
BOOST, PSI	0	6.0	6.0
EXHAUST, PSI	0	6.5	13.5
PMEP	0	(0.5)	(7.5)
ΔFMEP	0	(0.5)	(7.5)
ΔIMEP	0	33.8	35.3
ΔBMEP	0	33.3	27.8
BMEP, CALC.	125	174.3	166.8
% GAIN BMEP CALC.	0	23.6	20.0

* BENDIX 70 C.I.D. @ 1.1:1 SPEED RATIO

Fig 8 Turbocharger/supercharger comparison; 1.6ℓ SI engine

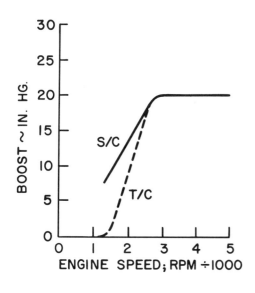
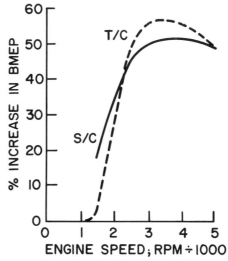

TURBOCHARGED			
RPM	1500	3000	5000
BMEP N.A.	119	129	128
IMEP N.A.	131	152	163
INTERCOOLER EFFECTIVENESS	0.52	0.47	0.41
MANIFOLD DENSITY RATIO	1.027	1.478	1.458
BOOST, PSI	0.4	10.0	10.0
EXHAUST, PSI	0.8	10.5	21.7
PMEP	(0.4)	(0.5)	(11.7)
ΔFMEP	(0.4)	(0.5)	(11.7)
ΔIMEP	3.5	72.7	74.7
ΔBMEP	3.1	72.2	63.0
BMEP, CALC.	122.1	201.2	191
BMEP, EXP.	120.4	205.6	192
% GAIN BMEP CALC.	2.6	56	49.2

SUPERCHARGED*			
RPM	1500	3000	5000
BMEP N.A.	119	129	128
IMEP N.A.	131	152	163
INTERCOOLER EFFECTIVENESS	0.52	0.47	0.41
MANIFOLD DENSITY RATIO	1.204	1.478	1.458
BOOST, PSI	4.0	10.0	10.0
EXHAUST, PSI	0.0	0.1	0.5
PMEP	4.0	9.9	9.5
S/C HP.	1.8	8.35	15.9
FMEP S/C	(7.9)	(18.3)	(20.9)
ΔFMEP	3.9	8.4	11.4
ΔIMEP	26.7	72.7	74.7
ΔBMEP	22.8	64.3	63.3
BMEP, CALC.	141.8	193.3	191.3
% GAIN BMEP CALC.	19.2	49.8	49.5

* BENDIX 90 C.I.D. @ 1:1 SPEED RATIO

Fig 9 Turbocharger/supercharger comparison; 2.0ℓ SI engine

C432/84

SAE 841286

Fuel economy opportunities with an uncooled DI diesel engine

W R WADE, BME, MSME, P H HAVSTAD, BSME, MSME, MBA, E J OUNSTED,
F H TRINKER, BSME, MSME and I J GARWIN, BSME
Ford Motor Company, Dearborn, Michigan, USA

SYNOPSIS An experimental, uncooled, single cylinder DI Diesel engine with ceramic coated cylinder head and valves, a heat insulated steel topped piston and a short, partially stabilized zirconia cylinder liner in the area above the piston rings provided 4 to 7% improvement in fuel consumption at operating conditions typical of the EPA CVS driving cycle for light duty vehicles relative to the baseline water-cooled engine. An engine simulation model was used to explain the measured improvements in fuel consumption. Generally, the uncooled engine provided reductions in HC, NOx and particulate emissions. The trend of lower HC and NOx emissions was explained by the reduction in the amount of premixed combustion resulting from the shortened ignition delay period. The trend of lower particulate emissions was explained by an increase in the diffusion combustion rate. Reduced rates of pressure rise for the uncooled engine were projected to result in lower noise levels.

1 INTRODUCTION

The primary reason for applying the light duty Diesel engine to passenger cars and light trucks is improved fuel economy relative to the gasoline engine. The fuel economy improvement of the Diesel engine relative to the gasoline engine was approximately 29% in 1978 when the popularity of Diesel engines was increasing in the United States. However, this improvement has subsequently declined to approximately 13% in 1984 (1).* This decline has resulted from the introduction of fast burn, low friction gasoline engines with electronic controls. As a result, opportunities to improve the fuel economy of the light duty Diesel engine by reductions in engine friction and by conversion from indirect injection to direct injection combustion systems are receiving widespread attention (1,2,3). Some of these features are already being incorporated in production Diesel engines. The Ford 2.5L DI Diesel, which went into volume production in early 1984 in England for the Transit light truck, was the first high speed, automotive Diesel engine to use direct fuel injection.

The reduction in heat losses from the combustion chamber provides another opportunity to improve the fuel economy of the Diesel engine. In research programs on uncooled heavy duty Diesel engines operating near full load, turbocompounding has generally been applied at the same time heat losses from the combustion chamber were reduced (4,5). Studies by Komatsu and Cummins have projected that a turbocompound, uncooled Diesel engine should provide a 13 to 26% improvement in fuel economy, including the effect of the reduction in cooling fan power, relative to a water-cooled,

---
*Numbers in parenthesis designate references at end of paper.

turbocharged Diesel engine. Uncooled operation alone was projected in the Komatsu study to provide a 5% improvement in fuel economy (4).

The study described in this paper was conducted to determine the benefits of an uncooled, light duty, direct injection (DI) Diesel engine operating at speeds and loads typical for passenger cars and light trucks. For the light duty Diesel engine, turbocompounding was not considered because of marginal improvements in fuel economy at light loads, packaging constraints and cost. In this study, a single cylinder DI Diesel engine, which was designed to operate without cooling, was built and tested. Improvements in measured fuel consumption from this engine relative to the baseline water-cooled engine were analyzed and compared to the improvements in fuel consumption that were estimated for uncooled operation from an engine simulation model. Exhaust emissions of the uncooled and the water-cooled Diesel engines were compared and a detailed analysis of the Diesel combustion process was used to explain the observed changes in emissions.

2 EXPERIMENTAL ENGINE DESIGNS

The baseline, single cylinder engine configuration used in this investigation was an experimental, water-cooled, high speed, direct injection Diesel engine with an 80 mm bore and an 88 mm stroke. A schematic of this engine is shown in Fig 1a. The AVL type of high speed, direct injection combustion system in this engine consisted of a helical intake port, a compression ratio of 21:1, a valve covered orifice (VCO) multi-hole injector and a re-entrant combustion bowl in the piston. The cylinder head was made from high strength, compacted graphite (CG) cast iron. A conventional aluminum, auto-thermatic piston with production type piston rings was used. The engine lubricant was a conventional mineral based oil formulated for use in Diesel engines.

Techniques for reducing heat rejection in an engine are illustrated in Fig 2. For each case shown in the figure, temperature profiles through the combustion gas, cylinder wall and outside environment were calculated. The first case shows the water-cooled baseline where a major temperature drop occurred in the combustion gas as a result of the low combustion chamber surface temperature of the water-cooled wall. Eliminating the water cooling resulted in a 93% reduction in heat losses for the simplified steady state conditions illustrated in the example shown in the figure. For actual transient operation, the reduction in heat losses would be less than that shown for the steady state example. With the elimination of water cooling, the temperature of the cast iron increased from 200°C to 875°C. In this case, the major temperature drop occurred in the ambient air outside the combustion chamber wall. A zirconia coating was considered as a means to provide a thermal barrier to reduce the temperature of the cast iron. However, a 1.0 mm thick zirconia coating provided only a 25°C drop in temperature, even though the thermal conductivity of zirconia is low. As a result, the temperature of the cast iron was not changed significantly from the previous case and the heat loss was practically unchanged. Because of its effectiveness, operation without water cooling was the primary technique used for reducing heat rejection in this experimental investigation.

To evaluate the improvements in fuel consumption that could be obtained with reduced heat rejection, the baseline engine was modified to operate without water coolant circulated through the cylinder block and head. The modifications to this engine are shown in Fig 1b. The compacted graphite cast iron cylinder head incorporated oil cooled valve guides and local oil cooling near the injector tip. A 1.0 mm thick zirconia coating was applied to the cylinder head face and the valve heads. A solid, partially stabilized zirconia insert was added to the cylinder block in the area above the top ring reversal location. An articulated piston featuring a steel top with a reduced heat flow path to the piston rings and an aluminum skirt was used in place of the original aluminum piston (6). Two directed oil jets were used to provide cooling to the aluminum skirt of the articulated piston. The re-entrant combustion bowl was retained in the articulated piston. The combustion systems of the uncooled and water-cooled engines were identical. Coking problems occurred when the conventional mineral based oil was used in the uncooled engine. To resolve this problem, a synthetic ester based oil with higher temperature capability was used during the uncooled engine tests. To obtain the test results reported in this paper, the uncooled engine was operated for nearly 200 hours at conditions ranging from light loads and low speeds up to maximum power at 4500 RPM.

3 FUEL CONSUMPTION RESULTS

The baseline water-cooled and the uncooled DI Diesel engines were tested at 3 speed/load conditions which simulated the operating conditions of an engine over the EPA CVS* cycle for light duty vehicles. The 3 CVS speed/load operating conditions, shown in Table 1, include a low speed-light load condition, a moderate speed-medium load condition and a higher speed-heavy load condition. In the uncooled engine, the higher combustion chamber wall temperatures reduced the volumetric efficiency. To compensate for this reduction, a modest boost pressure was applied to restore the engine air flow to the level of the baseline water-cooled engine. Whenever boost pressure was applied, exhaust back pressure was increased so that the net differential pressure across the engine was equal to that of the baseline water-cooled engine. By matching the air flow to the level of the baseline water-cooled engine, only minor changes in air/fuel ratio occurred between the uncooled and water-cooled engines as a result of changes in specific fuel consumption. With this test procedure, the fuel consumption and emissions of the uncooled and water-cooled engines could be compared at nearly equal air/fuel ratios. To provide a compromise between NOx emissions and fuel consumption, start of combustion timing was set at top dead center (TDC) in both engines.

A comparison of the measured indicated specific fuel consumption (ISFC) data for the uncooled and water-cooled engines is shown in Fig 3. The improvements in fuel consumption of the uncooled engine ranged from 4% at the heavy load condition to 7% at the light load condition.

The ISFC data for the single cylinder engines shown in Fig 3 were determined using conventional techniques. The indicated power associated with the measured fuel flow was obtained by adding the measured power required to motor the engine 30 seconds after the termination of firing to the measured brake power of the firing engine. The ISFC data derived in this manner are useful for making relative comparisons of engine configurations. These ISFC data can also be used to estimate the brake power and associated brake specific fuel consumption (BSFC) of a multi-cylinder engine by subtracting the estimated power required to motor the multi-cylinder engine from the indicated power. However, the indicated mean effective pressure (IMEP) obtained by this technique does not represent the true IMEP that would be determined by integrating the cylinder pressure of the firing engine over volume. This difference occurs because the integral of pressure over volume during the motoring compression and expansion strokes is not zero, particularly for a high compression ratio Diesel engine(7). The ISFC

---

* CVS is used in this paper to refer to the urban driving cycle specified by the Federal emission test procedure. CVS is an abbreviation for the constant volume sampling technique specified by this procedure.

data for the single cylinder engine were not reported on a true IMEP basis because of possible errors in measuring cylinder pressure, particularly in an uncooled engine operating at elevated temperatures.

Reductions in heat rejection increased combustion chamber surface temperatures and exhaust port material temperatures. Chromel-alumel thermocouples were installed in strategic locations of the uncooled engine structure to measure these temperatures. Typically, two thermocouples were located along a heat transfer path so that temperature gradients could be determined. These temperature gradients were subsequently used to estimate surface temperatures. Extrapolated combustion chamber surface temperatures and measured exhaust port material temperatures for the uncooled and water-cooled engines are shown in Fig 4 for the 3 CVS speed/load operating conditions and the rated power condition. The increases in material temperatures resulting from the elimination of the water cooling were largest at the highest engine loads. The highest surface temperatures occurred in the partially stabilized zirconia insert in the area above the piston rings as a result of the low thermal conductivity of this material. The surface temperature of this component increased by 290°C to a maximum level of 425°C. Along the length of the cast iron cylinder, elimination of water-cooling increased surface temperatures by less than 100°C. Increases in the temperature of the cylinder at the half-stroke location were small as a result of the cooling effect of the intake charge and the shielding of this part of the cylinder from the combustion gases when the piston was above this location.

4 FUEL CONSUMPTION PREDICTIONS

4.1 Diesel Engine Model

To explain the experimental test results from the uncooled engine, a Diesel engine simulation model was used to estimate the improvements in fuel consumption that would result from fully insulated, uncooled operation of a light duty, Diesel engine. In this model, commonly termed an "emptying and filling" model, the thermodynamic, heat transfer and fluid flow processes affecting the fuel economy of a Diesel engine were included, as shown in Fig 5. During one engine cycle, the mass and energy transfers across the boundaries of a control volume, which included the volume enclosed by the piston, cylinder head, and cylinder wall, were calculated. The mass transfer was calculated assuming quasi-steady, compressible, isentropic flow between the control volume and infinitely large plenums. The energy transfer was calculated by determining the P·V work, convection and radiation heat transfer, combustion heat release and the energy flow through the valves. The convective heat transfer coefficient was calculated using the Woschni correlation (8). The radiation heat transfer was calculated using the Stephan-Boltzman equation (9, 10). The flame emissivity, which consisted of the sum of the emissivities of the combustion soot and the infrared active gases, was calculated from Watson's correlation which is based on the concentration of the infrared active gases (11). The rate of heat release was calculated from Watson's correlation which describes Diesel combustion with premixed and diffusion combustion terms (12). Since ignition delay was described as a function of temperature, shorter ignition delays and less premixed combustion were predicted for the uncooled engine. The thermodynamic properties of the intake air, combustion and exhaust gases were calculated from curve fits (13, 14).

In order to predict component temperatures of the cylinder head, cylinder, piston, and valves, a wall heat transfer sub-model was developed. A resistor network, shown in Fig 6, was used to determine the heat flow paths in the engine. This network was described by the component geometries, thermal conductivities and heat transfer coefficients. In the engine simulation model, component wall temperatures were initially assumed and the resulting heat transfer rates were subsequently calculated. These heat transfer rates were then used in the wall heat transfer sub-model to calculate a new value of wall temperature resulting from these heat flows. By iteration between the engine simulation model and the wall heat transfer sub-model, wall temperatures which resulted in equal heat transfer rates from the gas to the wall and through the wall could be calculated.

Calibration of the model was performed by adjusting the three coefficients in the Woschni heat transfer correlation for breathing, compression/expansion, and combustion until agreement within 4% was obtained with experimental baseline, water-cooled, single cylinder engine data for volumetric efficiency, motoring IMEP, and ISFC, respectively, over the entire speed and load operating range. These comparisons were based on true values of IMEP that would be determined by integrating cylinder pressure of the firing engine over volume. Experimental values of IMEP, derived from torque measurements on the engine during firing and motoring conditions, were corrected to true values of IMEP by subtracting the integral of pressure over volume during the motoring compression and expansion strokes, as described in Table 2.

4.2 Adiabatic Engine Predictions

The engine simulation model was first used to project the reduction in indicated specific fuel consumption that could be obtained with a fully insulated engine design. Fully insulated, uncooled operation was simulated by elevating the combustion chamber wall temperatures to the average gas temperature to reduce the net cycle heat loss to zero, applying a modest boost pressure to compensate for the reduction in volumetric efficiency in order to maintain equal IMEP levels at equal air/fuel ratios, and reoptimizing the injection timing for minimum fuel consumption. Optimum start of combustion timings for the fully insulated uncooled engine were several degrees advanced relative to the optimum values for the water-cooled engine. The slight advance in start of combustion timing was required to compensate for the longer combustion period of the uncooled engine. The longer combustion period resulted from the reduced amount of fuel burned in the rapid premixed combustion mode and the larger amount of fuel burned in the slower diffusion combustion mode. The projected reductions in ISFC for fully insulated operation varied from 3 to 14% over a range of engine speeds and loads, as shown in Fig 7. Reductions in ISFC varied from 3 to 12% in the operating region for a typical light duty Diesel engine over the CVS cycle, as shown in the figure.

Although the "heat rejected to the coolant" from the combustion gases, which equals about 20% of the fuel energy, would be eliminated in the fully insulated, uncooled engine, only modest reductions in indicated specific fuel consumption were projected. The following fundamental principles, illustrated in Fig 8, control the improvements in thermal efficiency that can be obtained in a fully insulated, uncooled DI Diesel engine:

1. The thermodynamic cycle efficiency limits the conversion of recovered heat to work, as shown in Fig 8a.

2. Part of the heat loss occurs late in the power stroke or exhaust stroke where it cannot be recovered as work without auxiliary devices. As shown in Fig 8b, the heat lost to the coolant, as a percent of the fuel energy has been estimated to be 6% late in the power stroke and 6% in the exhaust stroke. Only 8% of the fuel energy, which is lost to the coolant near TDC, is available for conversion to work.

3. The surface of the cylinder walls cannot follow the instantaneous gas temperature so that the temperature difference between the gas and the wall during combustion near TDC precludes total elimination of heat loss during this period, as shown in Fig 8c. The large instantaneous heat loss near TDC will moderate the improvements in thermal efficiency that can be achieved. Since the overall cycle heat loss can approach zero, the heat lost to the combustion chamber walls near TDC will be added to the air during the induction and compression strokes of the next cycle.

4. Part of the "heat rejected to the coolant" results from engine friction, as shown in Fig 8d, and is not available for conversion to work.

Further insight into the reasons for the projected modest reductions in ISFC was obtained by examining detailed calculations from the engine simulation model. The changes in gas and wall temperatures, heat losses and energy balances for the baseline water-cooled and fully insulated, uncooled engines are shown in Fig 9. Bulk gas temperatures and wall temperatures for the water-cooled and fully insulated, uncooled engines are shown in Fig 9a as functions of crank angle. For the fully insulated, uncooled engine, temperatures are shown for the surfaces of the piston, cylinder head, valves, and liner which are exposed continuously to the gases as well as for the surfaces of the cylinder wall which are exposed only periodically to the gases. The significant observations from Fig 9a are:

1. The higher wall temperatures of the fully insulated, uncooled engine increased the gas temperatures throughout the engine cycle.

2. Since the gas temperatures were higher than the wall temperature during combustion, heat transfer from the gas occurred during combustion and a portion of the expansion stroke, even though the net cycle heat transfer was zero.

3. The high temperature walls of the fully insulated, uncooled engine transferred heat to the gas during the induction stroke and a portion of the compression stroke.

The rate of heat loss, or heat transfer rate, and the cumulative heat loss for the water-cooled and fully insulated, uncooled engines are shown in Figs 9b and 9c respectively. Only modest reductions in the rate of heat loss are shown for the fully insulated, uncooled engine. During the compression and expansion cycles, the cumulative heat transfer was reduced by about 30%, as shown in Fig 9c. Negative heat loss during the intake stroke and part of the compression stroke is shown for the fully insulated, uncooled case in Fig 9b, indicating that the intake charge was being heated by the high temperature walls.

Energy balances for the baseline water-cooled and the fully insulated, uncooled Diesel engines are shown in Fig 9d. The net cycle heat transfer losses were eliminated in the fully insulated, uncooled engine and most of these losses were converted into exhaust energy. About 5% of the heat transfer losses was converted into increased work. The break down of the indicated work into compression and expansion work was included to illustrate the significant effects that intake charge heating has on increasing compression work and that reduced heat rejection has on increasing expansion work. Since the net indicated work available from the cycle is equal to the difference between the expansion work and compression work, the increased compression work resulting from the intake charge heating significantly diminished the increase in net work for an adiabatic engine.

The predicted reductions in ISFC for the fully insulated, uncooled engine were previously shown in Fig 7 to be a function of speed and load. These reductions in ISFC were larger at higher engine speeds and lighter loads with leaner air/fuel ratios. Detailed calculations from the engine simulation model provided insight into the reasons for these trends. The effects of air/fuel ratio on several engine parameters affecting the reduction in ISFC are shown in Fig 10. For the baseline water-cooled engine, the heat losses were nearly constant over the range of loads. However, for the fully insulated, uncooled engine, both gas and wall temperatures increased with increases in load. At higher loads, compression work increased more rapidly than the expansion work so that the increase in net work, or reduction in ISFC, with reduced heat losses was not as large.

The effects of engine speed on several engine parameters affecting the reduction in ISFC are shown in Fig 12. At lower speeds, the heat losses from the water-cooled engine were larger because more time was available for heat transfer. However, in the fully insulated, uncooled case, more intake charge heating occurred at lower speeds when more time was available. Therefore, at low speeds, the compression work increased more rapidly than the expansion work and the increase in net work, or reduction in ISFC, with reduced heat losses was not as large.

4.3 Partially Insulated Engine Predictions

Fuel consumption projections for partially insulated engines were made by using combustion chamber wall temperatures ranging from the level for the water-cooled engine to the level for the fully insulated, uncooled engine. The reductions in heat loss and the corresponding projected reductions in ISFC are shown in Fig 12. For the water-cooled engine operating at low speed and light load, about 15% of the total heat loss was projected to be due to radiation. As the total heat loss was reduced, the radiation heat transfer remained nearly constant. With 100% reduction in heat loss, the convection heat transfer was negative and cancelled the radiation heat transfer. Negative convection heat transfer indicated that the engine structure would be heating the gases in the engine. The projected reductions in ISFC were proportional to the reductions in heat loss and reached a maximum reduction of 5% for the fully insulated configuration at the low speed and light load condition.

The engine simulation model was used to estimate the fuel consumption improvements for the specific uncooled engine design tested. For this case, the cylinder surface temperatures measured in the experimental uncooled engine were used in the model. The piston, cylinder head and valves were initially assumed to be fully insulated. Reduced amounts of insulation with corresponding lesser reductions in heat losses were also analyzed. The projected reductions in ISFC as a function of the reduction in heat loss for this engine configuration are shown in Fig 12. When the engine was fully insulated except for the cylinder, heat losses were reduced by only 60%. However, the reduction in ISFC was approximately 10% for this case whereas only a 5% reduction was obtained for the fully insulated engine. This larger reduction in ISFC resulted from a reduction in the compression work since the intake charge was not heated as much by the lower temperature cylinder walls. The heat losses during combustion near top dead center where most of the combustion chamber was insulated remained low so that the increase in expansion work was nearly the same as for the fully insulated, uncooled engine.

As a result, a larger increase in net work was obtained. At the 1250 RPM light load condition, the 4% reduction in ISFC, which was measured on the experimental uncooled engine, was in the range predicted by the engine simulation model for a 25% reduction in heat loss, as shown in Fig 12.

5 EMISSIONS

5.1 Test Results

The measured emissions from the uncooled and water-cooled single cylinder DI Diesel engines are shown in Fig 13. Also shown in the figure are the manifold pressures required to restore the air flow of the uncooled engine to the level of the water cooled engine. By matching the air flows, nearly constant air/fuel ratios were obtained so that the effect of uncooled engine operation on emissions could be isolated. Generally, the uncooled engine provided reductions in HC, NOx and particulate emissions. At the light and medium loads, HC emissions were 8 to 50% lower relative to the water-cooled engine. Particulate emissions were 38 to 56% lower at the light and medium loads. HC and particulate emissions increased at the heavy load condition, perhaps as a result of non-optimized fuel spray and air swirl patterns for the higher temperature, uncooled operation. NOx emissions were 2 to 25% lower relative to the water-cooled engine.

5.2 Combustion Analysis

A detailed analysis of the combustion process was conducted to explain the changes in emissions that resulted from reductions in heat losses from the single cylinder DI Diesel engine. This analysis was based on the cylinder pressure versus crank angle data. The cylinder pressure data was used to calculate the apparent rate of heat release, defined as the actual rate of heat release due to the combustion of the fuel minus the heat loss from the combustion gases.

A comparison of the apparent rates of heat release, AROHR, for the uncooled and baseline water-cooled DI Diesel engines operated at the three CVS speed/load conditions, shown in Fig 14, illustrate two major changes that occurred in the uncooled engine. First, the ignition delay was reduced in the uncooled engine so that, for the same start of combustion timing, the start of injection timing was retarded. As a result, less fuel was injected during the ignition delay period and less time was available for fuel-air premixing. The second major change that occurred in the uncooled engine was a major reduction in the initial premixed combustion spike which resulted from the reduced ignition delay period.

The effects of the uncooled engine on the Diesel combustion process were further analyzed by comparing the normalized cumulative apparent heat release to the normalized cumulative available heat release calculated from the fuel injection rate and the heating value of the fuel. From this analysis, the following combustion performance parameters, which are described in detail in References 1 and 15, were developed:

Ignition delay, ID, which is defined as the crank angle degrees between the start of injection (available heat release) and the start of combustion (apparent heat release).

Premixed combustion fraction, PCF, which is defined as the fraction of the total apparent heat release which occurs at the end of the premixed combustion mode.

Diffusion combustion index, DCI, which is defined as the ratio of the average apparent rate of heat release in the diffusion mode to the average available rate of heat release, excluding the heat release in the premixed mode.

The three combustion performance parameters for the uncooled and water-cooled DI Diesel engines are shown in Fig 15 as functions of air/fuel ratio. As a result of the intake charge heating and subsequent increase in gas temperatures near the end of the compression stroke at the time of injection, the ignition delays of the uncooled engine were reduced by approximately 50% relative to the water-cooled engine. Likewise, the premixed combustion fraction of the uncooled engine was reduced by approximately 50% relative to the water-cooled engine. Although the actual amount of fuel which burned in the premixed mode remained relatively constant with variations in load, the premixed combustion fraction decreased at richer air/fuel ratios as a result of the larger amount of total fuel injected at the higher loads. Previous studies (16-18) have shown that a significant portion of the HC and NOx emissions in a Diesel engine are related to the amount of fuel associated with the premixed combustion mode. Therefore, the general trend of lower HC and NOx emissions with the uncooled engine, shown in Fig 13, can be explained by the reduction in the premixed combustion fraction for this engine. At light loads, the decrease in NOx production due to the reduction in the amount of premixed combustion was greater than the increase in NOx production due to the higher combustion gas temperatures so that a net reduction in NOx emissions was obtained.

The diffusion combustion index of the uncooled engine was slightly higher relative to the water cooled engine as a result of the higher temperature conditions which promoted more rapid combustion rates. In both engines, the diffusion combustion index decreased at leaner air/fuel ratios where lower temperature conditions tend to reduce combustion rates. A previous study(19) showed that, at a specific air/fuel ratio, particulate emissions tend to decrease when combustion is completed earlier in the expansion stroke. Larger values of the diffusion combustion index indicate that combustion is completed earlier in the expansion stroke. Therefore, the general trend of lower particulate emissions with the uncooled engine, shown in Fig 13, can be explained by the increases in the diffusion combustion index. Higher values of the diffusion combustion index indicate faster combustion rates which tend to result in more complete oxidation of the particulates formed in the early stages of combustion.

6 NOISE

In addition to improving fuel economy and reducing exhaust emissions, the uncooled engine was projected to result in lower noise levels. Noise levels of a Diesel engine are influenced by the maximum rate of pressure rise. A comparison of the maximum rates of pressure rise for the uncooled and water-cooled DI Diesel engine is shown in Fig 16. Significant reductions in the rate of pressure rise were obtained with the uncooled engine. The reduced rates of pressure rise are a direct result of the reduced ignition delay and premixed combustion fractions of the uncooled engine. The reduced rates of pressure rise of the uncooled engine were, therefore, projected to result in lower noise levels.

7 CONCLUSIONS

1. An experimental, uncooled DI Diesel engine with ceramic coated cylinder-head and valves, a heat insulated steel topped piston, and a short, partially stabilized zirconia cylinder liner in the area above the piston rings provided a 4 to 7% improvement in fuel consumption at part load operating conditions typical of the EPA CVS driving cycle for light duty vehicles relative to the baseline water-cooled engine.

2. An engine simulation model projected 3 to 12% improvement in fuel consumption over the speed and load operating region of the CVS cycle for fully insulated, uncooled operation. At the 1250 RPM light load condition where a 5% reduction in fuel consumption was projected for fully insulated, uncooled operation, approximately a 10% reduction in fuel consumption was projected for an engine which was fully insulated everywhere except in the cylinder liner. The additional reduction in fuel consumption with this engine configuration was due to the lower compression work resulting from the lower temperature cylinder walls while the expansion work remained nearly constant because of the insulated combustion chamber near TDC. The 4% reduction in ISFC, which was measured on the experimental uncooled engine at the 1250 RPM light load condition, was in the range predicted by the engine simulation model for a 25% reduction in heat loss.

3. Generally, HC, NOx and particulate emissions were reduced in the uncooled engine. The trend of lower HC and NOx emissions with the uncooled engine was explained by the reduction in the amount of premixed combustion resulting from the shortened ignition delay period. The trend of lower particulate emissions was explained by the increase in the diffusion combustion rate.

4. Reduced rates of pressure rise measured on the uncooled engine were projected to result in lower noise levels.

REFERENCES

(1) W. R. Wade, J. E. White, C. M. Jones, C. E. Hunter and S. P. Hansen, "Combustion, Friction and Fuel Tolerance Improvements for the IDI Diesel Engine," SAE Paper No. 840515, March 1984.

(2) W. R. Wade and C. M. Jones, "Current and Future Light Duty Diesel Engines and Their Fuels," SAE Paper No. 840105, February, 1984.

(3) R. Cichocki and W. Cartellieri, "The Passenger Car Direct Injection Diesel - A Performance and Emission Update," SAE Paper No. 810480, 1981.

(4) K. Toyama, T. Yoshimitsu, T. Nishiyama, T. Shimauchi and T. Nakagaki, "Heat Insulated Turbocompound Engine," SAE Paper 831345, Sept. 1983.

(5) W. Bryzik and R. Kamo, "TACOM/Cummins Adiabatic Engine Program," SAE Paper No. 830314, March 1983.

(6) L. Elsbett and M. Behrens, "Elko's Light Duty DI Diesel Engines with Heat Insulated Combustion System and Component Design," SAE Paper No. 810478, 1981.

(7) D. Lancaster, R. Krieger and J. Lienesch, "Measurement and Analysis of Engine Pressure Data," SAE Paper No. 750026, 1975.

(8) G. Woschni, "A Universally Applicable Equation for the Instantaneous Heat Transfer Coefficient in the Internal Combustion Engine," SAE Paper No. 670931, 1967.

(9) P. Flynn, M. Mizusawa, O. Uyehara and P. Meyers, "An Experimental Determination of the Instantaneous Potential Radiant Heat Transfer Within an Operating Diesel Engine," SAE Paper No. 720022, 1972.

(10) G. Sitkei, *Heat Transfer and Thermal Loading in Internal Combustion Engines*, Akademiai, Kiado, Budapest 1974, p. 47-58.

(11) M. Kamel and N. Watson, "Heat Transfer in the Indirect Injected Diesel Engine," SAE Paper No. 790826, 1979.

(12) N. Watson, M. Marzouk and A. Pilley, "A Combustion Correlation for Diesel Engine Simulation," SAE Paper No. 800029, 1980.

(13) R. Krieger and G. Borman, "The Computation of Apparent Heat Release for Internal Combustion Engines," ASME Paper No. 66-WA/DGP-4, 1966.

(14) N. Watson, "Filling and Emptying Models for System Optimization," Presented at *Turbocharging the Internal Combustion Engine*, Troy, MI, 1982.

(15) W. R. Wade and C. E. Hunter, "Analysis of Combustion Performance of Diesel Fuels," CRC Workshop on Diesel Fuel Combustion Performance, Atlanta Georgia, September 1983.

(16) G. Greeves, I. M. Khan, C. H. T. Wang, I. Fenne, "Origins of Hydrocarbon Emissions from Diesel Engines," SAE Paper No. 770259, 1977.

(17) R. C. Yu, T. W. Kuo, S. M. Shahed, and T. W. Chang, "The Effect of Mixing Rate, End of Injection, and Gas Volume on Hydrocarbon Emissions from a D.I. Diesel Engine," SAE Paper No. 831294, 1983.

(18) V. K. Duggal, T. Priede and I. M. Khan, "A Study of Pollutant Formation Within the Combustion Space of a Diesel Engine," SAE Paper No. 780117, 1978.

(19) H. Hiroyasu, M. Arai, and K. Nakanishi, "Soot Formation and Oxidation in Diesel Engines," SAE Paper 800252, 1980.

TABLE 1

Test Conditions For The Water Cooled and Uncooled Single Cylinder DI Diesel Engines

RPM	Water Cooled Multi Cylinder Reference Conditions BMEP Bar	Single Cylinder Conditions[1] IMEP Bar
1000	1.03	2.8
1250	2.55	4.3
1700	5.17	7.0

[1] IMEP values derived from torque measurements on the engine during firing and motoring conditions

TABLE 2

Derivation of Experimental Values of IMEP

Torque Based IMEP:

$$IMEP_T = BMEP_T + FMEP_M$$

Where:

    T = Torque based measurement
    M = Torque based measurement on a motoring engine

Corrected IMEP:

$$IMEP_C = BMEP_T + FMEP_M - IMEP_M$$

Where:

    C = Corrected values

$IMEP_M = \int PdV$
(Derived from cylinder pressure measurements on a motoring engine)

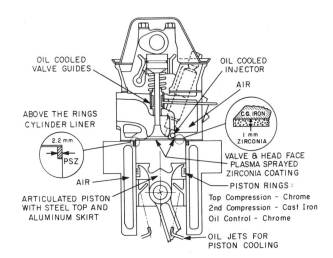

Fig 1b   Experimental uncooled DI diesel engine

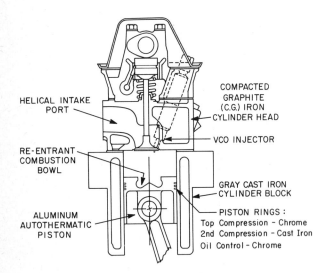

Fig 1a   Experimental baseline water-cooled DI diesel

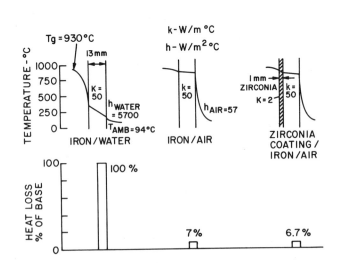

Fig 2   Techniques for reducing heat loss in an engine

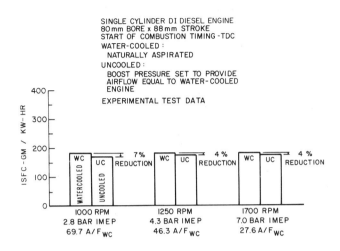

Fig 3 Indicated specific fuel consumption at the 3 CVS speed/load operating conditions for the uncooled and baseline water-cooled DI diesel engines

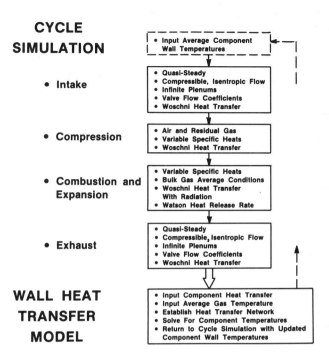

Fig 5 Schematic of the diesel engine simulation model

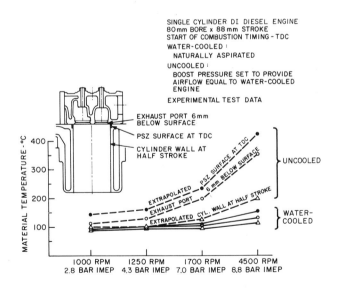

Fig 4 Combustion chamber surface temperatures and exhaust port material temperatures for the uncooled and baseline water-cooled DI diesel engines as a function of operating conditions

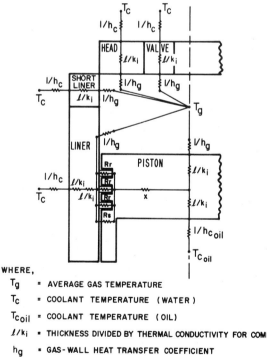

WHERE,

T_g = AVERAGE GAS TEMPERATURE
T_c = COOLANT TEMPERATURE (WATER)
$T_{c_{oil}}$ = COOLANT TEMPERATURE (OIL)
ℓ/k_i = THICKNESS DIVIDED BY THERMAL CONDUCTIVITY FOR COMPONENT, i
h_g = GAS-WALL HEAT TRANSFER COEFFICIENT
h_c = WALL-COOLANT HEAT TRANSFER COEFFICIENT (WATER)
$h_{c_{oil}}$ = WALL-COOLANT HEAT TRANSFER COEFFICIENT (OIL)
R_r, R_s = RING AND SKIRT THERMAL RESISTANCE

Fig 6 Diesel engine wall heat transfer resistor network

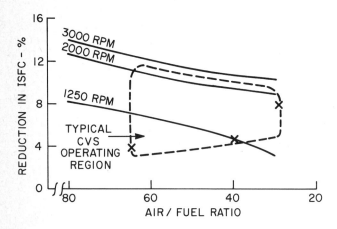

Fig 7 Calculated improvement in indicated specific fuel consumption of a fully insulated DI diesel engine relative to a baseline water-cooled DI diesel engine as a function of air/fuel ratio and engine speed

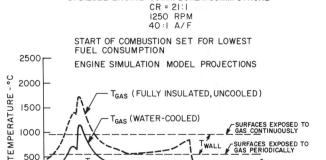

Fig 9a Calculated gas and wall temperatures for the uncooled and baseline water-cooled DI diesel engines

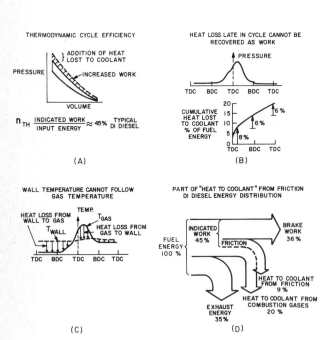

Fig 8 Illustration of the fundamental principles controlling the improvement in thermal efficiency of an adiabatic DI diesel engine

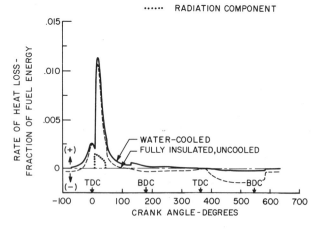

Fig 9b Calculated rates of heat loss for the uncooled and baseline water-cooled DI diesel engines

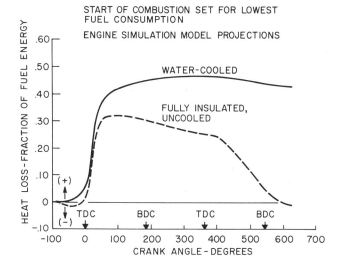

Fig 9c Calculated cumulative heat losses versus crank angle for the uncooled and baseline water-cooled DI diesel engines

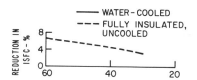
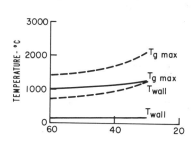
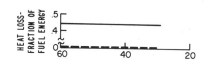
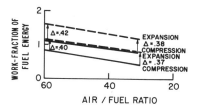

Fig 10 Calculated reduction in ISFC, gas and wall temperatures, heat loss, and compression and expansion work versus air/fuel ratio for the uncooled and baseline water-cooled DI diesel engines

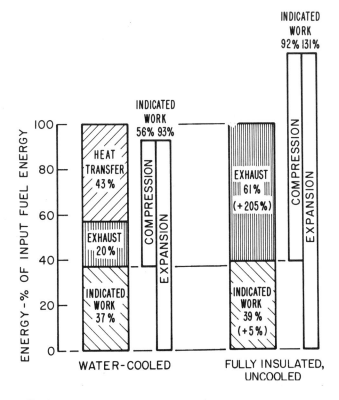

Fig 9d Calculated energy balances for the uncooled and baseline water-cooled DI diesel engines

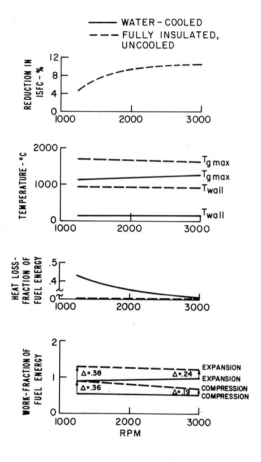

Fig 11 Calculated reduction in ISFC, gas and wall temperatures, heat loss, and compression and expansion work versus engine speed for the uncooled and baseline water-cooled DI diesel engine

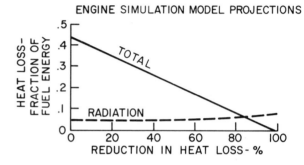

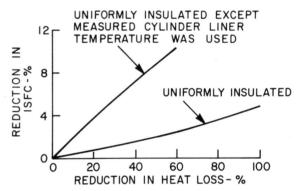

Fig 12 Calculated reductions in indicated specific fuel consumption versus percent reduction in heat loss for a DI diesel engine which is (a) insulated everywhere and (b) insulated everywhere except in the cylinder liner which operated at less than 100°C over the water-cooled operating temperature

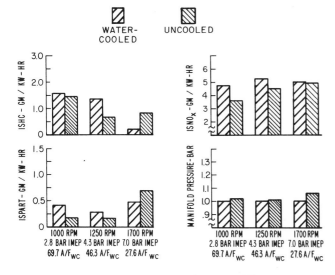

Fig 13 HC, NOx and particulate emissions and manifold pressure for the uncooled and baseline water-cooled DI diesel engines at the 3 CVS speed/load operating conditions

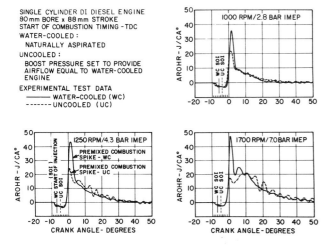

Fig 14 Apparent rate of heat release for the uncooled and baseline water-cooled DI diesel engines at the 3 CVS speed/load operating conditions

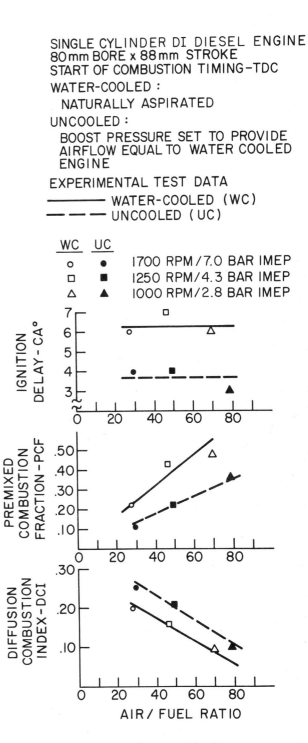

Fig 15 Ignition delay, premixed combustion fraction and diffusion combustion index versus air/fuel ratio for the uncooled and baseline water-cooled DI diesel engines

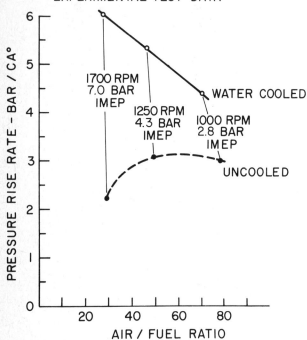

Fig 16 Maximum pressure rise rate as a function of air/fuel ratio for the uncooled and baseline water-cooled DI diesel engines

C440/84

SAE 841287

An appraisal of advanced engine concepts using second law analysis techniques

R J PRIMUS, MSME, BS, PE, K L HOAG, MSME, BSME, P F FLYNN, PhD, ME, BS, MBA, PE and M C BRANDS, BSME, MBA
Cummins Engine Company Inc, Columbus, Indiana, USA

SYNOPSIS The Diesel engine, as an automotive power plant, is assessed in terms of the Second Law of Thermodynamics. The Second Law analysis provides additional insights concerning the distribution of engine thermodynamic losses. The results of these analyses are contrasted with First Law results. Promising concepts for modification of the diesel cycle to improve fuel economy are identified and analyzed in terms of the Second Law. These concepts include turbocharging, charge air cooling, turbocompounding, the implementation of a bottoming cycle, and the use of insulating techniques. The results provide an indication of which processes within the vehicle power plant must be changed to improve fuel economy.

1 INTRODUCTION

An appraisal of the processes that affect vehicle fuel economy must include an analysis of all the processes that occur from the introduction of fuel to the delivery of tractive effort. The processes involved must convert the chemical potential of the fuel into mechanical work for vehicle propulsion. While the impact of mechanical losses in the vehicle drive train can be accurately cataloged with First Law analysis techniques, a similar accounting of process losses within the power plant cannot be obtained without applying Second Law analysis techniques.

Traditional First Law analysis of the thermodynamic processes within the vehicle power plant treats all forms of energy equally, irrespective of the ability of the system to utilize the energy to perform useful work. These First Law analyses yield only minimal insights regarding working fluid thermodynamic state degradations and process changes that might reduce these state degradations.

Second Law evaluation of the processes within the power plant addresses the fact that all forms of a unit of energy are not necessarily converted to the same quantity of useful work. By evaluating the elements of the thermodynamic system in terms of available energy (Second Law), succinct statements about the maximum achievable performance of the vehicle power plant can be made. Second Law comparisons of engine processes allow a quantification of the relative size of the various losses within the power plant. Furthermore, process changes in the system may be easily assessed in terms of their impact upon system losses. This "management of losses" approach to automotive power plant evaluation helps establish priorities to guide engine development for improving vehicle fuel economy.

Availability analysis, performed by direct application of the Second Law, is by no means a new technique (1)*. This type of analysis has been used for many years for evaluating stationary systems (2,3), but few have applied this form of analysis to automotive engines (4,5,6).

The formulation of an approach to Second Law analysis of automotive Diesel engines has been described by Flynn (6). The technique described in Ref. 6 used a control volume approach to formulate a generalized availability balance. The availability balance formulation was then applied to the specific types of thermodynamic control volumes encountered in automotive power plants.

The availability balance formulation, as described by Flynn, will first be reviewed and summarized. The Second Law analysis technique will then be applied to the basic Diesel cycle of a naturally aspirated engine. These results will be compared to a First Law analysis of the same thermodynamic system to compare the perspectives gained from the two analytical procedures. The Second Law analysis will then be used to assess the alterations of available energy distributions resulting from the application of cycle modification techniques. The techniques to be addressed are: (i) turbocharging, (ii) charge air cooling, (iii) turbocompounding, (iv) insulating, and (v) cycle bottoming.

The intent of this study is to present a Second Law evaluation of these thermodynamic cycles to better understand their potential

---
*Numbers in parentheses refer to references listed at the end of this paper.

for improving vehicle fuel consumption when used as automotive power plants. For this analysis, the definitions of thermodynamic system boundaries do not include the mechanical drive system of the vehicle. Although the specific examples used in the analysis are typical of heavy-truck power plants, the conclusions drawn from these calculations provide general assessments of the thermodynamic concepts at design (rated) operation.

2 NOTATIONS

A	availability (cal)
C	empirical constant
dM	change in mass (g)
K	particulate concentration per unit length (1/m)
L	optical length (m)
h	specific enthalpy (cal/g)
LHV	lower heating value (cal/g)
M	mass (g)
P	pressure (Pa)
Q	heat transfer (cal)
q	heat flux (kW/m^2)
S	entropy (cal/K)
s	specific entropy (cal/g/K)
T	temperature (K)
u	specific internal energy (cal/g)
v	specific volume (m^3/g)
W	work (cal)
ε	apparent emissivity
σ	Stephan-Boltzman constant (kW/m^2-K^4)

Subscripts

destroyed	destroyed within the control volume
fuel	associated with liquid fuel
gas	associated with gas mass transfer
heat transfer	associated with heat transfer
in	entering the control volume
o	at reference state
out	leaving the control volume
R	radiant
stored	stored within the control volume
W	wall
work	associated with work

3 BASIS OF SECOND LAW ANALYSIS

The First Law of thermodynamics leads to the concept of energy conservation. It states that energy is conserved in every device or process. Energy may exist in many different forms and may be converted from one form to another within a thermodynamic system, but energy is neither created nor destroyed.

A measure of the potential for energy to perform useful work is availability. The concept of availability is derived from the Second Law of thermodynamics and, unlike energy, availability is a non-conservative quantity. Availability may be destroyed. When destruction occurs the thermodynamic system looses some of its potential to perform useful work.

To properly evaluate energy-use strategies for vehicle power plants both energy and availability must be considered.

3.1 Availability - a review

The availability of a system in a given state is defined as the maximum useful work which can be obtained from the combination of the system and the surrounding atmosphere as the system goes from that state to equilibration with the atmosphere while exchanging heat only with the atmosphere. Availability is a property of the system but its value depends not only on the properties of the system, but also on the properties of the atmosphere.

Calculation of the availability of a given system is accomplished through application of both the First and Second Laws. By constructing an energy equation (First Law) for the system to describe the transition from its initial state to that of equilibration with the atmosphere and introducing the definition for entropy (Second Law), the maximum work of the system (availability) can be quantified. Specific equations for the calculation of availability for various processes are given in the next section.

The impact of inefficiencies in a system or process may be quantified by performing a detailed accounting of the availability at various locations in the system or states during the process. To facilitate the accounting process the concept of an availability balance is introduced.

3.2 The availability balance

A generalized control volume for availability analysis is given in Figure 1. Since availability is not a conserved quantity, a destruction term must be introduced to formulate an expression that maintains equality. Note that the availability destruction has been depicted as a quantity which diminishes the available energy of the control volume. This is not intended to depict a physical process but rather dissipation of available energy within the system. It is presented in this fashion for compatibility with the desired sign convention. The destruction term, defined in this manner, is always a positive value.

Writing the "balance" equation for the control volume produces:

$$\Sigma A_{in} = \Sigma A_{out} + A_{stored} + A_{destroyed} \qquad (1a)$$

Equation (1a) may be used to solve for the destruction term, producing:

$$A_{destroyed} = \Sigma A_{in} - \Sigma A_{out} - A_{stored} \qquad (1b)$$

The transfer and storage terms must be independently evaluated before the destruction term can be quantified.

Availability may be transferred across the boundary by work, heat, or mass transfer. These quantities are analogous to the work, heat, and mass transfers that appear in a First Law balance.

The availability associated with work is equal to the amount of work crossing the boundary. Therefore,

$$dA_{work} = dW \qquad (2)$$

Work crossing a system boundary is by definition available.

The availability associated with heat transfer crossing the system boundary is:

$$dA_{heat\ transfer} = dQ\,(1 - T_0/T) \qquad (3)$$

Where T is the temperature of the system boundary at which heat transfer occurs.

Two types of mass transfer occur in a Diesel engine. There are gases of varying composition, and liquid fuel. The availability associated with gas flow across the boundary is given by:

$$dA_{gas} = dM_{gas}[(h-h_0) - T_0(s-s_0)] \qquad (4)$$

The availability of the liquid fuel entering the system is the Gibb's free energy of combustion at reference conditions (the temperature and pressure entering the system). For liquid fuels of a $C_N H_{2N}$ composition, the relationships of Moran (8) and Rodriguez (10) can be combined to show the thermochemical availability of the fuel to be approximated by a value of 1.0317 times the lower heating value of the fuel (6).

For the direct-injection Diesel engine, after entering the system at reference conditions, the fuel is heated and pressurized before entering the combustion chamber. The heat addition to the fuel results in a very small increase in the availability of the fuel. For an engine at normal operating temperatures the availability increase is approximately .0008 times the lower heating value of the fuel.

The availability of the fuel is further augmented by the PV work supplied during introduction into the combustion chamber. This change in availability is roughly .0014 times the lower heating value of the fuel.

By combining these three factors, the thermomechanical availability of the liquid fuel as it crosses the boundary of the combustion chamber is given by:

$$dA_{fuel} = dM_{fuel}[1.0338 * (LHV_{fuel})] \qquad (5)$$

The storage of availability within a control volume is assessed by:

$$a_c = (u-u_0) - T_0(s-s_0) + P_0(v-v_0) \qquad (6)$$

$$dA_{stored} = d(M_c\,a_c) \qquad (7)$$

It is assumed that the contents of the control volume ae homogeneous and in chemical equilibrium.

The above equations allow the various terms of the availability balance to be evaluated provided values for the work, heat transfer, mass flows, and properties of the control volume are known. These are quantities which can be determined from a conservation of mass and energy (First Law) model of the thermodynamic system.

One property required in an availability balance which is generally not calculated in a First Law model is entropy. In these situations values for entropy must be determined from an external formulation given two other property values and the composition of the working fluid in the control volume. One such formulation, which is based upon the Krieger and Borman (7) equations for calculating working fluid properties, is given in Reference 6.

By substituting equations (2) through (7) into equation (1b) a general availability balance equation for a control volume is obtained. All resulting terms on the right hand side of the equation may be evaluated using values from a First Law model of the control volume and a formulation for entropy. Therefore, the instantaneous availability destruction within the control volume can be calculated. Integration of the availability balance equation over a process allows the availability destruction resulting from the process to be quantified.

For example, the availability destruction resulting from throttling across a valve may be determined by defining control volume boundaries around the valve and its receiving chamber and integrating the availability balance equation from valve opening to valve closing. For an intake valve the receiving chamber is the engine cylinder and for an exhaust valve the receiving chamber is the exhaust manifold. As in a First Law balance, when backflow occurs, care must be taken to insure that the availability term for the gas flow is evaluated at the appropriate conditions.

Detailed discussions of the application of this Second Law analysis approach to intake, compression, combustion and exhaust processes may be found in Reference 6.

4 EVALUATION OF POWER PLANT CONCEPTS

The First Law model used to determine the energy and property values used in the Second Law analysis was based upon the work of Borman (11). Substantial empirical data have been incorporated into this model to establish accurate correlation between simulation and actual engine operation. These empirical data were obtained from detailed in-cylinder measurements taken on turbocharged, aftercooled and turbocompound engines.

The Second Law calculations were produced by a companion program which post-processes the data generated by the First Law model. The model uses the First Law energy and property values in conjunction with an entropy calculation routine to perform an availability balance for each element of the system.

The availability analysis of the power

plants were performed on an indicated basis since mechanical friction can be adequately addressed by First Law calculations.

The calculated values of availability are directly dependent upon the selection of reference dead states (8,9). For the following discussions, a reference dead state of thermal and mechanical equilibrium with the atmosphere was selected. This reference dead state selection provides a single standard of comparison to be used for all thermodynamic systems to be discussed within this paper. Values of 100 kPa for atmospheric pressure and 300 K for atmospheric temperature were used for the reference dead state of each thermodynamic system.

4.1 Naturally aspirated engine

The baseline power plant for this study is an in-line 6 cylinder direct-injection Diesel engine with a 14-liter displacement (140 mm bore x 152 mm stroke). A typical naturally aspirated rating of 185 kW at 2100 rpm was selected for the initial application of Second Law analysis. The pertinent performance parameters for this engine rating are summarized in the Appendix.

The results from the First and Second Law analysis of this naturally aspirated engine are given in Table I. This table shows that 25.1% of the fuel energy leaves the combustion chamber in the form of heat transfer. According to the Second Law balance, the heat transfer accounts for 21.4% of the fuel availability.

The difference between the energy associated with the heat transfer and the availability associated with the heat transfer is best explained by referring to equation (3). As stated earlier, the ambient temperature (T_o) is fixed at 300 K. The control volume used in the availability balance is defined such that the boundaries crossed by the instantaneous heat transfer (dQ) are at the instantaneous bulk gas temperature (T). The relationship between energy and availability presented in equation (3) describes the fact that the amount of availability associated with heat transfer will asymptotically approach the amount of energy transferred as the temperature at the control volume boundaries (bulk gas temperature) approaches infinity.

In realistic, physically constrained systems, the availability associated with heat transfer will always be less than the energy. For the naturally aspirated engine, the instantaneous gas temperatures and heat transfers produce an integrated value for available energy which is about 15% smaller than the heat transfer.

The table shows that 34.6% of the fuel energy (First Law) is contained in the exhaust products. But, only 20.4% of the available energy of the fuel (Second Law) remains in the exhaust products. The ratio of these two quantities shows that only about 59% of the exhaust energy could be accessed by ideal thermodynamic devices. The exhaust gas leaves the system in a high-temperature, ambient-pressure state and, therefore, has high entropy (relative to the P_o, T_o reference state). Referring to equation (4), the entropy difference between the exit and reference conditions significantly reduces the available energy associated with products of combustion that leave the system.

Table I Comparison of First and Second Law analysis for a 14-liter naturally aspirated engine at 185 kW at 2100 rpm

	First Law (% Fuel Energy)	Second Law (% Fuel Availability)
Indicated Work	40.3	39.1*
Combustion Loss	-	15.9
Cyl. Heat Transfer	25.1	21.4
Int. Val. Throttling	-	0.7
Exh. Val. Throttling	-	2.5
Exh. to Ambient	34.6	20.4

*Note that the indicated work for the Second Law balance is a lower percentage than for the First Law. This occurs because the availability of the fuel is 1.0317 times the fuel energy (6).

The quantity referred to as combustion loss in Table I is determined by constructing an availability balance for the combustion chamber and integrating equation 1b over the combustion portion of the cycle. The resulting change in the magnitude of the destruction term represents the deviation in the combustion process from a completely reversible process. The Second Law balance indicates that the magnitude of the irreversibility associated with the combustion and mixing processes is such that 15.9% of the availability associated with the fuel is dissipated. Had the combustion process occurred at a higher air-fuel ratio, the combustion availability dissipation would have been greater due to increased mixing and lower resulting bulk gas temperatures (6). This point will be addressed in a later discussion.

4.2 The turbocharged engine

To study the effect of turbocharging on the engine system, the analysis was performed for a typical turbocharged-engine rating of the baseline 14-liter engine. The selected operating point was 220 kW at 2100 rpm. The pertinent performance parameters for this engine operating point are tabulated in the Appendix.

A comparison of the Second Law results for the 185 kW naturally aspirated and the 220 kW turbocharged ratings is presented in Table II. The comparison of First Law quantities in the Appendix shows that the turbocharged engine has considerably better brake thermal efficiency (39.2%) than its naturally aspirated counterpart (33.9%). By turbocharging, the combined heat transfer and exhaust availabilities are reduced from 41.8% to 31.7% (a difference of 10.1 percentage points) while the combustion and added turbomachinery availability losses increase from 15.9% to 21.4% (a difference of 5.5

percentage points).

Table II Comparison of Second Law analysis of naturally aspirated and turbocharged ratings of a 14-liter Diesel engine (% fuel availability)

	Naturally Aspirated	Turbocharged
Indicated Work	39.1	43.9
Combustion Loss	15.9	19.2
Cyl. Heat Transfer	21.4	17.6
Int. Val. Throttling	0.7	0.7
Exh. Val. Throttling	2.5	2.3
Loss in Compressor	–	1.4
Loss in Turbine	–	0.8
Exh. to Ambient	20.4	14.1

The in-cylinder conditions associated with the combustion of a relatively rich mixture (A/F = 17.9) in the naturally aspirated engine leads to high combustion chamber heat transfer. Also, combustion of fuel late in the expansion stroke leads to high exhaust-gas energy which is, for the naturally aspirated engine, rejected directly into the atmosphere. By improving the time distribution of heat release and expanding the exhaust gases through a turbine, the turbocharged engine displays a significant reduction in heat transfer and exhaust gas available energies relative to fuel input.

It is interesting to note that, while the in-cylinder conditions in the turbocharged case have led to a relative reduction in heat transfer and exhaust availability, the availability loss associated with combustion has increased. This increased combustion availability loss is due to the relatively higher mixing and lower resulting bulk gas temperatures associated with the lean-mixture (air-fuel ratio = 28.0) combustion of the turbocharged engine versus the relatively rich (air-fuel ratio = 17.9) operation of the naturally aspirated engine. The rich mixture maintains high in-cylinder bulk gas temperatures during combustion; therefore, conversion of chemical potential to heat occurs at a relatively high effectiveness. For the turbocharged situation, the high-temperature products of combustion are mixed with the excess air, causing the fuel conversion to produce lower bulk temperatures. Therefore, higher combustion availability losses are associated with higher air-fuel ratios (6).

A slight reduction is observed in the exhaust valve throttling loss with the addition of the turbocharger. Even though the mass flow through the valves has increased by over 50%, the expansion ratio across the exhaust valves is significantly reduced by the back pressure imposed upon the cylinder by the addition of the turbocharger. The decreased expansion ratio has more of an effect than the increased mass flow, resulting in a slightly lower exhaust valve throttling loss for the turbocharged situation.

When turbocharging is applied, a trade-off between combustion availability loss and exhaust to ambient loss is occurring. The higher air-fuel ratio operation of the turbocharged engine increases mixing losses and uses a greater portion of the chemical potential of the fuel to heat excess air. This leaner combustion produces lower bulk gas temperatures. Therefore, for the same fueling, the combustion availability loss increases. The lower in-cylinder bulk gas temperature translates into a lower cylinder exhaust temperature, which reduces the specific availability of the gas leaving the cylinder. In addition to this, the turbocharger transfers available energy from the cylinder exhaust to the inlet air. This transfer process further reduces the amount of available energy which the system exhausts to ambient.

The remaining discussion will be held on a 14-liter engine rating of 300 kW at 2100 rpm. The comparisons are made at a fixed peak cylinder pressure of 12,500 Kpa and an air-fuel ratio of 30.0. Injection timing is held constant, and the desired peak cylinder pressure is maintained by adjusting the compression ratio. The air-fuel ratio is held constant by changing the turbine nozzle area.

4.3 Charge air cooling

The addition of a charge air cooler to the turbocharged engine is becoming increasingly popular on heavy-duty engines due to the resulting fuel consumption improvement. This improvement is seen in spite of the fact that the charge air cooler removes energy and availability from the cycle in the form of heat transfer.

Figure 2 provides an indication of the level of performance improvement associated with cooling of the intake charge. Fitting a charge cooler to a given engine without reoptimization causes an increase in the air-fuel ratio. The lower-temperature charge is more dense, allowing a greater mass of air to enter the cylinder. A reduction in peak cylinder pressure accompanies the lower charge temperature since it leads to a reduced pressure rise during combustion. The change in performance resulting from the reduction in charge-air temperature is indicated by the upper curve in Figure 2. Reoptimizing the engine to restore the cylinder pressure and air-fuel ratio to their original values, by altering the compression ratio and turbine nozzle area, will further improve performance, as indicated by the lower curve.

Results from Second Law analysis of the three highlighted points on the curves in Figure 2 are summarized in Table III. The pertinent performance parameters for these three cases are given in the Appendix. It should be understood that the non-aftercooled operating condition shown here is being used for comparison purposes only. Such an engine does not typify current production engines because of the high required compressor pressure ratio. The upper portion of Table III provides a detailed availability balance for each of the cases under discussion. In the lower portion of the table, various terms were summed together in an attempt to clarify the interactions among the physical processes being affected.

Table III Availability balance summarizing the effect of charge-air cooling (% fuel availability)

	CASE 1 Turbocharged	CASE 2 Charge Temp = 340 K	CASE 3 Charge Temp. = 340 K
Indicated Work	43.14	43.60	44.89
Combustion Loss	19.11	21.94	20.83
Heat Transfer	15.05	12.68	13.08
Intake Valve Throttling	0.85	1.09	1.02
Exhaust Valve Throttling	2.49	2.73	2.80
Charge Air Cooler Heat Transfer	0.00	1.09	1.02
Loss in Compressor	1.62	1.74	1.57
Loss in Turbine	1.89	1.84	1.46
Exhaust to Ambient	15.85	13.37	13.75
Indicated Work	43.14	43.60	44.89
Cyl. + Charge Air Cooler Heat Transfer	15.05	13.73	13.88
Combustion + Exhaust Loss	34.96	35.51	34.58
Valve + Turbocharger Losses	6.85	7.40	6.85

Table IV Availability balance for a turbocompound Diesel engine at various power-turbine pressure ratios turbocharger efficiency = 0.50; power-turbine efficiency = 0.80 (% fuel availability)

POWER TURBINE PRESSURE RATIO	1.00	1/25	1.50	1.75	2.0
Indicated Work + Power Turbine	44.41	45.20	45.40	46.24	44.83
Combustion Loss	20.15	20.33	20.35	20.35	20.34
Cylinder Heat Transfer	13.70	13.95	14.23	14.54	14.87
Intake Valve Throttling	0.94	0.92	0.82	0.72	0.62
Exhaust Valve Throttling	2.69	1.83	1.32	1.00	0.80
Charge Air Cooler Heat Transfer	0.67	0.92	1.06	1.18	1.29
Loss in Compressor	1.61	1.66	1.68	1.68	1.68
Loss in Turbine	1.65	1.83	1.95	2.01	2.06
Loss in Power Turbine	0.00	0.63	1.13	1.62	2.09
Exhaust to Ambient	14.35	12.73	12.07	11.66	11.42

Table V Availability balance demonstrating the effects of insulated operation on the turbocompound Diesel engine (% fuel availability)

	COOLED	UNCOOLED	INSULATED
Indicated Work + Power Turbine	45.40	45.93	47.08
Combustion Loss	20.35	19.16	17.68
Cylinder Heat Transfer	14.23	12.14	8.24
Intake Valve Throttling	0.82	0.93	1.18
Exhaust Valve Throttling	1.32	1.46	1.70
Charge Air Cooler Heat Transfer	1.06	1.18	1.31
Loss in Compressor	1.68	1.73	1.81
Loss in Turbine	1.95	1.87	1.73
Loss in Power Turbine	1.13	1.40	1.30
Exhaust to Ambient	12.07	14.20	17.97

The effects of adding the charge-air cooler to a given engine configuration without reoptimization is typified by comparing Cases 1 and 2 in Table III. Combustion with the reduced temperature charge has, associated with it, greater losses due to the cooler bulk gas temperature at which conversion of the chemical potential of the fuel occurs. The cooler temperature is also responsible for a reduction in exhaust gas losses. The latter is not sufficient to offset the combustion losses, as indicated in the lower portion of Table III.

A slight increase in throttling losses is seen across both the intake and exhaust valves. This is due to the increased mass flow through the valves and the increased density of the charge resulting from the lower temperature. Finally, although the charge-air cooler produces an additional loss of available energy it can be seen from the table that the availability associated with the cylinder heat transfer has decreased significantly.

The charge-air cooler reduces the cylinder gas temperature throughout the cycle, which results in a reduction in cylinder heat rejection. The total engine heat rejection (cylinder and charge air cooler) increases, as indicated in the Appendix, but the availability transfer decreases. The energy transfer in the charge air cooler has little availability associated with it since the process removes heat from the working fluid at a relatively low temperature.

Case 3 in Table III represents the aftercooled engine in Case 2 with the following alterations in hardware:

1. increased compression ratio (to restore original peak cylinder pressure),
2. increased turbine nozzle area (to restore the air-fuel ratio to that of Case 1).

Combustion losses have been reduced by decreasing the air-fuel ratio; however, the losses are still higher than in the turbocharged engine (Case 1). Intake valve throttling has been reduced slightly from that of Case 2 due to the reduced air flow. The exhaust valve throttling has increased over Case 2 because of the increased pressure ratio across the exhaust valve associated with the lower air-fuel ratio (12). However, performance gains were achieved through a reduction in turbocharger losses. The reduced airflow requirement has brought about a reduction in compressor pressure ratio and thus turbine pressure ratio reducing both compressor and turbine losses.

4.4 Turbocompounding

Turbocompounding has been demonstrated as a viable means of improving Diesel engine efficiency (13). The concept involves the expansion of Diesel exhaust gas through a low-pressure turbine which is linked directly to the crankshaft. In the examples presented here, the power turbine is located downstream of a conventional turbocharger turbine. The engine also includes a charge-air cooler maintaining the incoming air at a temperature of 370 K.

The turbocompound Diesel engine operates with a significantly higher exhaust manifold pressure than the turbocharged or aftercooled Diesel engines. As the level of turbocompounding increases, the exhaust manifold pressure becomes increasingly higher than the intake manifold pressure; thus, the power output from the reciprocator is reduced as the output of the power turbine increases. There will thus be an optimum power-turbine pressure ratio for any given configuration or operating condition.

Figures 3 and 4 present the change in net brake specific fuel consumption of a turbocompound engine as a function of power-turbine pressure ratio. These two figures represent cases with combined turbocharger efficiencies (product of compressor, turbine and shaft efficiencies) of 50 and 60 percent, respectively. In each figure, curves for power-turbine efficiencies of 70, 80, and 90 percent are given. Power-turbine efficiency, as used here, is the product of the power-turbine adiabatic efficiency and the mechanical efficiency of the drive system to the crankshaft. The dashed line in each figure indicates the change in optimum power-turbine pressure ratio with power-turbine efficiency. As is indicated in the figures, the optimum power-turbine pressure ratio increases with improved power-turbine efficiency as well as with improved turbocharger efficiency.

Second Law analysis has been applied to the engine configuration represented by the middle curve of Figure 3 (turbocharger efficiency = 50%; power-turbine efficiency = 80%) at various power-turbine pressure ratios. The analysis results are summarized in Table IV. As the power-turbine pressure ratio increases, the working fluid temperature increases causing an increase in the availability associated with the cylinder heat transfer. The increased back pressure lowers the pressure ratio across the valves, thus reducing throttling losses. By maintaining a fixed air-fuel ratio with the increasing back pressure, the compressor power requirement increases, which leads to greater losses in the turbocharger. Furthermore, the increased compressor power leads to a higher compressor discharge temperature, which implies that heat rejection at the charge air cooler must be greater to maintain the same intake manifold temperature.

The deviations in the various loss mechanisms from Table IV are plotted as a function of power-turbine pressure ratio in Figure 5. The power-turbine losses, total heat transfer (cylinder + charge air cooler), and turbocharger losses increase linearly with power-turbine pressure ratio. The valve losses and the available energy removed with the exhaust gases decrease rapidly at first and taper off significantly as the pressure ratio across the power turbine is further increased. Initially, the pressure ratio across the exhaust valve is relatively high, producing significant throttling losses. Increasing the power-turbine pressure ratio causes more of the expansion process to occur in the turbomachinery and reduces the expansion at the exhaust valves (12). However, as the power-turbine pressure ratio is increased further, the pressure ratio across the exhaust valve becomes less significant and the reduction in throttling and exhaust losses become less than the increase in heat transfer and turbomachinery losses.

Referring again to Figure 5, the optimum power-turbine pressure ratio corresponds to the point at which the sum of all the loss curves is minimal. This will occur when the turbomachinery (power-turbine and turbocharger) and total heat transfer (in-cylinder and charge air cooler) losses are increasing at the same rate at which the valve and exhaust losses are decreasing. Improvements in power-turbine efficiency reduce the slope of the power-turbine loss curve. Exhaust availability losses are also reduced as power-turbine efficiency is improved. The net result is a shift in the optimum to a higher power-turbine pressure ratio.

A similar shift in optimum power-turbine pressure ratio is observed when the turbocharger efficiency is increased. An increase in turbocharger efficiency not only reduces the turbocharger loss but also reduces the exhaust to ambient loss. Unfortunately, the increase in turbocharger efficiency is somewhat offset by an increase in the exhaust valve throttling losses, as discussed by Primus (12). As the turbocharger efficiency is improved, less exhaust pressure is needed to drive the turbocharger to supply the same intake manifold pressure. With increased turbocharger efficiency, the exhaust valve throttling losses become more sensitive to changes in power-turbine pressure ratio. The net result is a shift in the optimum power-turbine pressure ratio to a higher value.

4.5 Insulating the turbocompound engine

Insulating the Diesel engine, as a possible means of improving efficiency, has recently been treated with great interest (14). The premise has been that substantially reduced heat rejection leads to either improved work output or increases in exhaust energy (where conversion to useful work can be more readily accomplished). It has been found that any potential gains must come by accessing available energy in the exhaust gas, and turbocompounding has been utilized as a means of extracting this energy.

Before proceeding to an analysis of the insulated engine, a brief digression is in order to discuss the modeling of insulated engines. Diesel engine simulation models have evolved to a high degree of sophistication. As in the case of the insulated engine concept, the simulations are often used as predictive tools, giving engineers an indication of the performance potential of a given design prior to conducting costly experiments.

However, inherent in this exercise of the model is the danger of extrapolation beyond the realm for which its process models have been validated. Many of the submodels included in simulations are not exact representations of physical processes, but empirical correlations based on past data. Among the submodels falling into this category are those for heat transfer, friction, and combustion or heat release. Insulating the engine will affect the processes in all of these submodels.

Experience with insulated engines has demonstrated that early performance predictions were optimistic. Therefore, it was felt that, before presenting a Second Law analysis of the insulated engine, it would be necessary to take steps to bring the simulation results more in line with those observed experimentally.

The authors have chosen to make the modifications in the form of an enhanced radiant heat transfer term, believing that radiation is generally underestimated in simulations. Its contribution to the driving forces of heat transfer from the combustion chamber will be quite sensitive to even modest changes in combustion temperatures.

It is recognized that the modifications have not been based on fundamental measurements of the heat transfer mechanisms, but they do represent an educated speculation as to the relative significance of the radiant process. The model changes have allowed the alignment of analytical results with those observed experimentally.

A radiant heat flux term of the following form was included during the combustion portion of the cycle:

$$q_R = \sigma \epsilon (T_W^4 - T_R^4) \qquad (7)$$

The radiant heat flux expression as given in equation (7) is dependent on expressions for the emissivity (ϵ) and radiant temperature (T_R) as a function of crank angle. The expression chosen to represent the variation in emissivity throughout the process is of the form:

$$\epsilon = C (1 - \exp[-KL]) \qquad (8)$$

This expression provided a reasonable correlation to the experimental data for apparent emissivity obtained by Flynn (15).

Radiant temperatures versus crank angle were calculated by using a multi-zone combustion model (16). The flame was divided into zones of constant equivalence ratio. It was assumed that the onset of soot (the prevalent radiating species) occurs at an equivalence ratio of 1.5 and that the flame could be considered optically thick at richer equivalence ratios. Therefore, the radiant temperature seen by the walls was that generated by the zone in the combustion-model that corresponded to an equivalence ratio of 1.5. The driving temperature for the radiation term in the heat transfer model was the temperature of this rich zone as it varied throughout the combustion and expansion processes.

Implicit in the radiant heat flux model described above is the assumption that, at any moment in time, all exposed combustion chamber surfaces see the radiating flame over the full field of view. While it is recognized that this condition may not be correct for all combustion chamber surfaces it was deemed appropriate for the cylinder head and piston face.

Using the simulation with the model described above, the insulated engine will now be considered. One of the first points which must be clarified concerns the constraints under which the insulated engine is simulated. As the heat transfer from a reciprocator is reduced, the in-cylinder temperature and pressure increase. If the effect of insulation is assessed at constant compression ratio and injection timing, the peak cylinder pressure will rise with increasing insulation. This produces more substantial performance gains than would be seen at constant peak cylinder pressure.

Engine designs are often constrained by peak cylinder pressure and it was felt that comparison should be done at conditions which are optimal for each engine under this constraint. Therefore, both insulated and cooled engines should be considered at the same peak cylinder pressure. The difference in insulated-engine performance with and without the peak cylinder pressure constraint is depicted in Figure 6. All further analysis presented assumes a constant peak cylinder pressure of 12500 kPA.

In Figure 7, the performance potential of the insulated engine is summarized. The solid curve was obtained using the revised simulation with the radiant model described above. The sensitivity of the results to this radiant model are indicated by the dashed curves. Also indicated in this figure is the insulation thickness required for a given reduction in heat transfer. The thicknesses given are those required using monolithic zirconia (k = 2.49 W/m-K) on the piston and cylinder head surfaces. The reduction in heat transfer diminishes with increasing insulation thickness. Liner insulation was found to have a negligible impact on performance and, therefore, was not considered in this analysis.

A Second Law analysis of the insulated engine is summarized in Table V, and the corresponding performance parameters are presented in the Appendix. For each of the cases given, the engine was turbocompounded with a power-turbine pressure ratio of 1.50. Turbocharger efficiency was 50 percent and power-turbine efficiency 80 percent. The first data set given is for the cooled turbocompound engine which was previously presented in Table IV. The second column of data, labeled "uncooled", results from the removal of the cooling fluid from the engine. No additional steps toward insulation have been taken. The Appendix shows that this provides an approximate 24 percent reduction of in-cylinder heat transfer. The data for the insulated engine, given in the third column of Table V, results from applying 5.08 mm of insulation to the piston and cylinder head face. This provides an approximate 61 percent reduction of in-cylinder heat transfer.

The values for the available energy associated with cylinder heat transfer, presented in Table V, are also reduced. However, the percentage reductions are considerably different than those obtained in the First Law Analysis. For the uncooled engine, the availability associated with the cylinder heat transfer has been reduced by only 15 percent versus the 24 percent reduction on an energy basis. Similarly the insulated engine displays a 42 percent reduction in available energy associated with heat transfer compared to the energy reduction value of 61 percent. The availability associated with heat transfer, as given by equation (3), is proportional to the temperature at which the transfer occurs. This points to the fact that insulating the engine has greatest effect during portions of the cycle where temperatures are relatively low reducing the availability associated with this heat transfer. Insulation is not as effective during combustion and the expansion stroke when temperatures are high, and only a slight improvement in power output during the expansion process is seen. This improvement is more than offset by an increase in compression work due to the more rapid heating and thus rapid pressure rise in the insulated engine. The net result of insulation is a reduction in power output during the closed portion of the cycle.

Looking further at Table V it is seen that the effect of insulation is to increase throttling at both the intake and exhaust valves. Intake throttling losses are increased due to the higher pressure ratio across the intake valve required to overcome the increased heating of the charge air during intake. Exhaust throttling losses also increase since the higher exhaust gas temperature allows the turbine inlet pressure to be reduced while still providing the additional energy required to the compressor for the increased intake manifold pressure. Although the valve throttling losses have increased, the cylinder work during the gas exchange portion of the cycle has been significantly improved.

In summary, insulating the engine allows a reduction in exhaust manifold pressure while providing an increased intake manifold pressure. This has a favorable impact on reciprocator power output during the gas exchange portion of the cycle which more than offsets the reduction in power during the closed portion of the cycle discussed previously.

Note that, even with the addition of the power turbine, the available energy associated with the exhaust to ambient is still sizable and has increased with insulation. The power turbine is an expansion device which can only extract work from the exhaust gas through pressure equilibration with the ambient. The power turbine has no capacity for work extraction through temperature equilibration of exhaust products and the ambient. Therefore, a significant amount of available energy is not accessed by the power turbine. This fact points to the need for some form of heat recovery device to utilize the additional exhaust energy resulting from the insulated engine.

4.6 Bottoming cycles

No attempt will be made in this paper to address the details of any particular bottoming cycle. As was demonstrated in the previous section, the availability remaining in the exhaust gas increases significantly with insulation of the turbocompound Diesel engine. Additional efficiency improvements can be made by utilizing a portion of this high-temperature, ambient-pressure available energy of the exhaust gas.

In Figure 8, the performance potential of cycle bottoming is indicated as a function of the degree of insulation of the turbocompound Diesel engine. The results are given for various bottoming cycle efficiencies, defined as the percentage of the total exhaust energy actually converted to useful work. Additional performance gains available through the use of bottoming cycles must be weighed against cost and package-size constraints for the given application.

5 CONCLUSIONS

Various advanced concepts for the improvement of diesel engine performance have been assessed in terms of the Second Law of thermodynamics. This method holds the potential for gaining new insights concerning possible manipulations of the various processes within the engine to reduce fuel consumption. The following general conclusions may be drawn from the work presented herein.

1. By permitting an improvement in the time distribution of heat release and expanding the exhaust gases through a turbine, the turbocharged diesel engine displays a significant reduction in heat-transfer and exhaust-gas available energy loss compared to a naturally aspirated version of the same engine.

2. When increasing the air-fuel ratio by turbocharging, a trade-off is made between combustion availability loss and the available energy contained in the system exhaust gases.

3. Reduction of total availability associated with heat transfer is directly responsible for the performance gains associated with charge air cooling the turbocharged engine. Turbocharger losses may also be reduced by reoptimizing the charge air cooled engine.

4. The Second Law clearly presents the effects of various engine parameters on the optimum power-turbine pressure ratio of a turbocompound engine.

5. Performance improvements with the insulated engine are a result of an increase in the ratio of intake manifold pressure to exhaust manifold pressure brought about by the increase in exhaust temperature. This more than offsets a power reduction in the closed portion of the cycle which is caused by a substantial increase in compression work overshadowing a small improvement in the expansion process.

6 ACKNOWLEDGEMENT

The authors wish to thank the management of Cummins Engine Company for their permission to publish this paper and for making available the necessary resources to carry out the work.

REFERENCES

1. Bruges, E. A., *Available Energy and the Second Law Analysis*, Butterworths Scientific Publications, London, 1959.

2. Shapiro, H. N. and Kuehn, T. H., "Second Law Analysis of the Ames Solid Waste Recovery System," Energy, August/September, 1980, pp. 985-991.

3. Brzustowski, T. A. and Golem, P. J., "Second Law Analysis of Energy Processes Part I: Exergy-An Introduction," Trans. Can. Soc. Mech. Eng., 1976-1977.

4. Foster, E. F. and Myers, P. S., "Factors Affecting Heavy Duty Diesel Fuel Economy", ASME Paper 82-DGEP-5, 1982.

5. Clarke, J. M., Letters and Comments, Mechanical Engineering, March, 1983.

6. Flynn, P. F., Hoag, K. L., Kamel, M. M., and Primus, R. J., "A New Perspective on Diesel Engine Evaluation Based on Second Law Analysis", SAE Paper 840032, 1984.

7. Krieger, R. B. and Borman, G. L., "The Computation of Apparent Heat Release for Internal Combustion Engines", ASME Paper 66-WA/DGP-4, 1966.

8. Moran, M. J., *Availability Analysis: A Guide to Efficient Energy Use*, Prentice Hall, Inc., Englewood Cliffs, New Jersey, 1982.

9. Wepfer, W. J. and Gaggioli, R. A., "Reference Datums for Available Energy", *Thermodynamics: Second Law Analysis*, Am. Chem. Soc. Symposium Series No. 122, Am. Chem. Soc., Washington, D.C., 1980.

10. Rodriguez, S. J., "Calculation of Available Energy Quantities", *Thermodynamics: Second Law Analysis*, Am.

Chem. Soc. Symposium Series No. 122, Am. Chem. Soc., Washington, D.C., 1980.

11. Borman, G. L., "Mathematical Simulation of Internal Combustion Engine Processes and Performance Including Comparisons with Experiment", PhD Thesis, Mechanical Engineering Dept., University of Wisconsin, 1964.

12. Primus, R. J., "A Second Law Approach to Exhaust System Optimization", SAE Paper 840033, 1984.

13. Brands, M. C., Werner, J. R., Hoehne, J. L., Kramer, S., "Vehicle Testing of Cummins Turbocompound Diesel Engine", SAE Paper 810073, 1981.

14. Kamo, R., Bryzik, W., "Cummins-TARADCOM Adiabatic Turbocompound Engine", SAE Paper 810070, 1981.

15. Flynn, P., Mizusawa, M., Uyehara, O., Myers, P., "An Experimental Determination of the Instantaneous Potential Radiant Heat Transfer within an Operating Diesel Engine", SAE Paper 720022, 1972.

16. Shahed, S. M., Chiu, W. S., Lyn, W. T., "A Mathematical Model of Diesel Combustion", Paper C94/75, Proc. Inst. of Mech. Engineers (1975), pp. 119-128.

APPENDIX

Engine Simulation Results for Naturally Aspirated and Turbocharged Diesel Engines Corresponding to Table II of the Text

	Naturally Aspirated	Turbocharged
Engine Speed (rev/min)	2100	2100
Fuel Rate (g/s)	12.6	13.0
Cylinder Power (kW)	185	220
Cylinder Ind. Work (kcal/min)	3125	3623
Cylinder Heat Transfer (kcal/min)	1943	1812
Cylinder Exh. Energy (kcal/min)	2680	2566
Air Flow (kg/min)	13.5	21.8
A/F	17.9	28.0
Int. Man. Temp. (K)	302.8	367.6
Exh. Man. Temp. (K)	1025	813.2
Int. Man. Press. (kPa)	95.8	167.3
Exh. Man. Press. (kPa)	104.1	155.8
Peak Cyl. Press. (kPa)	9468	11,722
Compression Ratio	15.3	16.0
Compressor Efficiency	–	.72
Compressor Pressure Ratio	–	1.68
Turbine Efficiency	–	.70
Turbine Pressure Ratio	–	1.56
BSFC (g/kW-hr)	247	214

Engine Simulation Results for Aftercooling Study Corresponding to Table III in Text

	Turbocharged	Turbocharged Aftercooled	Turbocharged Aftercooled (Reoptimized)
Engine Speed (rev/min)	2100	2100	2100
Fuel Rate (g/s)	16.92	16.92	16.92
Cylinder Power (kW)	294	299	303
Cylinder Ind. Work (kcal/min)	4683	4756	4813
Cylinder Heat Transfer (kcal/min)	2056	1723	1758
Cylinder Exh. Energy (kcal/min)	3982	4091	4020
Aftercooler Heat Trans. (kcal/min)	0.0	559	450
Air Flow (kg/min)	30.44	32.50	30.47
A/F	30.0	32.02	30.0
Int. Man. Temp. (K)	429	340	340
Exh. Man. Temp. (K)	870	792	811
Int. Man. Press. (kPa)	256.2	227.2	212.0
Exh. Man. Press. (kPa)	218.6	211.6	192.4
Peak Cyl. Press. (kPa)	12,500	12,040	12,500
Compression Ratio	13.90	13.90	14.85
Compressor Efficiency	.720	.720	.720
Compressor Pressure Ratio	2.57	2.28	2.13
Turbine Efficiency	.700	.700	.700
Turbine Pressure Ratio	2.18	2.11	1.92
BSFC (g/kW-hr)	210	207	201

Engine Simulation Results for Turbocompound Study
Corresponding to Table IV in Text

	Turbocharged Aftercooled	Turbocompound Optimum Press. Ratio	Turbocompound
Engine Speed (rev/min)	2100	2100	2100
Fuel Rate (g/s)	16.92	16.92	16.92
Cylinder Power (kW)	299	270	240
Power Turbine Power (kW)	0.0	38	63
Cylinder Ind. Work (kcal/min)	4762	4345	3915
Cylinder Heat Transfer (kcal/min)	1858	1967	2070
Cylinder Exh. Energy (kcal/min)	4012	4024	4139
Aftercooler Heat Trans. (kcal/min)	304	442	516
Air Flow (kg/min)	30.45	30.45	30.45
A/F	30.00	30.00	30.00
Int. Man. Temp. (K)	370	370	370
Exh. Man. Temp. (K)	832	861	894
Int. Man. Press. (kPa)	227	258	276
Exh. Man. Press. (kPa)	203	335	459
Peak Cyl. Press. (kPa)	12,500	12,500	12,500
Compression Ratio	14.50	13.70	13.30
Power Turbine Press. Ratio	1.0	1.50	2.00
Compressor Efficiency	.720	.720	.720
Compressor Pressure Ratio	2.28	2.59	2.77
Turbine Efficiency	.700	.700	.700
Turbine Pressure Ratio	2.03	2.23	2.29
Power Turbine Efficiency	–	0.80	0.80
BSFC (g/kW-hr)	203	198	201

Engine Simulation Results for Insulated Study
Corresponding to Table V in Text

	Cooled	Uncooled	Insulated
Engine Speed (rev/min)	2100	2100	2100
Fuel Rate (g/s)	16.92	16.92	16.92
Cylinder Power (kW)	270	271	274
Power Turbine (kW)	38	42	46
Cylinder Ind. Work (kcal/min)	4345	4370	4410
Cylinder Heat Transfer (kcal/min)	1967	1498	772
Cylinder Exh. Energy (kcal/min)	4024	4287	4975
Aftercooler Heat Trans (kcal/min)	442	483	528
Air Flow (kg/min)	30.45	30.45	30.45
A/F	30.0	30.0	30.0
Int. Man. Temp. (K)	370	370	370
Exh. Man. Temp. (K)	861	918	995
Int. Man. Press. (kPa)	258	267	277
Exh. Man. Press. (kPa)	335	328	318
Peak Cyl. Press. (kPa)	12,500	12,500	12,500
Compression Ratio	13.70	13.15	12.50
Power Turb. Press. Ratio	1.50	1.50	1.50
Comp. Efficiency	.720	.720	.720
Comp. Pressure Ratio	2.59	2.69	2.79
Turbine Efficiency	.700	.700	.700
Turbine Pressure Ratio	2.23	2.18	2.12
Power Turb. Efficiency	0.80	0.80	0.80
BSFC (g/kW-hr)	198	195	190

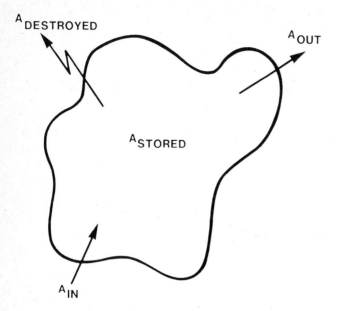

Fig 1 Generalized control volume for availability analysis

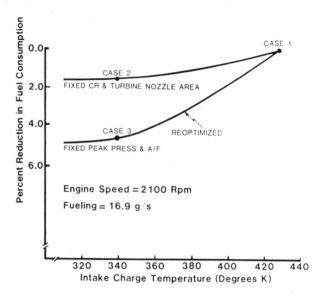

Fig 2 The effect of charge-air cooling on the performance of a turbocharged diesel engine

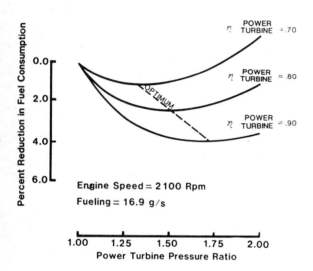

Fig 3 The effect of power-turbine pressure ratio on engine system efficiency; turbocharger efficiency = 0.50

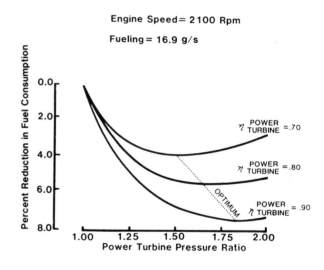

Fig 4 The effect of power-turbine pressure ratio on engine system efficiency; turbocharger efficiency = 0.60

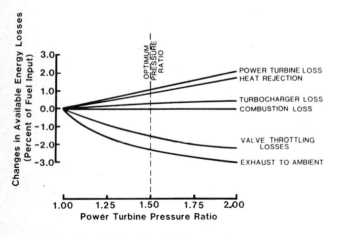

Fig 5 Sensitivity of available energy losses to power-turbine pressure ratio

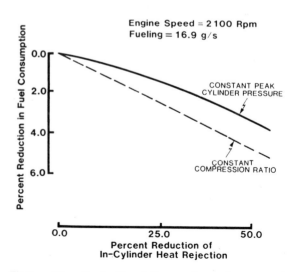

Fig 6 The effect of insulation on the performance of a turbocompound diesel engine

© IMechE/SAE 1984 C440/84

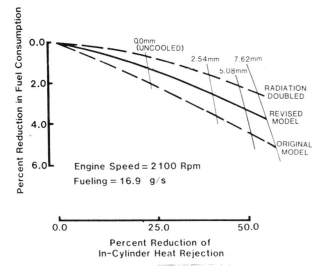

Fig 7   Sensitivity of performance prediction to radiant heat transfer model

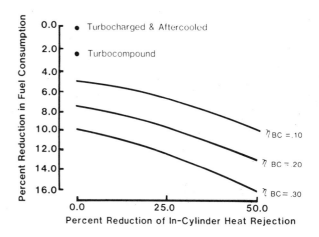

Fig 8   The effect of cycle bottoming on the efficiency of an insulated turbocompounded diesel engine

© IMechE/SAE 1984 C440/84

C421/84

SAE 841288

Fuel economy, emissions and noise of multi-spray light duty DI diesels—current status and development trends

K P MAYER, Dr techn, Dipl-Ing
AVL LIST Ges m b H, Graz, Austria

SYNOPSIS The current status and development trends of light duty diesel engines equipped with a multi-spray direct injection combustion system are described. Fuel consumption maps of a naturally aspirated and a turbocharged engine show that competitive performance and best fuel economy can be achieved. As a means for the reduction of combustion noise a "split injection device" is installed in the high pressure part of the fuel system. It permits a reduction of the amount of fuel injected during the ignition delay and consequently reduces the rate of pressure rise and maximum cylinder pressure. In addition, also a reduction of exhaust emissions is experienced. Advanced piston technology is presented as a feature for a further improvement of performance and fuel economy.

1 INTRODUCTION

Fuel economy is the only reason for the use of diesel engines in light duty vehicles and passenger cars. Such engines obtain the best fuel economy with the multi-spray direct injection combustion system (airborne mixing) as traditionally used in heavy duty vehicles. Therefore, AVL is making a major effort to develop this type of combustion system for light duty high speed DI (HSDI) diesel engines (1),(2),(3). The progress made in the development of other types of light duty HSDI diesels was reported earlier several times as in ref.(4),(5),(6),(7).

The conventional multi-spray direct injection combustion system when applied to light duty high speed diesel engines has a number of disadvantages such as a lower high speed power potential, higher exhaust emissions and higher combustion noise. It was therefore necessary to develop technologies which permit these difficulties to be overcome without sacrificing the inherent fuel economy advantage of the multi-spray DI combustion system.

This paper is intended to present an update of the current status of the multi-spray DI combustion system applied to light duty diesel engines and to describe two examples of the many new technologies currently evaluated for the HSDI diesel.

2 DESIGN FEATURES OF MULTI-SPRAY HSDI DIESELS

In contrast to the traditional direct injection combustion systems used in heavy duty diesels, the light duty version of such engines incorporates the following features:

a) A high compression ratio in the range from 19 to 22 is a necessity for light duty DI diesels to control hydrocarbon emissions and noise. The high compression ratio also provides for excellent cold starting and reduces exhaust odour and cold idle noise which are unacceptable with light duty vehicles. It also makes the DI diesel more tolerant to lower quality diesel fuels.

b) Shallow re-entrant combustion bowls with a turbulence rim in combination with modest swirl levels generated by efficient helical intake ports are used for improved air-fuel mixing and for the acceleration of the combustion. This bowl design enables the operation of the engines at competitive BMEP and smoke levels despite the use of the high compression ratio.

c) Distributor pumps or small pump injectors are primarily used with HSDI diesels. The distributor pumps are essentially of conventional design but modified for higher injection pressures and low dead volume. They are either mechanically or electronically controlled. Small low cost pump injectors with electronic controls give very promising engine test results and are currently under development.

d) The injectors are equipped either with VCO (valve covers orifices) or minimized sac volume nozzles in order to reduce hydrocarbon emissions.

e) Fast glow plugs penetrating into the combustion bowl are installed for cold starts below -15 deg.C. There is no penalty with respect to performance or smoke when glow plugs are used.

In addition to these design features, new technologies are under development to further advance the multi-spray DI diesel as described later on in this paper.

3 PERFORMANCE

A typical fuel consumption map of an advanced naturally aspirated DI-diesel engine of 2.1 litre displacement is shown in fig. 1a. The full load curve was established at a smoke level of around 3.5 Bosch. With this criterion, a maximum BMEP of 8 bar at peak torque and 6.5 bar at rated speed are achieved. The latter corresponds to a rated specific power of 24 kW/litre displacement. Besides a low specific fuel consumption at full load the part-load fuel consumption is even more important for HSDI diesels used in vehicles. For example, at an engine speed of 2000 rpm and a BMEP of 2 bar which is typically the center of the load schedule for the US urban driving cycle this engine shows a specific fuel consumption of 295 g/kWh. The minimum fuel consumption in the island is 218 g/kWh.

The emissions and fuel economy in the US cold-hot urban driving cycle (FTP 75) were evaluated in a vehicle of 2625 lbs inertia weight. The engine was equipped with an exhaust gas recirculation (EGR) system in order to meet the future US NOx- standard of 1.0 g/mile. The following cycle emissions in g/mile were obtained with and without EGR using US no. 2 diesel fuel (Phillips Reference Fuel):

		with EGR	w/o EGR
US urban	HC	.26	.36
driving cycle	CO	1.25	1.45
(FTP75), emis-	NOx	.75	1.60
sions in g/mile	Part.	.11	.10

The fuel economy of this vehicle, which incorporates a standard 5-speed transmission system is

Cycle	with EGR	w/o EGR
Urban (FTP75) (mpg)	46.2	45.5
Highway (mpg)	60.3	60.4
Composite (mpg)	51.6	51.2

It has to be noted that the particulate emissions are well below the future US-limit of 0.2 g/mile. This is the result of the very low smoke obtainable with multi-spray DI diesels at part load (8).

The slightly better fuel economy as well as the lower HC-emissions of the engine with EGR can be explained by the earlier injection timing which is used with EGR. HC emissions are also reduced with EGR due to the increased intake charge temperatures.

Fig. 1b presents the fuel map of an advanced turbocharged HSDI diesel of 1.7 l displacement. It shows on the one hand that a good full load performance can be achieved over the whole speed range and on the other hand the advantage in specific fuel consumption in comparison to other systems especially in the part load range. At 2000 rpm and 2 bar the specific fuel consumption is about 300 g/kWh which is only slightly higher than for the naturally aspirated HSDI diesel described before. The minimum fuel consumption in the island is 221 g/kWh.

The turbocharged engine described above was developed to meet the European emission standards. In a vehicle of 1150 kg test weight a combined emission of HC + NOx = 8 g/test was measured. The full load curve was established for a smoke level of around 3.5 Bosch.

Modern HSDI-diesel engines have to meet the contradictory demands of fuel economy, lowest exhaust emissions and lowest combustion noise. This calls for a "two-dimensional" dynamic injection timing pattern which can be best realized by use of electronic controls. The timing requirements referred to engine load and speed are shown in fig. 2. Optimum fuel economy would require a simple speed advance, whereas low combustion noise would require a retardation. Retardation of timing would also mean a reduction of NOx emissions. But since cycle emissions are only measured at part-load, a retardation is mainly required in this range. Over-retarded timing at medium speeds and light loads, however, creates increased hydrocarbon emissions. Therefore, as a compromise, a light load advance is provided.

4 SPLIT INJECTION AS MEANS TO REDUCE COMBUSTION NOISE

4.1 General

The conventional multi-spray DI diesel engine, especially in its naturally aspirated version, suffers from high combustion noise. This is due to the instantaneous combustion of the fuel mixed with air during the ignition delay, causing a sudden pressure rise in the power cylinder. According to (9), however, the combustion noise is decisively determined by the maximum value of the pressure rise (dp/dt) and the speed of pressure rise (d^2p/dt^2). Therefore, a reduction of combustion noise would require a reduced fuel quantity injected during the ignition delay period.

One of the possibilities to reduce the fuel quantity which becomes combustible within the ignition delay and then burns instantaneously is staged or pilot injection. Therefore, in the past numerous proposals were made for devices generating pilot injection. However, most of such systems failed to be introduced into production because of their complexity, poor reliability or other disadvantages involved.

One possibility to reduce the combustion noise of multi-spray direct injection diesel engines without penalizing other advantages consists in the control of the initial rate of fuel discharge by the so-called "split injection" technology. The split injection technique reduces the amount of fuel injected during the ignition delay period and thus reduces the initial rate of cylinder pressure rise. This technique achieves a considerable reduction of combustion noise and also a reduction of gaseous emissions over the entire engine speed and load range.

4.2 Design features of the split injection device (SID)

Several designs are currently under development to achieve the proper shaping of the initial rate of fuel discharge.

The mechanical split injection device, in its simplest form, is a retraction valve which is operated hydraulically. It can be installed in any convenient location in the high pressure part of the fuel system between the fuel pump and the nozzle.

As shown in Fig. 3, this device consists of a piston in a barrel which is held against its seat by means of a spring. The piston of the retraction valve is connected, via the seat bore, to the high pressure side of the fuel system, and its opening pressure is adjusted to be slightly greater than that of the nozzle. Fuel leaking past the piston is fed into the fuel leak-off system.

4.3 Operation

An example of the operation of the split injection device is shown in fig.4 in comparison to a normal injection. The needle lift, the line pressure and the injection rate are plotted for an engine speed of 3000 rpm.

The piston remains against its seat until the fuel system pressure reaches its opening pressure (point A in fig.4). Once the piston has lifted off its seat, the fuel pressure then acts upon the full area of the piston causing it to move rapidly back against its stop. In so doing, it retracts a fixed volume of the pump delivery (equivalent to Piston Area x Stroke). This produces a rapid decay in the line pressure and a momentary reduction in the fuel delivery (point B in fig. 4). The injection continues when the fuel pressure has risen sufficiently. At the end of injection the piston returns to its seat when the line pressure falls below its closing pressure. For equal fuelling split injection extends the injection duration. This is compensated in an actual engine by advancing the injection timing accordingly.

The effect of the split injection device depends upon its opening pressure, the volume of fuel it retracts and its position in the fuel system. It can be matched therefore to any development requirement.

4.4 Effect of split injection

Optimum results are obtained by split injection when at low idle, as well as low engine speeds and light loads the pre-injection is completely separated from the main injection (split injection). At medium and high engine speeds it is necessary to avoid a separation between the initial and the main injection because of the resulting increase in smoke. At these operating conditions the split injection device is therefore adjusted to only briefly slow down the initial rate of fuel discharge as shown in fig. 4.

In the following, several results are presented which were obtained on single- and 4-cylinder DI diesels as examples for the effect of split injection.

- Rate of heat release (ROHR)

The first peak of the rate of heat release which occurs at the start of combustion and which is typical for conventional multi-spray DI diesels, as well as the maximum value of the pressure rise decisively influence the combustion noise. Both parameters can be controlled by split injection as the following example demonstrates.

Fig. 5 shows the time history of cylinder pressure needle lift and the rate of heat release at 1800 rpm and full load conditions for a single cylinder research DI diesel engine (85.0 mm bore, 92.0 mm stroke) operating with a normal injection equipment as well as with an injection equipment providing split injection.

The normal injection with a nearly rectangular needle lift diagram, although retarded for low NOx levels, provides a steep pressure rise after the start of combustion. The respective ROHR diagram shows the typical steep gradient at the beginning of combustion ending in a high peak. A clear improvement is achieved with injection rate shaping (dashed curves in fig. 5) where the needle closes slightly after a first complete opening thus splitting up the injection into an initial and a main injection but without distinct separation. This reduces the quantity of combustible fuel at the beginning of combustion. Split injection, in general, reduces the NOx emissions. Therefore, for equal NOx emissions split injection permits a more advanced timing of the (pre-)injection in comparison to the normal injection.

Fig. 5 shows that with split injection used as an injection rate shaping device the cylinder pressure rise and the first peak of the rate of heat release are reduced considerably.

An even smoother combustion, this means a further reduction of pressure rise, can be obtained at this engine speed and load when a small pilot quantity is injected ahead of and separated from the main injection (dash-dotted lines). It can be clearly seen that in this case the combustion of the pilot quantity starts already before the main injection begins, thus decreasing the maximum pressure rise. The first spike in the ROHR diagram which exists with normal injection is reduced considerably.

- Combustion noise

As already mentioned earlier, at low idle optimum results are obtained when the fuel is injected in two injections of about equal quantities. As an example the third octave cylinder pressure spectra are shown in fig.6 for low idle conditions of a 1.7 litre 4-cylinder DI/NA engine. It can be seen clearly that a significant reduction of noise level up to 10 dB can be obtained especially in the range between 200 and 2000 Hz, a frequency range which is decisive for the excitation of the engine structure and consequently for the airborne noise radiated from the engine surface.

The cylinder pressure frequency spectra of a single cylinder research engine (85.0 mm bore, 92.0 mm stroke) for a medium engine speed of 1800 rpm and full load are shown in fig. 7. In this case also the influence of injection timing on the level of the cylinder pressure spectra with and without split injection is demonstrated. Without split injection the cylinder pressure spectrum level can be only reduced by 1 - 3 dB between 500 and 5000 Hz when using a retarded timing. However, by use of split injection a decrease of up to 17 dB can be achieved.

The effect of noise reduction by split injection is also experienced when the engine is operated under transient conditions. If the engine is accelerated, the temperature of the cylinder charge is lower than under steady operation comparing equal engine speeds and loads thus causing an increased ignition delay and in the case of multi-spray DI diesels a larger quantity of combustible mixture at the start of combustion (9). Thus, the cylinder pressure rise and the combustion noise are increased significantly, a problem which can also be alleviated by split injection.

- Vehicle noise

The improvement in noise of a passenger car which is equipped with a multi-spray HSDI diesel engine incorporating a split injection device is up to 3 dB(A) in drive-by noise (cf. also (10)).

It has to be noted that these dB(A) values do not reflect the improvement of the noise characteristics. Since the noise reduction is achieved in critical frequency ranges a substantial improvement of the subjective noise characteristic is obtained.

- Emissions

Fig. 7 shows also the effect of split injection on emissions measured under steady test bed conditions. As mentioned above, the same NOx-levels as with normal injection can be obtained by an earlier injection timing. The hydrocarbon emissions are reduced by up to 50 percent at full load and at part load. Smoke levels increase slightly if the rest of the combustion system remains unaltered. However, the reduction of hydrocarbons would permit a small decrease of the compression ratio, which eliminates the smoke increase. Also the swirl level which is decisive for the quality of combustion with airborne mixing has to be matched to the new conditions. In essence, a slightly lower swirl level than with normal combustion will be required.

In addition to the reduction of hydrocarbon emissions, obtained with split injection, a reduced sensitivity of HC emissions due to retarded injection timing was experienced. Fig. 8 shows as an example the HC-NOx trade-off curves again for the single cylinder research engine with and without split injection at low idle conditions: the parabola shaped trade-off curve as measured with normal injection is shifted to lower HC-levels with the application of split injection and the upper branch is flattened out so that the HC-emissions remain nearly unchanged when the timing is changed for NOx levels between 150 and 300 ppm.

The BSFC is not influenced by split injection. Sometimes a tendency for a slight improvement may be experienced which could be explained by the earlier timing.

The SID helps also to improve cycle emissions. In the following table as an example the emissions in the US CVS cold-hot driving cycle in g/mile and the fuel economy in mpg are presented for a vehicle of 2750 lbs. inertia equipped with a naturally aspirated 2.2 l DI diesel operated without EGR with and without split injection at the same timing schedule. This example shows that hydrocarbon emissions are equal with and without SID but NOx emissions are reduced by about 25 percent.

	HC g/mile	CO g/mile	NOx g/mile	Part. g/mile	FE mpg
w/o SID	0.38	1.24	1.80	0.19	43.4
with SID	0.38	1.18	1.36	0.18	43.4

5 FURTHER COMPONENT DEVELOPMENT

For a further improvement of performance, fuel economy and emissions of the multi-spray light duty DI diesel engine, the development work is currently also concentrating on improved piston designs. One promising feature is the minimization of parasitic volumes in the combustion space. This can be mainly achieved for a conventional piston design by a reduced top land height and smaller top land clearances. Although a high position of the first compression ring as well as low top land clearances create a number of difficulties (e.g. higher top ring and groove temperatures) there exist technologies today by which these problems can be overcome, like scuff resistant coating on the top ring periphery, fibre reinforced piston grooves, etc.(11).

An example of the gain which can be obtained by a reduced dead volume is presented for a 1.6 litre naturally aspirated engine in fig. 9. In the lower speed range, not only the smoke limited full load BMEP is increased, but also the specific fuel consumption is reduced by up to 8 g/kWh. In the higher speed range, the performance is approximately equal. However, in the part load range the reduction of specific fuel consumption is even more important. BSFC-improvements of up to 5 percent are experienced.

Investigations were also made with light weight articulated pistons, because this type of piston incorporates a number of advantages which are worth looking into despite its higher cost. Without a weight penalty, the piston top can be manufactured of steel with an aluminium carrier. Steel has a lower thermal expansion and a lower coefficient of thermal conductivity than aluminium. This permits the first compression ring to be installed in a higher position, and allows the piston to be designed for a tighter fit and less bump clearance. At low engine speeds and loads this results in a substantial decrease of the dead volume in a running engine.

With such piston designs improvements are mainly experienced in the part load range. Because of the lower coefficient of thermal conductivity of steel in comparison to aluminium the temperature of the combustion chamber becomes higher. This helps to reduce hydrocarbon emissions. However, a slight increase of NOx emissions is experienced. Nevertheless, steel top pistons are advantageous for the control of cycle emissions because of their faster temperature response under transient conditions.

As a consequence of the smaller wall thickness of the piston top, the temperatures of the bottom surface are higher causing higher oil temperatures. A positive effect of these higher oil temperatures is mainly seen at light loads where oil temperatures are undesirably low and therefore a temperature increase helps to reduce engine friction.

By use of the articulated design the piston slap is eliminated which results in a reduction of the engine noise.

The possible advantages obtainable by such advanced piston technology were experienced in the course of initial investigations. Considerably more development work has still to be performed on improved piston designs and they represent a key for further improvements such as friction reduction, minimization of blow-by, lower HC emissions, better cold starting, lower noise and better fuel economy.

6 CONCLUSIONS

1. The multi-spray direct injection principle gives the best fuel economy of all combustion systems currently under development for light duty diesel engines.

2. In order to meet the requirements of high speed light duty application it was necessary to make a number of design changes to the conventional multi-spray diesel combustion system such as high compression ratio, re-entrant combustion bowl, VCO or minimized sac volume nozzles.

3. Fuel injection systems which meet the requirements of multi-spray HSDI diesels are currently under development and include both modified distributor pumps and small low-cost pump injectors. Both systems also offer electronic control possibilities.

4. Multi-spray DI diesel combustion systems with electronically controlled FIE are capable of meeting the

US Federal emission standards proposed for 1987 in vehicles up to 3625 lbs. with reserves. Particulate emissions are currently well below .2 g/mile for vehicles with inertia weight of 2625 lbs. Development work is in progress to further lower the particulate emissions in order to avoid particulate traps with heavier vehicles.

5. The major disadvantage of the multi-spray HSDI-system - its high combustion noise - could be eliminated by the application of a "split injection device" installed in the fuel system. This device permits the amount of fuel injected during the ignition delay period to be minimized and provides a shaping of the rate of fuel injection in such a manner that the combustion noise is reduced without any performance penalties. Improvements in drive-by noise of up to 3 dB(A) can be achieved but more importantly this technique greatly improves the subjective noise character of DI diesels at all operating conditions including low idle. Additionally split injection helps to improve gaseous emissions.

6. Improvements of low end torque and specific fuel consumption can be obtained with advanced piston designs with minimized dead volume. Investigations are in progress using advanced piston designs which are expected to offer further advantages in cycle emissions, part load fuel consumption and noise.

7. In addition to the current component development described in this paper a number of additional new technologies and new designs are being evaluated to further optimize the obvious potential of light duty DI diesels with multi-spray combustion system especially in view of further improvements of the vehicle fuel economy, emissions and noise.

REFERENCES

(1) CARTELLIERI, W. and SCHUKOFF, B.: Die direkte Einspritzung für leichte Dieselmotoren. FISITA Congress, Budapest, Paper 3.3.5, 1978.

(2) CARTELLIERI, W. Direct Injection for Light Duty Diesel Engines. SIA Congress, Paris, April 2-3, 1980.

(3) CICHOCKI, R. and CARTELLIERI, W. The Passenger Car Direct Injection Diesel - A Performance and Emissions Update. SAE Paper 810 480, 1981.

(4) NEITZ, A. and D'ALFONSO, N. The M.A.N Combustion System with Controlled Direct Injection for Passenger Car Diesel Engines. SAE Paper 810 479, 1981.

(5) ELSBETT, L. and BEHRENS, M. Elko's Light Duty D.I. Diesel Engines with Heat Insulated Combustion System and Component Design. SAE Paper 810 478, 1981.

(6) MONAGHAN, M.L. The Best High Speed Direct Injection Diesel System for Light Duty Application. I Mech. E Conference, London, Paper C 105/82, October 5-7, 1982.

(7) MONAGHAN, M.L. A Wall Wetting Direct Injection Diesel for Passenger Cars. FISITA-Congress, Vienna, Paper No., 845040, May 6-11, 1984.

(8) CARTELLIERI, W. and TRITTHART, P.: Particulate Analysis of Light Duty Diesel Engines (IDI & DI) with Particular Reference to the Lube Oil Particulate Fraction. SAE Paper No. 840 418, 1984.

(9) THIEN, G.E. Sources of Noise in the Diesel Engine VDI-Berichte, 1983, no. 499, pp. 107 - 119.

(10) RÖNITZ, R. Chancen der Direkteinspritzung bei schnellaufenden PKW-Dieselmotoren für die Großserie. FISITA-Congress, Vienna, Paper 845 090, May 6-11, 1984.

(11) MUNRO, R. Fuel Economy - Its Influence on Piston Design Features. SAE Paper 830067, 1983.

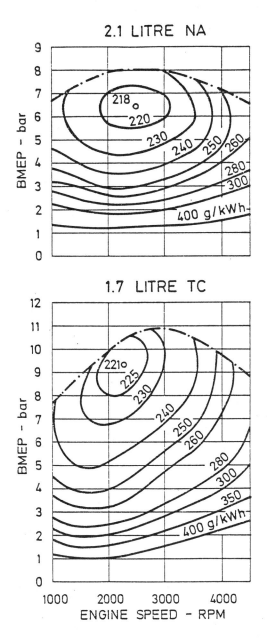

Fig 1 Fuel consumption maps of a 2.1 litre naturally aspirated and a 1.7 litre turbocharged multi-spray light duty DI diesel engine

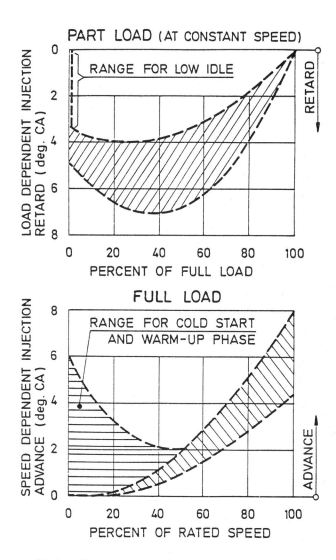

Fig 2 Requirements on injection timing dependent on engine speed and load for a low emission multi-spray DI diesel engine

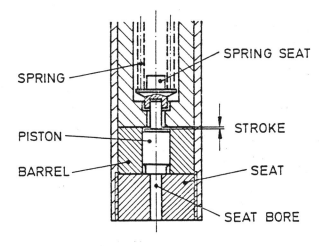

Fig 3 Split injection device (SID)

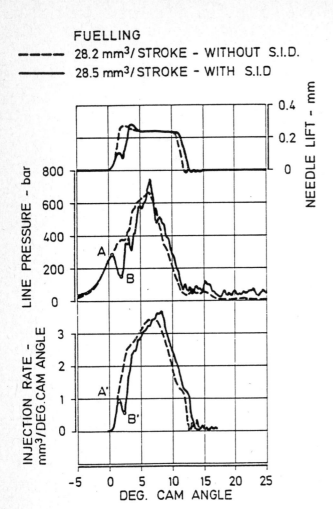

Fig 4 Effect of split injection on injection rate (injection rate shaping)

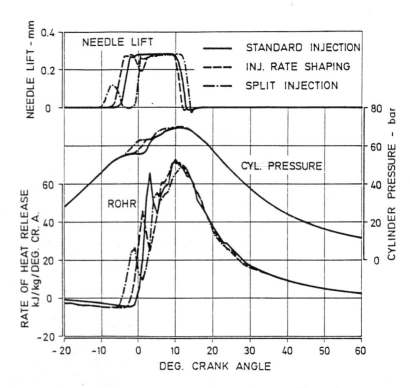

Fig 5 Effect of split injection on the rate of heat release (single cylinder research engine)

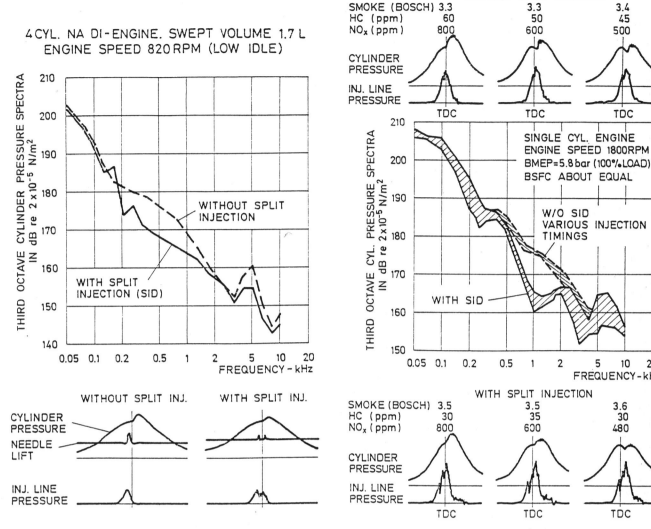

Fig 6 Third octave cylinder pressure spectra for low idle with and without split injection

Fig 7 Third octave cylinder pressure spectra and gaseous emissions at variable timing conditions — comparison between injection systems with and without SID

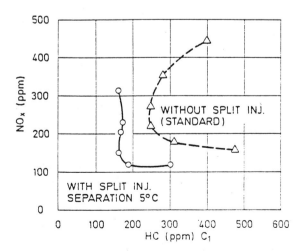

Fig 8 HC — NOx trade-off for low idle with and without split injection

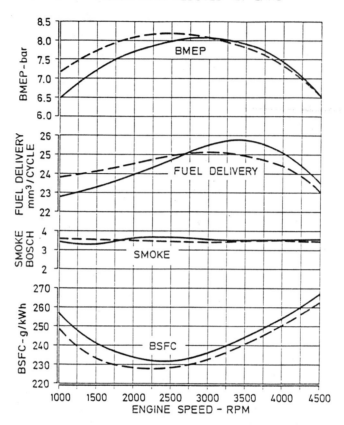

Fig 9 Improvement of full load performance by reduced dead volume

C434/84

Knock protection—future fuels and engines

D L SUTTON, BSc, CEng, MIMechE, MSAE and D WILLIAMS
The Associated Octel Company Limited, Engine Laboratory, Bletchley, Milton Keynes

SYNOPSIS As compression ratios are maximised for improved economy, engine-fuel inter-relationships become critical. Fuel composition effects are examined in conventional and high compression spark ignition engines.

1 INTRODUCTION

Because of the increased costs of crude oil, fuel economy legislation, and market place competition, improvements in vehicle fuel economy are being sought continuously. Attention to improvement is being concentrated in areas such as vehicle weight, drag and drive line efficiency. Considerable improvements have been made in car drive line efficiency and further improvements are to be expected.

Between 1945 and 1980 the combined efforts of automotive engineers and fuel technologists succeeded in raising the compression ratio of spark ignition engines from typically 6:1 to 9.5:1. This was made possible as high octane fuels became available which were able to resist engine knock; the spontaneous ignition of the unburned fuel/air charge ahead of the normally advancing flame front in the combustion chamber. Knock can quickly give rise to abnormal engine loads resulting in engine damage.

This trend towards higher ratios results from an unalterable law of thermodynamics which says that an engine's thermal efficiency is directly related to the compression ratio at which it operates. The operating cycle used by the majority of present day engines was first demonstrated by Dr. Nikolaus August Otto in 1876 and its efficiency can be expressed as follows:-

$$\eta = 1 - \frac{1}{r}^{(n-1)}$$

where η = thermal efficiency
    r = compression ratio
    n = ratio of specific heats

Fig.1 shows the effect of change of compression ratio on the maximum work available from the process by creating a larger expansion for the liberation of the chemical energy.

In actual fact the full advantage of compression ratio increases cannot be realised, the short fall being a result of other factors, for example a fall in the ratio of specific heats due to working with mixtures of air and fuel, energy expenditure in chemical dissociation, friction and pumping losses etc. Nevertheless, efficiency is improved for increases in geometric compression ratio up to around 14:1. The compression ratio of production model engines available in Europe has reached 10 - 12:1, and the use of higher ratios is limited, mainly because of knock.

The laboratory procedures for evaluating antiknock quality were first proposed in 1931 and are based on the Waukesha CFR fuel rating engine. The octane scale dates from 1926. Two methods were initially evolved:-

Method F1 or the Research method
Method F2 or the Motor method.

Since their introduction the suitability of these methods has been discussed continuously. Other procedures such as the R100 ON method or the Distribution Octane Number method have been used to evaluate the antiknock properties of particular fuel fractions. None of these laboratory procedures entirely predict the antiknock quality of fuels when operated in road vehicles. Because of this, fuel technologists find it necessary to test future fuels in current car designs then construct mathematical models to help predict future fuel blend quality. This paper describes tests carried out using current and future design engines with a range of possible fuel formulations.

2 DRIVE LINE EFFICIENCY

The search for optimised designs of combustion chamber to give higher knock free performance is continuing and is reported in the literature, (1), (2) and (3).

High compression ratio engine designs which give high engine efficiency also give high engine torque and power. The engine torque improvement associated with an increase in compression ratio can be used to operate the car at a lower numerical axle ratio. In this way the car accelerating performance remains unchanged but the engine is required to deliver increased torque at lower rotational speeds for any given driving condition. The wider throttle openings associated with the higher torque reduce gas pumping losses and a reduction in speed lowers friction, thus yielding further gains in fuel economy for the high compression ratio engine.

The most reliable data available for vehicle fuel consumption comparisons is published from the ECE 15 test procedure.

To be able to compare fuel efficiency with different drive lines, factors of aerodynamic drag rolling resistance and vehicle weight must be taken into account. The constant speed test data comparisons would have to make allowances for the aerodynamic drag. However, the urban cycle with its low average speed (18.7 km/hour) presents a possibility for test data comparison of drive line efficiency.

The energy used in the various modes of the cycle has been calculated to be of the following order:-

    Acceleration energy - 45%
    Rolling friction   - 35%
    Aerodynamic drag   - 13%
    Transmission losses -   7%

(This data was derived during work by a CEC group studying vehicle fuel consumption measurements).

Acceleration and rolling friction are the largest energy demanding components of the cycle and both are strongly influenced by vehicle weight. Hence if vehicle weight is taken into account then a good comparison of drive line efficiency for a range of cars is possible.

For most European cars the ECE 15 cycle fuel consumption is known, together with the kerb weight.

Drive line efficiency can be expressed as:-

ECE 15 cycle consumption litres/100 km
Vehicle kerb weight in 100 kg units

This concept is not new and has been presented in a paper by Doran K. Samples and Richard C. Wiquist, (4).

Figure 2 shows the comparison of cars marketed within the United Kingdom to run on premium or regular fuels. The figure includes data from manual transmission saloon cars only. A large scatter of results is shown but by comparing the best in group examples the benefits possible from premium fuel designs are clear.

Table 1 shows data from selected European cars where premium and regular model options existed. The compounding effect of compression ratio and gearing can yield very high gains in drive line efficiency.

3 FUTURE FUEL FORMULATIONS

With the trend towards higher compression ratios and the operation of engines at close to full load for cruise economy at the higher speeds, spark knock protection becomes a limiting parameter. The engine designer has to be cautious if engine damage is to be avoided. Engines may be used in many countries and knowledge of fuel quality and variations in quality is essential before he can finalise the engine design.

Figure 3 gives an indication of octane quality of European gasoline over the last eight years (5). Four of the major car producing countries have been chosen for comparison. Premium grade gasoline is generally specified as 97 or 98 RON in these areas but in Germany a MON of 88 units is specified. England and Italy specify 86 MON, while France has no MON specification.

Table 1 Fuel economy improvements with various vehicles designed to run at high compression ratios

CAR	VW GOLF 1093 cc		BL CITY 998 cc		CITROEN VISA SUPER E	
UK SPECIFICATION	C	'E' Formel	1981	1982	1980	1982
Compression ratio	8:1	9.7:1	8.3:1	10.3:1	9.2:1	10.2:1
Top gear	4.41	3.248	3.44	2.95	3.7	2.67
0 - 60 mph secs.	16.8	16.8	18.7	18.5	15.4	15.1
Urban consumption litres/100 km.	9.3	= 25.8% 6.9	7.3	= 15.1% 6.2	8.3	= 15.7% 7.0
90 kph litres/100 km	7.0	= 25.7% 5.2	5.8	= 19% 4.7	5.8	= 12.1% 5.1
120 kph litres/100 km	9.2	= 22.8% 7.1	8.5	= 24.7% 6.4	7.6	= 13.2% 6.6
*Overall gain		24.8%		19.6%		13.7%

* Based on Euromix fuel consumption - one-third urban, one-third 90 km/h, one-third 120 km/h

Work by Bell and others (6) shows that Research octane quality reflects the fuel antiknock performance at the low and medium engine speed range (say 2000 rev/min). At speeds of typically 3500 rev/min and above the Motor octane quality becomes the important indicator. The engine designer will have to allow for the worst case fuel and accept a potential wastage of engine fuel efficiency when the engine is over provided with octane quality.

Knock sensor systems are now being featured on some engines and these systems are arranged to sense spark knock in the engine then reduce spark advance levels to limit the occurrence of knock. In this way engines will self-adjust to suit the fuel quality provided but the spark advance control required for lower quality fuel will always incur penalties of engine fuel efficiency and power.

To maximise the quantity of gasoline obtained from every barrel of crude oil, refinery features are being introduced to process and crack the heavy fractions of the crude oil supply into lighter more useful products. The fuel components cracked from the heavy oil fractions are short of hydrogen (unsaturated) and frequently contain olefinic hydrocarbons (hydrocarbons with multiple bonds between some carbon atoms). Olefinic hydrocarbons blended into gasoline give good RON characteristics but their MON (high speed) antiknock properties are poor. This means that for a given premium RON quality the MON quality will fall (high sensitivity).

Oxygenates such as methanol or MTBE are now being considered as high octane blending components and their inclusion can effectively extend the gasoline supply. Most oxygenates give this high sensitivity characteristic to gasoline. To improve the understanding of future possible fuel formulations a programme of tests was carried out using five current European cars. A total of 104 different fuels were tested, selected to include the most likely commercial characteristics.

The main variables examined were:-

1. Olefin level   0%   -   15%

2. Lead level   0, .15, .4, g/litre

3. Lead alkyl type   Tetraethyl lead
   Tetramethyl lead
   Mixtures

4. Oxygenate type   Methanol
   (0%, 5%, 10% volume)   Ethanol
   MTBE
   Oxinol (50/50 methanol-TBA)
   TBA

A summary of the gasoline product data for the 104 fuels is shown in Table 2 and the Research and Motor octane quality of all the fuels is tabulated in Appendix 1.

The cars used for the test programme are listed in Table 3, and all testing was carried out on a single roll dynamometer. Repeated accelerating tests and then constant speed tests were made on all the fuels. For every fuel and test condition the knock limited spark advance (KLSA) was established and then a primary reference fuel to give equivalent spark knock determined.

Table 2 Gasoline product quality data

Test	Non-olefinic gasolines	Olefinic gasolines
Specific gravity. 60/60°F	0.762 - 0.781	0.732 - 0.765
Vapour pressure, bars	0.52 - 0.80	0.57 - 0.82
Distillation, recovery @ 70°C	18 - 40	27 - 48
@ 100°C	42 - 56	50 - 69
Octane numbers, Research	96.8 - 97.7	96.7 - 97.6
Motor	85.4 - 89.2	84.0 - 87.1
R100°C	81.7 - 97.0	90.5 - 97.7
Olefin content, % vol.	–	15.0

Despite the wide variations in the R100°ON fuel properties (see Table 2, the octane number of each fuel in the boiling range to 100°C) only small changes in KLSA were recorded during the accelerating tests. It was concluded that the cars used for testing were equipped with carburetter and intake systems which minimised fuel mixture fractionation to the cylinders. To achieve current levels of gaseous control designers must target for homogeneous and consistent fuel air supplies to all cylinders of an engine. Figure 4 shows some data from test car 3 under accelerating conditions. The KLSA for two test fuels and two PRF is shown in the top part of the graph. Also marked is the manufacturer's spark timing. The lower part of Figure 4 shows the manufacturer's spark setting deducted, giving a value of reserve, or spark knock protection (SKP). Both test fuels gave similar SKP at speeds to 2500 rev/min but the lower motor octane quality of one fuel caused a rapid deterioration of protection as the engine speeds increased above 3000 rev/min. This trend was recorded for all cars and the low MON fuels caused the most critical knock speed to be in the higher speed ranges. Because of this the constant speed tests serve best to compare the antiknock quality of the fuels. The constant speed tests were all carried out at the maximum octane requirement speed of each car being tested (worst case). At these constant speed conditions a strong relationship between MON and spark knock protection was found, see Figure 5.

To confirm our test consistency, one test fuel (trailer fuel) was tested in every car on every day to give a control on the consistency of the tests. Results were discarded where changes of weather or car condition influenced the trailer fuel results. When looking at the results shown in Figure 5 it must be remembered that all the fuels were close to 97 RON and the MON change is due to fuel composition effects. Generally most of the non olefinic fuels gave adequate spark knock protection and a MON quality of 86 units would give an average of six degrees of safety. The olefinic fuels, however, having lower MON quality gave less

Table 3 Technical details of test vehicles

COUNTRY OF ORIGIN	BRITAIN	SPAIN	BRITAIN	FRANCE	GERMANY
CAR TYPE (No.)	3-DOOR HATCHBACK (1)	3-DOOR HATCHBACK (2)	3-DOOR HATCHBACK (3)	4-DOOR SALOON (4)	2-DOOR COUPE (5)
Swept volume (cm^3)	998	1117	1597	1647	1921
Bore/stroke (mm)	64.59/76.2	73.96/64.95	79.96/79.52	79/84	79.5/77.4
Compression ratio	10.3:1	9.1:1	9.5:1	9.3:1	10.0:1
No. of cylinders	4	4	4	4	5
Valve gear	OHV-Rocker/Tappet	OHV-Rocker/Tappet	OHC-Toothed belt drive	OHV-Rocker/Tappet	OHC-Toothed belt drive
Max. power (BHP (DIN) @ rpm)	46 @ 5500	53 @ 5700	79 @ 5800	79 @ 5500	114 @ 5900
Max. torque (kg/m @ rpm)	7.47 @ 3300	8.16 @ 3000	12.7 @ 3000	12.5 @ 3000	15.7 @ 3700
Top gear (km/h @ 1000 rpm)	27	25.4	30.88	28.8	N/A
Transmission	4-speed manual	4-speed manual	4-speed manual	4-speed manual	5-speed manual
Fuel system	Single carburettor, constant vacuum	Single carburettor, down-draught	Single carburettor, constant vacuum	Single carburettor, down-draught	Single carburettor, twin barrel compound
Max. kerb wt. (kg)	747	727	765	940	1260

protection than their non-olefinic counterparts. A number of the test cars would not run without knock on many of the fuels and on average 86 MON quality would give four to five degrees of engine safety.

Different slopes are shown for the two fuel families in Figure 5 and it is interesting to note that for a unit reduction of MON fuel quality, a knock sensor equipped car would action less spark retard in olefinic fuels than in non-olefinic blends. In an effort to understand the influences of the various fuel components the results were grouped as shown in Figure 6. The vertical axis represents spark knock protection (SKP) while the base axes show the blend proportions. The corner of the diagrams showing zero oxygenate, and zero lead level, are shown in broken lines. These are extrapolated points for ease of viewing. Insufficient octane quality was available in the existing base hydrocarbon blends to allow 97 RON fuel quality to be made. The diagrams show that olefinic fuels blended to 97 RON quality using oxygenates to offset reductions in lead alkyl levels are likely to give high speed knock in current cars. All the oxygenate types tested gave similar characteristics. During this work no adjustments were made to the car carburetters, and fuel containing oxygenates was operated at the normal settings for hydrocarbon fuels.

4 FUEL EFFECTS ON FUTURE ENGINES

From the tests described so far it can be seen that at constant speed MON is a good guide for assessing the amount of spark knock protection. Future engines are likely to operate with the highest practical compression ratio and with lean air/fuel ratio settings.

The normal practice for motor manufacturers is to adjust full throttle maximum power at WMMP (Weakest Mixture for Maximum Power). If modern engines were designed to run at stoichiometric conditions at full throttle then appreciable fuel saving could be realised with some trade off in maximum power.

Figure 7 shows this comparison, which was recorded on a 2.2-litre four-cylinder petrol injection engine. Not all engine designs will react in exactly this way as imperfect fuelling or mixture distribution between cylinders can mask this effect.

A reduction in KLSA also occurs, showing that for a given fuel, less spark advance can be applied at the engine when at stoichiometric fuelling. Lean operation engines would have less spark knock protection on a given fuel and hence be at potentially higher risk for engine damage on poor octane quality fuels.

Many manufacturers are moving towards the stoichiometric operation for best full load efficiency.

The four-cylinder 2.2-litre engine had previously been modified by the Engine Laboratory (2) to run at high compression and be capable of operating with lean mixtures. A fuel injection system and breakerless ignition were fitted to improve engine running stability.

To explore fuel effects on this high compression lean burn engine a series of fuels listed in Table 4 were tested at constant speed engine conditions. The fuels were arranged in complementary pairs with different octane quality. The groupings of fuels in Table 4 were as follows:-

Fuels   Common parameters
A - E   High olefin content, high lead
B - F   High olefin content, low lead
C - G   Low olefin content, high lead
D - H   Low olefin content, low lead

These fuels were compared on a knock limited spark advance (KLSA) basis and also by matching the KLSA of the fuel to a Primary Reference Fuel (PRF) to give similar knock at wide open throttle conditions.

EON at 2500 rpm = .86 RON + .204 MON
EON at 4500 rpm = .232 RON + .873 MON

Less dependence on MON at stoichiometric conditions was indicated by the following equations:-

EON at 2500 rpm = .911 RON + .135 MON
EON at 4500 rpm = .402 RON + .67 MON

Table 4 Test gasoline samples - Inspection data

SAMPLE CODE	A	B	C	D	E	F	G	H
Specific gravity, 60/60°F	0.7632	0.7457	0.7609	0.7663	0.7304	0.7325	0.7406	0.7606
Vapour pressure, bar	0.42	0.41	0.48	0.53	0.58	0.55	0.48	0.52
FIA analysis,								
Saturates, % vol	37.2	57.2	52.1	47.7	55.4	52.3	63.6	55.3
Olefins, % vol	22.2	22.0	0.6	0.5	23.8	23.8	0.8	0.9
Aromatics, % vol	40.6	20.8	47.3	51.8	20.8	23.9	35.6	43.8
Lead content, g/litre	0.35	0.15	0.39	0.15	0.38	0.16	0.40	0.17
Lead type	TEL	TEL	TML	TEL	PM30	TEL	TEL	PM90
Distillation,								
IBP, °C	36	40	36	35	34	34	36	35
10% @ °C	54	63	49	50	48	49	49	49
30% @ °C	75	83	64	78	62	63	59	65
50% @ °C	102	99	98	106	84	82	79	102
70% @ °C	124	112	128	132	102	107	115	127
90% @ °C	146	144	148	150	140	139	146	147
FBP, °C	179	188	178	178	172	170	172	174
Octane numbers,								
Research	99.6	98.2	99.9	99.7	96.7	96.7	96.0	97.0
Motor	87.2	86.1	92.1	90.4	85.8	85.1	89.8	89.3
R100°C	93.5	94.9	87.2	85.4	91.5	91.2	76.7	86.7
Sensitivity	12.4	12.1	7.8	9.3	10.9	11.6	6.2	7.7
Delta R100°C	6.1	3.3	12.7	14.3	5.2	5.5	19.3	10.3

GLOSSARY:
R100°C = The Research octane number of hydrocarbons having a boiling point < 100°C in any given fuel
Sensitivity = Research octane number - Motor octane number
Delta R100°C = Research octane number - R100°C
TEL = Tetraethyl lead
TML = Tetramethyl lead
PM30 = A mixture of 30% TML and 70% TEL (example)

Engine octane quality values could be allocated to each test fuel and engine condition in terms of primary reference fuel (PRF) and were as shown in Table 5.

At 2500 rpm all fuels gave antiknock quality in terms of the primary reference fuels higher than their respective RON, showing 'engine appreciation'. At 4500 rpm the appreciation effect also occurred (EON-RON) but to a smaller degree. This is unusual as tests with conventional engines with compression ratios of 9.1, say, tend to give low-medium speed fuel ratings close to the fuel RON and depreciate fuels toward the MON at higher engine speeds (6).

Judging from these results the PRF ratings follow a different pattern in the high compression engine. From Table 5 it can be seen that the engine octane requirements exceed 101 units but tests with full boiling range fuels show that the engine will operate with 98 RON.

Prediction equations from this particular series of tests were developed and results indicated that the EON values had a strong correlation with RON at 2500 rpm. Equations to describe the fuel performance at WMMP conditions can be derived:-

Table 5 Engine octane number, primary reference fuels

SAMPLE CODE	CFR ENGINE RATING		ENGINE OCTANE NUMBER (EON)			
	RON	MON	WMMP		STOICHIOMETRIC	
			2500 rpm	4500 rpm	2500 rpm	4500 rpm
A	99.6	87.2	103.5	100.6	102.4	99.5
B	98.2	86.1	101.7	96.6	100.9	96.2
C	99.9	92.1	104.1	102.7	102.4	101.5
D	99.7	90.4	104.4	102.6	102.7	101.0
E	96.7	85.8	101.6	97.5	100.4	97.0
F	96.7	85.1	99.8	95.8	98.9	94.3
G	96.0	89.8	100.6	99.9	99.6	98.1
H	97.0	89.3	102.4	102.1	101.0	99.4
PRF engine octane requirement			101.3	95.6	101.4	96.3

Additional analyses to identify the effects on the prediction equations of other gasoline parameters such as olefin content and lead levels were also undertaken. These showed that the additional fuel parameters had no marked effects on the prediction equations for EON at 2500 rpm and essentially replaced the effect of the MON variable at 4500 rpm test conditions.

As discussed earlier, it was found that with the high compression ratio engine tested, primary reference fuels can over-estimate the engine octane requirement. To avoid this problem an alternative method of predicting RON and MON requirements for this particular series of fuels was devised. It does not depend on the EON primary reference rating scale but takes the engine requirement from a knowledge of the knock limited spark advance readings for fuels in a particular group.

Figure 8 demonstrates the method used. The RON and MON values for each test fuel can be illustrated together with the KLSA values at a given engine condition. By studying the change in KLSA over each pair of fuel comparisons and knowing the optimum ignition timing the octane requirement of the engine condition can be derived. This procedure avoids the need to compare fuels on a PRF standard and several repeated tests at each condition can confirm the validity of each point. Several ways to compare engine fuel compositions thus become possible. Tables 6 and 7 give the results of this method for the fuel groups A to H at 2500 rpm and 4500 rpm.

For many years discussion has centred around a lead bonus effect. This means that a leaded fuel will give better antiknock performance than an unleaded equivalent in a real engine, in fact more than would be predicted by the conventional test method of RON and MON.

Table 6 Table of derived fuel octane quality requirements, 2500 rev/min wide open throttle

FUEL BLENDS (refer Table)	FUEL CHARACTERISTICS LEAD g/l	OLEF.% VOL	DERIVED ENGINE OCTANE REQUIREMENT RON	MON
WMMP ENGINE CONDITIONS				
A-E	0.35-0.38	>20	96.2	85.6
B-F	0.15-0.16	>20	97.7	85.8
C-G	0.39-0.40	<1	96.5	90.1
D-H	0.15-0.17	<1	95.5	88.7
STOICHIOMETRIC ENGINE CONDITIONS				
A-E	0.35-0.38	>20	98.0	86.4
B-F	0.15-0.16	>20	98.7	86.4
C-G	0.39-0.40	<1	98.1	91.0
D-H	0.15-0.17	<1	97.3	89.4

Table 7 Table of derived fuel octane quality requirements, 4500 rev/min wide open throttle

FUEL BLENDS (refer Table)	FUEL CHARACTERISTICS LEAD g/l	OLEF.% VOL	DERIVED ENGINE OCTANE REQUIREMENT RON	MON
WMMP ENGINE CONDITIONS				
A-E	0.35-0.38	>20	95.5	85.3
B-F	0.15-0.16	>20	96.3	84.7
C-G	0.39-0.40	<1	91.2	86.9
D-H	0.15-0.17	<1	91.0	86.8
STOICHIOMETRIC ENGINE CONDITIONS				
A-E	0.35-0.38	>20	95.7	85.4
B-F	0.15-0.16	>20	98.5	86.3
C-G	0.39-0.40	<1	95.1	88.8
D-H	0.15-0.17	<1	93.6	88.0

Throughout the test results a lead bonus effect occurs with olefinic fuels but not with non-olefinic fuels. (Compare results A-E with B-F, and also C-G with D-H in Tables 6 and 7). When the lead levels in gasoline are reduced then other high octane blending components have to be introduced to replace the lost lead antiknock effect. Depending on the base fuel and the way that lost octane quality is restored by other components then a fuel inferior or superior to the leaded original can result.

The effect of including olefinic components in the fuels can also be identified. The worst case tested to illustrate this effect is 4500 rev/min (higher engine speed). If olefinic fuel blends are used the Research octane quality to satisfy the engine becomes greater than that for non-olefinic fuels. The increase in Research octane quality required can offset the reduction in Motor octane quality associated with higher olefinic fuels. (Compare results B-F with D-H in Table 7).

From earlier in this paper it can be seen that fuel savings can be made by arranging engines to operate at conditions approaching stoichiometric air/fuel ratio when at full load, although this stoichiometric operation means a minor loss of power. The high compression ratio engine operating at stoichiometric air/fuel ratio will normally have a full throttle power advantage over its 9:1 counterpart operating at WMMP (optimum fuelling). To take full advantage of the fuel economy gains possible, using high compression lean burn, the octane quality should be maintained at the highest practical level. For example compare the derived engine octane requirement for any fuel type at the two air/fuel ratio settings shown on Tables 6 and 7. Stoichiometric engine operation requires fuel of 1 to 2.5 ON units higher than the case for WMMP operation.

5 CLOSURE

1 Recent changes to car engine design and compression ratio, coupled with carefully selected gearing arrangements, have yielded significant improvements in fuel economy.

2 If oxygenates are used to offset octane quality reduction when lead in gasoline is reduced a reduction in Motor octane quality will occur. Current vehicles will experience a significant deterioration in spark knock protection unless careful control of Motor octane quality is maintained.

3 Tests with an experimental high compression ratio engine show that Primary Reference Fuels over-estimate the octane requirement of the engine. Similarly, comparisons with Primary Reference Fuels can give misleading quality standards to full boiling range fuels compared in high compression ratio engines.

4 Engines can be made to operate successfully at high compression lean burn conditions using 98 RON fuels. Using this approach, appreciable fuel economy improvements are possible.

5 A method of deriving octane requirement for comparison of fuel quality can be used which avoids the use of Primary Reference Fuels.

With high compression lean burn engines of the type tested, Research Octane quality must be maintained at the highest predicted level for best engine efficiency. This is of increasing importance with the use of high olefinic fuels having reduced lead levels.

REFERENCES

(1) GRUDEN, D. and HOEHSMAN. G.H. Field experience acquired with high compression thermo-dynamically optimised Porsche (TOP) engines.
IN: XIX International Fisita Congress, November 8-12, 1982, Melbourne, Australia.

(2) SUTTON, D.L. Combustion chamber design for improved performance and economy with high compression lean burn operation.
Paper presented at the SAE Congress, Detroit, USA, February 1983.
(SAE Paper No.830336)

(3) DOWNS, D. The passenger car power plant: future perspectives.
IN: XIX International Fisita Congress, November 8-12, 1982. Melbourne, Australia.

(4) SAMPLES, Doran K. and WIQUIST, Richard C. TFC/IW
Warrendale, SAE 1978
SAE preprint 780937

(5) GROUP GASOLINE SURVEY, years 1976 to 1983 incl.
The Associated Octel Company Limited, Engine Laboratory, Bletchley.

(6) BELL, A.G. The relationship between octane quality and octane requirement.
Warrendale, SAE 1983.
SAE preprint 750935.

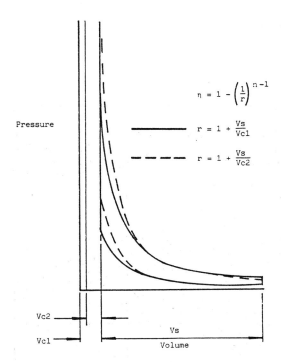

Fig 1 Otto cycle and compression ratio effect

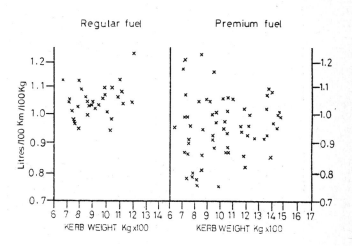

Fig 2 Comparison of fuel efficiency on urban cycle — Regular and Premium fuelled cars

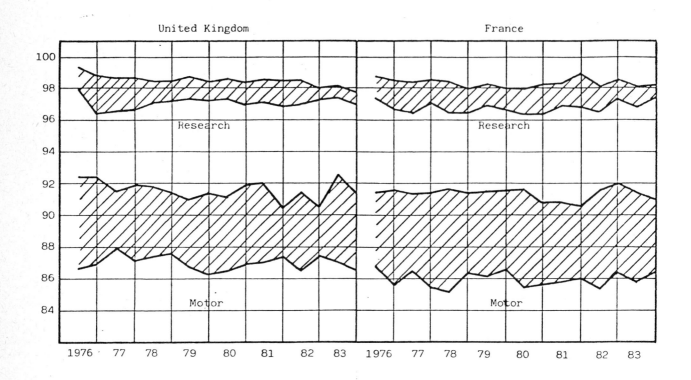
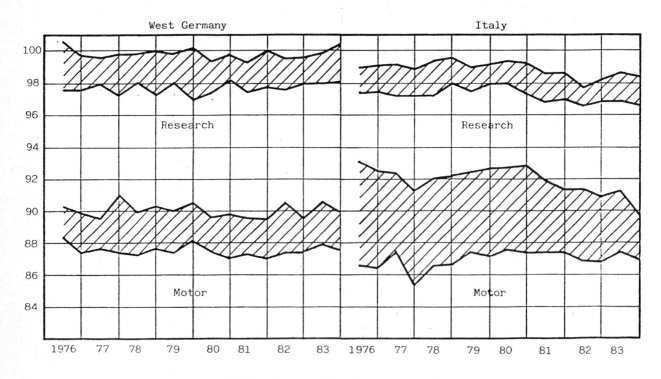

Fig 3 Octane quality — European gasolines

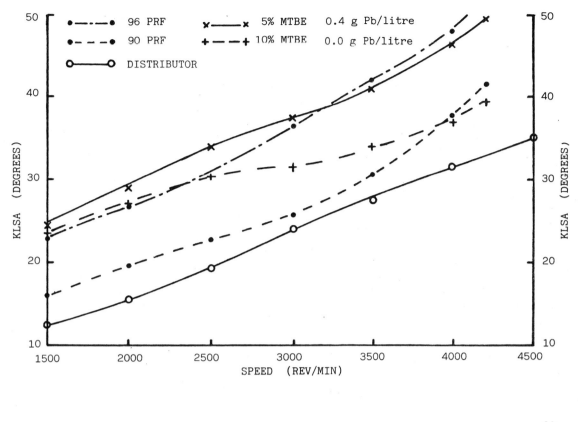
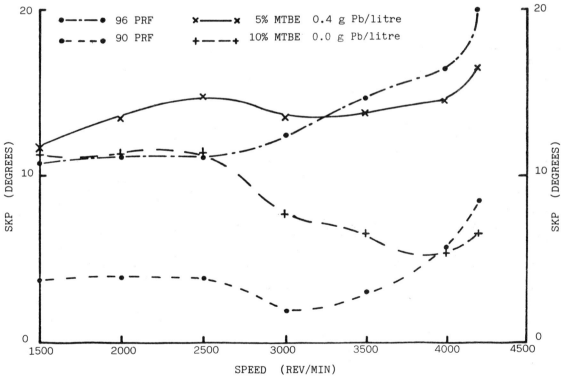

Fig 4 Knock limited spark advance and spark knock protection through full throttle acceleration

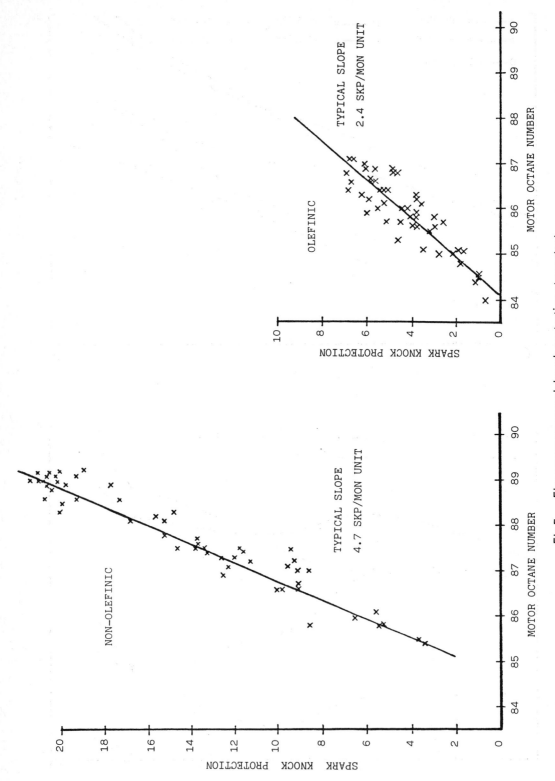

Fig 5 Five-car average spark knock protection at constant speed – non-olefinic and olefinic fuels

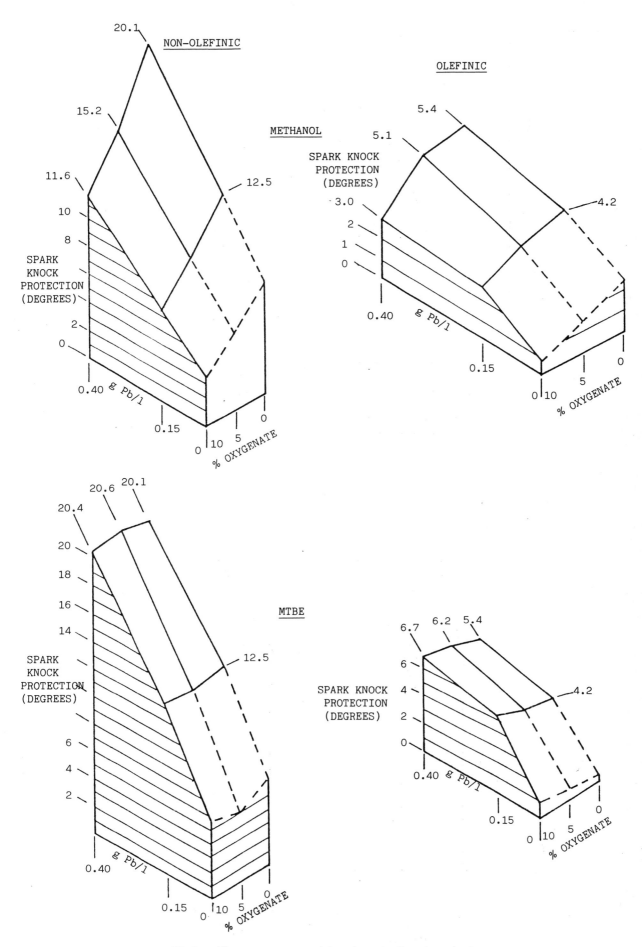

Fig 6 Five-car average spark knock protection at constant speed showing effects of oxygenate and lead additions

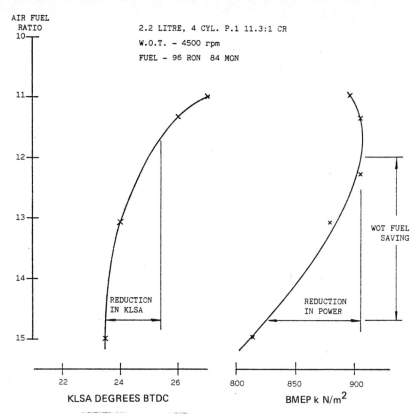

Fig 7　Effect of air/fuel ratio on engine torque and knock limited spark advance

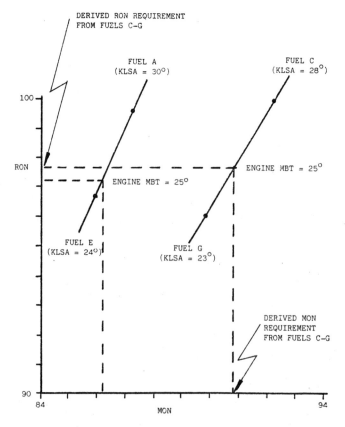

Fig 8　Method of deriving full boiling range fuel quality, RON and MON, for a given engine condition. 2500 r/min

LABORATORY OCTANE RATINGS
OLEFINIC AND NON-OLEFINIC BLENDS

APPENDIX 1

PARAMETER: AVERAGE RON OLEFINIC GASOLINES						
OXYGENATES	LEAD ALKYL (G/LITRE)					
	UNLEADED	0.15 TEL	0.15 TML	0.40 TEL	0.40 PM50	0.40 TML
NIL	-	97.2	-	97.2	97.1	97.1
5% METHANOL	-	97.1	-	97.2	97.0	96.9
5% ETHANOL	-	97.1	-	97.1	96.9	97.1
5% MTBE	-	97.2	-	97.0	96.8	97.0
5% 1:1 METHANOL + TBA	-	97.0	-	96.9	97.0	97.1
5% T-BUTANOL (TBA)	-	97.2	-	97.0	96.9	97.0
10% METHANOL	97.2	97.2	-	97.0	97.1	96.9
10% ETHANOL	97.1	97.1	-	96.9	97.0	96.9
10% MTBE	97.3	97.1	-	97.2	96.9	96.8
10% 1:1 METHANOL + TBA	97.1	97.1	-	97.2	96.9	97.0
10% T-BUTANOL (TBA)	97.4	97.2	-	97.5	97.3	97.0
PARAMETER: MON OLEFINIC GASOLINES						
NIL	-	86.0	-	86.4	86.6	86.8
5% METHANOL	-	85.1	-	85.7	85.9	86.3
5% ETHANOL	-	85.5	-	85.6	85.8	86.2
5% MTBE	-	85.3	-	86.3	86.2	86.6
5% 1:1 METHANOL + TBA	-	85.6	-	86.0	86.4	86.9
5% T-BUTANOL (TBA)	-	85.7	-	85.9	86.7	87.0
10% METHNOL	84.0	85.0	-	85.6	86.1	85.8
10% ETHANOL	84.5	85.0	-	85.1	84.8	85.7
10% MTBE	84.4	86.0	-	87.1	86.4	86.6
10% 1:1 METHANOL + TBA	84.5	85.8	-	86.9	86.8	86.1
10% T-BUTANOL (TBA)	85.1	86.4	-	86.8	86.9	87.1
PARAMETER: AVERAGE RON NON-OLEFINIC GASOLINES						
NIL	-	97.2	97.2	97.2	97.5	96.9
5% METHANOL	-	97.4	97.1	97.1	97.2	97.2
5% ETHANOL	-	97.4	96.9	97.5	97.5	97.1
5% MTBE	-	97.5	96.9	97.1	97.5	96.8
5% 1:1 METHANOL + TBA	-	97.1	96.9	97.1	97.4	96.8
5% T-BUTANOL (TBA)	-	97.0	96.9	97.4	97.6	97.0
10% METHANOL	97.2	97.3	-	97.2	97.3	97.2
10% ETHANOL	97.2	97.3	-	97.4	97.3	97.0
10% MTBE	97.0	96.9	-	97.1	97.1	97.1
10% 1:1 METHANOL + TBA	96.9	97.0	-	97.5	97.1	97.4
10% T-BUTANOL (TBA)	97.1	97.4	-	97.1	97.4	96.8
PARAMETER MON NON-OLEFINIC GASOLINES						
NIL	-	86.9	87.6	88.3	88.6	88.7
5% METHANOL	-	86.7	87.1	87.8	88.4	88.4
5% ETHANOL	-	86.6	87.0	87.7	88.1	88.3
5% MTBE	-	87.3	87.4	88.4	89.0	89.0
5% 1:1 METHANOL + TBA	-	86.6	87.2	88.1	88.5	88.7
5% T-BUTANOL (TBA)	-	87.0	87.5	88.6	88.7	88.6
10% METHANOL	85.4	86.0	-	87.4	87.5	87.5
10% ETHANOL	85.5	85.8	-	87.0	87.2	87.5
10% MTBE	86.1	87.1	-	88.3	88.5	89.2
10% 1:1 METHANOL + TBA	85.8	86.6	-	88.2	88.1	88.6
10% T-BUTANOL (TBA)	85.8	87.3	-	88.6	88.5	88.4

C424/84

SAE 841290

A cold-start track test procedure for evaluating fuel-efficient oils

G B TOFT, CEng, MIMechE
Shell Research Limited, Thornton Research Centre, Chester

SYNOPSIS A cold-start test procedure has been developed for use on a test track to show the benefits of fuel-efficient engine lubricants. Using this procedure the fuel economy benefits of a friction-modified SAE 15W/40 motor oil over a conventional SAE 20W/50 motor oil have been evaluated in a fleet of eight cars. These cars, which were chosen to be representative of the UK market, ranged in engine size from one to two litres.

Although some cars showed large benefits for the lower viscosity friction-modified oil, especially during the first mile from cold start, other cars showed little response to the oil change. Because of the limited number of tests in each car on each oil and the day-to-day variations in ambient conditions, the differences in fuel consumption measured in single cars were not significant at the 95% confidence level. When the eight cars were considered as a single fleet the fuel consumption benefit for the lower viscosity friction-modified oil became highly significant even for the early laps when repeatability is at its worst.

The benefits in fleet fuel consumption for the lower viscosity friction-modified oil varied from 5.95% during the first mile from a cold start to 1.94% when fully warmed up after eleven miles.

Experience with the test procedure has also shown the importance of using a low volatility fuel for making the fuel consumption measurements.

1 INTRODUCTION

As gasoline costs increase the fuel economy of engines becomes more and more important. Small but significant improvements in fuel economy can be obtained by reducing engine friction either by reducing viscosity or by the use of a friction modifier. In practice it is extremely difficult to measure small differences in fuel economy because of the large effect of uncontrolled variables such as ambient conditions, traffic conditions and driver behaviour.

A new test procedure which can be carried out on a test track has therefore been developed and using this procedure it has been possible to show significant differences in fuel economy between a conventional SAE 20W/50 grade oil and a friction-modified (FM) SAE 15W/40 grade.

The oils were tested in a fleet of eight cars ranging in capacity from 1.1 to 2.0 litres. The fleet was chosen to be representative of the UK car population at the time of planning the tests. The tests were run during the late winter months and thus the low ambient temperatures experienced probably maximised the benefit obtained by reducing viscosity.

2 TRACK TEST PROCEDURE

2.1 Test cars and preparation for test

The test fleet contained the following cars, which were numbered as shown:
(1) Ford Cortina 1.99 l
(2) Austin Metro 1.27 l
(3) Volkswagen Polo 1.09 l
(4) Renault 5 1.29 l
(5) Datsun Sunny 1.4 l
(6) Ford Escort 1.29 l
(7) Vauxhall Astra 1.3 l
(8) Talbot Alpine 1.44 l

Four of the cars (1, 3, 6 & 7) had overhead camshafts, the other four had overhead valve, pushrod engines.

With the exception of the Datsun all the cars were less than one year old when purchased and had completed mileages of between 5000 and 15000.

Before starting the test each car received a full diagnostic check and any faults found were rectified. New spark plugs, ignition contact points, oil and air filters were fitted and the ignition timing, idle speed and mixture strength were adjusted to the manufacturers recommended settings. Tyres and exhaust systems were checked and replaced where necessary to ensure that the test could be completed without having to replace these items.

2.2 Instrumentation and modifications

Each of the cars was fitted with thermocouples in the sump (thermocouple attached to end of replacement dipstick) and in the top water hose. The leads from these thermocouples and from tachometer connections on the ignition coil terminated in sockets mounted on a panel in the car. Temperature readings were obtained by plugging small battery-powered digital thermometers into the appropriate sockets in the panel.

Two sets of these thermometers together with electronic tachometers were moved from car to car when tests were carried out.

The fuel system was modified to enable 1 gallon (4.5 litres) fuel cans to be quickly connected for fuel consumption measurements. This entailed extending the fuel line from the tank to the fuel pump up into the rear luggage space, and then splitting the line and fitting 'quick-connect' connectors to each end so that when the two ends were connected together the car could run on its original fuel tank. When measuring fuel consumption the end connected to the fuel pump was plugged into the fuel can and another breather pipe being also plugged into the can, the open end of this pipe being passed to the outside of the car. When the 'quick-connect' connections are disconnected both ends automatically seal, thus avoiding any fuel loss or fuel movement in the fuel lines.

The fuel cans were located in wooden blocks attached to the luggage space floor by four magnets.

In order to control the rates of choke (in the cars with manual choke) and throttle deployment during the track tests and to improve test repeatability, choke and throttle stops were manufactured before the start of the test. The choke stops were simple spacers that were hooked over the shaft of the choke knob to prevent it returning to the fully closed position. Two spacers were used, one giving 0.8 of full choke and the other giving 0.5.

The throttle stop was designed to be universal and fully adjustable. It was clamped to the pedal shaft just above the foot pedal and had an adjustable stop that came into contact with the floor or bulkhead panel at the desired throttle position. The position of the throttle stop was adjusted so that when the car was fully warmed up only moderate acceleration rates were possible. Since these cars had to be driven on public roads as well as on the track the position of the stop had to allow reasonable maximum speeds. Once fitted and set up the throttle stops were not removed or adjusted for the duration of the test.

2.3 Procedure at the track

The track test procedure is described in detail in Appendix A. Briefly it consisted of parking the cars in open-fronted garages for an overnight soak prior to a fifteen lap, cold-start fuel consumption test using the one mile circuit at the test track, Figure 1. On cars fitted with manual chokes the choke deployment rate was controlled by using choke stops and the rate of acceleration was controlled by the use of throttle stops on all cars. Gear changes were made at set engine speeds, and cornering speeds were selected that did not require the use of the brakes at any point on the circuit other than at the start/finish line. The average speed for the driven part of the cycle was 42 km/h ±3 kph depending on car size and lap number from cold start (when the thirty second idle period is included this falls to 36.8 km/h).

To prevent carry-over of any effect of the friction modifier the cars were deconditioned after each run on the test oil by running for one hundred and fifty miles on the reference oil.

The order of testing varied from day to day as shown in the test design, 2.4. The ambient temperature and humidity was measured by using a wet and dry bulb hygrometer. These readings together with wind speed and direction and barometric pressure were recorded frequently throughout each test.

Tests were not commenced if the wind speed exceeded 20 mile/h (32 km/h) or if the humidity exceeded 93%. The tests were carried out in the months of February, March and April 1982. The mean temperature at the start of each test was 5.62°C.

2.4 Test design

A statistically designed sixteen day test was adopted using the eight cars and two drivers. The cars were numbered 1 to 8 as shown in 2.1 and driver A drove the odd numbered cars and driver B the even numbered. The oils were coded test oil (T) and a reference oil (R). Four cars were tested each day (two per driver) each starting from cold. The testing sequence is shown below.

Day	1	2	3	4	5	6	7	8	9	10	11	12	13	14	15	16
Cars tested in early morning	1 2	5 6	3 4	7 8	1 2	5 6	3 4	7 8	1 2	5 6	3 4	7 8	1 2	5 6	3 4	7 8
Oil	R	R	R	R	T	T	T	T	T	T	T	T	R	R	R	R
Cars tested in late morning	3 4	7 8	1 2	5 6	3 4	7 8	1 2	5 6	3 4	7 8	1 2	5 6	3 4	7 8	1 2	5 6
Oil	T	T	T	T	R	R	R	R	R	R	R	R	T	T	T	T

2.5 Oil change procedure

All tests were made on a fresh charge of the appropriate oil. After fifteen laps the oil and filter were changed. A charge of the next oil to be tested was used for flushing. On cars changing from reference to test oil, from test oil to test oil and from reference oil to reference oil the flushing run of three laps of the one mile circuit was followed by a final change to the next oil before parking the cars in the garages.

On cars changing from test oil to reference oil the de-conditioning run was carried out on the flushing charge of reference oil, the final charge of reference oil was added before parking these cars in the garages in the morning after the de-conditioning run.

2.6 Fuel consumption measurements

Fuel consumption on the test track was measured gravimetrically using 1 gallon (4.5 litre) cans weighed before and after driving each lap of the test circuit. The cans were weighed on a Sartorious 3802 MP electronic balance with a data print-out facility. This balance weighs to an accuracy of 0.1 g.

2.7 Choice of test fuel

During a previous test using the same car fleet but carried out during the months of July and August severe hot-fuel handling problems were experienced when running on a summer grade premium gasoline.

This led to a marked deterioration in the repeatability of the fuel consumption measurements.

To prove that the effects were due to hot-fuel handling problems, tests were carried out in two cars on a warm day using iso-octane (which has a very low volatility) as the test fuel. At the start of these two tests the fuel in the fuel line, pump and carburetter was the original test fuel but with iso-octane in the test can so that as the car warmed up the original fuel was quickly replaced by iso-octane; after about four laps the fuel being used was pure iso-octane.

The test results (illustrated by Figure 2) show that when iso-octane was used test repeatability was satisfactory (1% or better) whereas with the original test fuel repeatability was poor.

Despite the large differences in lap-to-lap fuel consumption on the original fuel the drivers did not notice any driveability problems.

Before starting the tests reported here a fuel of suitable volatility was blended. This gasoline had a RVP of 447 mbar and ASTM Distillation points of IBP 37.5°C, 10% 59.0°C, 50% 100.0°C, 90% 160°C and FBP 199.5°C. This fuel is now used for all lubricant fuel economy studies on the test track and on the chassis dynamometer where it appears to be satisfactory for all cars under the ECE 15 cycle driving conditions.

2.8 Choice of test oils

The SAE 15W/40 FM oil was a premium grade motor oil to which an ashless soluble friction modifier had been added. The friction modified oil was of API SF/CC and CCMC G2 performance level. Bearing in mind the importance of "effective" viscosity of the oil in the engine (1) the conventional SAE 20W/50 grade reference oil was chosen so that it had similar shear stability characteristics to the test oil. Figure 3 shows plots of dynamic viscosity versus shear rate for the two oils.

Other viscometric data for the two oils are shown below

Viscometric data

Oil	V_K 100°C, cSt	V_K 40°C, cSt	VI_E	V_d -18°C, poise
Test oil	13.91	100.1	141	41.2
Reference oil	18.66	152.6	138	78

Tests in the Thornton low velocity four ball friction machine (2) with centrifugal cast iron test pieces showed the coefficient of boundary friction (μ) of the test oil to be 20% less than that of the reference oil.

3 TEST RESULTS

As explained in the statistical Appendix (B) the results were analysed car by car and for the fleet as a whole. In each case separate analyses were made for lap 1, laps 1 to 5 and laps 11 to 15 to give a measure of cold start, short trip and fully warmed-up fuel consumption. The test results and results of the analyses are given in Tables 1 to 3 and are illustrated in the form of bar charts in Figures 4, 5 and 6.

The mean maximum oil temperature obtained on the test oil was 83.5°C which is 2.3°C less than the 85.8°C mean attained by the reference oil.

The overall results show that there is a significant difference (at 99.9% confidence level) in fuel economy between the two oils under all three conditions of test.

During lap 1, where the repeatability was at its poorest, a fuel economy benefit of 5.95% was seen.

During laps 1 to 5, (typical of short-trip motoring) the repeatability was better and the fuel economy benefit was 2.76%.

During laps 11 to 15 when the cars were fully warmed up the repeatability was good but the benefit for the test oil had dropped to 1.94%.

This is entirely consistent with what one would expect, i.e. that there is a greater difference during the cold start/warm-up period and a smaller difference under fully warmed-up conditions.

The picture for the individual cars is rather different in that because the number of data points is smaller none of the differences are significant at the 95% confidence level.

Similarly it is not possible to say categorically whether or not there are differences between cars in their response to the change in oil. Visual inspection of the results of two extreme cases, Figure 7 and Figure 8, suggests however that some cars may indeed be more responsive to change in viscosity/friction modification than others. Only further testing would prove this point one way or the other.

In an earlier field trial with 109 cars comparing a friction-modified SAE 10W/40 grade oil with a SAE 15W/50 reference oil a benefit of 1.9% (significant at the 95% confidence level) was seen for the 10W/40 FM oil.

4 COMPARISON WITH OTHER TEST PROCEDURES

A number of test methods for measuring the effect of engine oils on fuel economy have been described in the literature.

These include bench engine tests using the GFC procedure based on the Renault 20 TS engine (3), the ASTM five-car test procedure (4) and measurements made over long periods of time with large test fleets (2).

It is considered that this new procedure is to a certain extent complementary to these existing methods in that it stands somewhere between the bench tests and the ASTM five-car test in terms of both realism and cost. Furthermore initial results suggest that the overall values obtained in this test are consistent with the results of large-scale field trials.

5 SUMMARY AND CONCLUSIONS

(1) A carefully controlled track test has been developed to evaluate the effect of changes in engine oil on fuel economy in gasoline engines.

(2) Using an eight-car test fleet the observed differences in fuel consumption between a conventional SAE 20W/50 oil and an SAE 15W/40 friction-modified oil ranged from 5.95% under cold-start conditions to 1.94% under fully warmed-up conditions.

(3) Some of the cars tested appeared to show larger changes in economy than others but these differences were not statistically significant and further work would be needed to establish whether they were real or not.

(4) This test is seen to be complementary to existing test procedures and first indications are that the results obtained are consistent with the results of large-scale field trials.

(5) It was found that the use of an involatile fuel improved the repeatability of the measurements in this test and it is recommended that such fuels should be used for future fuel economy test work involving stop/start driving conditions.

ACKNOWLEDGEMENT

The author wishes to thank many of his colleagues in Shell Research Ltd. for the assistance in the development of this test procedure, in particular, Mr. P.J. Zemroch for his statistical evaluation of the test data, and Mr. G. Humphreys and Mr. L.E. Chew who carried out the track testing.

REFERENCES

(1) DOBSON, G.R. The prediction of fuel efficiency of engine oils. Paper No. EL/4/7 presented at CEC International Symposium on the Performance Evaluation of Automotive Fuels and Lubricants. Rome June 1981.

(2) WINTER, S.J., TOFT, G.B., MOORCROFT, D.W. and REDERS, K. Test Methods for the Evaluation of Fuel Economy Motor Oils. Erdöl & Kohle Erdgas Petrochemie, Nov. 1981, E2686 EX, p.492-496.

(3) G.F.C. Groupement français de coordination Comité Technique Lubrifiants Moteurs. Différences de Consommation de Carburant Liées au Lubrifiant Moteur. Méthode de Mesure à Chaud Sur Banc Moteur. G.F.C. L.007.83.

(4) The American Society for Testing and Materials. ASTM Fuel-Efficient Engine Oils Task Force. Report on the Pilot and Demonstration Programs February 1982.

APPENDIX A

Cold-start track test procedure for evaluating the effect of lubricant on fuel consumption during warm-up using the one mile circuit

Instructions to drivers

1. Record required data in garage before starting.

2. Start on full choke and immediately change to 0.8 choke.

3. Drive to start line.

4. Reduce choke to 0.5 and idle for 15 seconds. If necessary increase choke to maintain idle but return to 0.5 at 10 seconds.

5. Switch off and soak for 60 seconds. During

this period connect pre-weighed one gallon can into fuel system and record data.

6. Start engine at 60 seconds and idle for 30 seconds (still on 0.5 choke). Record idle speed.

7. At 1 minute 30 sec (i.e. end of 30 seconds idle) drive off in first gear changing gear at 2000 rev/min through second gear up to 30 mile/h (48 km/h) in third gear. The throttle must be on the stop for all accelerations. Maintain 30 mile/h in third gear to start of first corner on short circuit (marked by first cone). At this point lift off throttle completely and declerate down to 20 mile/h (32 km/h) (still in third gear) through the corner. Drive on the outside (to left) of cones. At last marker cone in corner accelerate (throttle on stop) back up to 30 mile/h, still in third gear.
Maintain this condition until first marker cone in second corner. At this point declerate (foot off throttle) to 20 mile/h. Maintain this speed (third gear) through the corner, driving on the outside (to left) of marker cones to last cone where accelerate (throttle on stop) to 2000 rev/min. AT 2000 rev/min change into top gear and continue accelerating (throttle on stop) up to 40 mile/h (64 km/h).
Maintain this speed, driving in centre of track until the 'H' marker board. At 'H' lift off throttle completely and change down into third gear as you pass to the right of the first marker cone for the third corner. Continue to decelerate, without braking, round corner driving on the inside (to the right) of the marker cones, down to 20 mile/h. Maintain this speed through the remainder of the corner. At the last marker cone commence acceleration (throttle on stop) up to 2000 rev/min where change up to top gear and adjust speed to 30 mile/h. Maintain this speed up to 'I' marker where lift off and decelerate, still in top gear (foot off throttle) to stop on start/finish line.

8. Immediately switch off engine and stop/reset the watch. The choke spacer is now removed and the fuel can be removed for weighing. Record data. Laps two onwards are driven without choke. The total number of laps to be driven will be 15.

APPENDIX B

Statistical analysis of test results

Each oil was tested four times in each car; thus there are eight data points for each of the fifteen laps in each car and with a total of sixty-four tests the total data points is nine hundred and sixty.

With only four tests on each oil in each car, test precision would have to be very good with excellent reproducibility in order to show significant consumption differences between the two oils in individual cars. When the fleet average on each oil is taken the comparison is based on thirty-two tests on each oil, and the chances of the measured differences in fuel consumption being significant is much higher.

Each test consisted of fifteen laps, and the fuel consumption was measured for each lap. However, it was not realistic to treat the data from each test as fifteen independent observations, as there was far more variation between tests than within tests. Therefore we conducted our analyses as follows:

(1) Fuel consumption for lap 1.
(2) Average fuel consumption for laps 1 to 5.
(3) Average fuel consumption for laps 11 to 15 (the fully warmed-up conditions).

Two first lap fuel consumption readings were rejected as outliers. The statistical test justifying this course of action is described in Appendix C.

Thus we considered there to be no results for these tests in the analyses of (1) and (2) above.

An analysis-of-variance was conducted for each of the three quantities described above, testing the effects of car and oil and for any car x oil interaction. Plotting the residuals against ambient temperature showed a marked negative relationship - that is fuel consumption decreased as ambient temperature increased. Therefore ambient temperature was used as a covariate in the analysis to improve the precision. We were, however, unable to find any similar relationship with other measured parameters (humidity, barometric pressure, wind speed etc.)

In the statistical analysis of the results when more than two means are being compared, for example when comparing cars, or comparing the differences between the two oils for different cars, the procedure is first to test for any overall differences using an F-test. If the F-statistic is significant, then the S.E. of differences may be used as a guide as to where the differences lie. A difference is significantly different from zero at P < 5% if, when divided by its S.E., the resultant value exceeds the critical value of the t-distribution, which in each case here equals 2.01. However, this use of the t-statistic should be used rather more as a guide than as a strict test, because it is possible to make many pair-wise comparisons, and therefore the significance level of 5% is not strictly applicable, as we will have selected the pairs for comparison after having seen the data.

Thus we used F-tests in this analysis, and found that the fuel consumption saving due to using the test oil rather than the reference oil, did not differ significantly from car to car (saving being measured as an absolute difference rather than as a percentage).

Before analysis the fuel consumption results were adjusted for ambient temperature. That is, the values are adjusted to what they would have been had all the experiments been conducted at the same temperature, namely the overall average temperature.

APPENDIX C

Rejection of outlying data points

We used the following statistic to test whether the two observations mentioned in Appendix B were outliers:

$$\tau^2 = \frac{\nu(S_c - S_m)}{2S_c}$$

where S_c is the residual sum of squares in our analysis-of-covariance table, S_m is the corresponding sum of squares after the two suspect points have been omitted from the analysis and ν is the number of degrees of freedom associated with S_c. This is an adaptation of a statistic for a single outlier given on page 241 of Outliers in Statistical Data by Barnett and Lewis. The distribution of τ^2 is approximately F with 2 and ν degrees of freedom, but as we have not chosen our 2 values a priori then we must test τ^2 as

$$\binom{n}{2} P(F_{2,\nu} > \tau^2)$$

where n is the total number of observations.

For the lap 1 data we found S_c = 17082.8 with 47 d.f., S_m = 3232.65 and hence τ^2 = 19.05. From tables of the F-distribution with 2 and 47 d.f., this value would be significant at P < 0.000001. Therefore the true significance, from above is

$$P < 0.000001 \times \frac{64 \times 63}{2} = 0.002, \text{ i.e. } P < 1\%.$$

Therefore the values concerned are indeed outliers and should be rejected.

Similarly for the lap 1 to 5 averages we found S_c = 971.16 and S_m = 373.63, therefore τ^2 = 14.459, which has a true significance of P < 5%, calculated as above. Once more, these values are outliers and should be rejected.

Table 1 Average fuel consumption for lap 1 (adjusted for ambient temperature, rejecting two outliers)

Car	Fuel consumption (g/lap) Reference oil	Test oil	Benefit for test oil	Diff, %	Standard error of diff, %	2X standard error	Limits Lower	Upper
Cortina	188.70	182.70	6.0	3.18	3.20	6.39	-3.21	9.57
Metro	130.10	119.50	10.6	8.15	4.63	9.27	-1.12	17.42
Polo	178.40	171.20	7.2	4.04	3.38	6.76	-2.72	10.80
Renault	137.00	132.10	4.9	3.58	4.40	8.80	-5.23	12.38
Datsun	188.00	164.10	23.9	12.71	3.21	6.41	6.30	19.13
Escort	288.50	265.00	23.5	8.15	2.09	4.18	3.97	12.33
Astra	209.40	199.90	9.5	4.54	2.88	5.76	-1.22	10.30
Alpine	160.20	157.80	2.4	1.50	3.76	7.53	-6.03	9.03
Mean	185.00[1]	174.00[2]	11.0	5.95	1.15	2.29	3.65	8.24

Fleet average[3] benefit for test oil = 5.95%

Significant at 99.9% confidence level

$$3 = \frac{1-2}{1} \times 100$$

The standard error of the difference in means for a single car is 6.03 g/lap and the standard error of the difference in the means averaged over all eight cars is 2.12 - g/lap.

Table 2 Average fuel consumption for laps 1 to 5 (adjusted for ambient temperature, rejecting two outliers)

Car	Fuel consumption (g/lap) Reference oil	Test oil	Benefit for test oil	Diff, %	Standard error of diff, %	2X standard error	Limits Lower	Upper
Cortina	131.69	129.09	2.60	1.97	1.55	3.10	-1.13	5.08
Metro	95.37	92.30	3.07	3.22	2.14	4.29	-1.07	7.51
Polo	112.83	110.40	2.43	2.15	1.81	3.62	-1.47	5.78
Renault	96.82	93.64	3.18	3.28	2.11	4.22	-0.94	7.51
Datsun	116.26	108.66	7.6	6.54	1.76	3.52	3.02	10.05
Escort	144.02	137.42	6.6	4.58	1.42	2.84	1.74	7.42
Astra	119.39	119.00	0.39	0.33	1.71	3.42	-3.10	3.75
Alpine	116.65	116.75	-0.1	-0.09	1.75	3.50	-3.59	3.42
Mean	116.63[1]	113.41[2]	3.22	2.76	0.62	1.23	1.53	4.00

Fleet average[3] benefit for test oil = 2.76%

Significant at 99.9% confidence level

$$3 = \frac{1-2}{1} \times 100$$

The standard error of the difference in means for a single car is 2.044 g/lap and the standard error of the difference in the means averaged over all eight cars is 0.720 - g/lap.

Table 3 Average fuel consumption for laps 11 to 15 (adjusted for ambient temperature)

Car	Fuel consumption (g/lap) Reference oil	Test oil	Benefit for test oil	Diff, %	Standard error of diff, %	2X standard error	Limits Lower	Upper
Cortina	104.32	102.65	1.67	1.60	1.06	2.11	-0.51	3.71
Metro	78.66	77.60	1.06	1.35	1.40	2.80	-1.45	4.15
Polo	89.38	86.66	2.72	3.04	1.23	2.47	0.58	5.51
Renault	79.71	77.45	2.26	2.84	1.38	2.77	0.07	5.60
Datsun	88.44	85.07	3.37	3.81	1.25	2.49	1.32	6.30
Escort	90.03	87.04	2.99	3.32	1.22	2.45	0.87	5.77
Astra	89.62	90.39	-0.77	-0.86	1.23	2.46	-3.32	1.60
Alpine	99.67	98.99	0.68	0.68	1.11	2.21	-1.53	2.89
Mean	89.98[1]	88.23[2]	1.75	1.94	0.43	0.87	1.08	2.81

Fleet average[3] benefit for test oil = 1.94%

Significant at 99.9% confidence level

$$3 = \frac{1-2}{1} \times 100$$

The standard error of the difference in means for a single car is 1.102 g/lap and the standard error of the difference in the means averaged over all eight cars is 0.390 - g/lap.

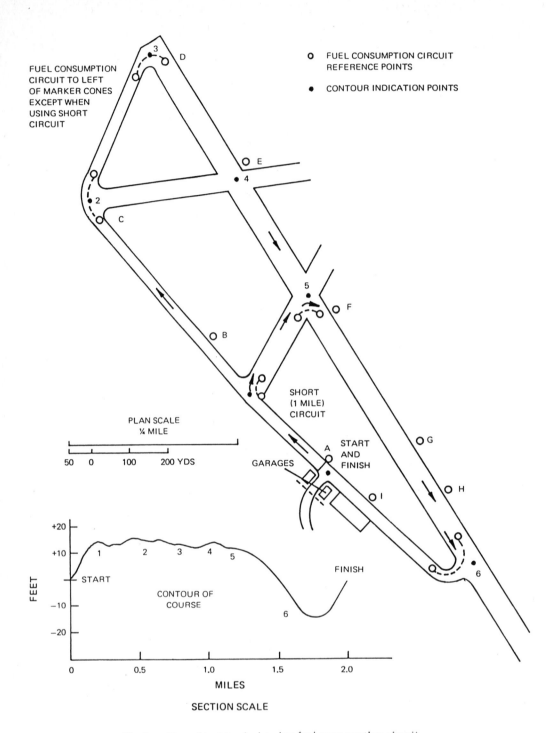

Fig 1 Plan of test track showing fuel consumption circuit reference points and circuit contour

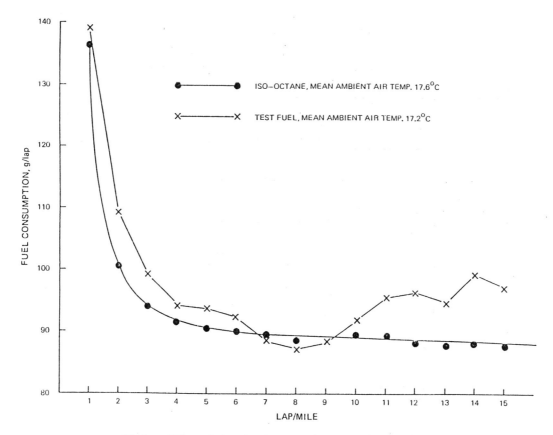

Fig 2 Effect of changing from premium to iso-octane test fuel on lap-to-lap fuel consumption from a cold start in the Talbot Alpine running on reference oil at similar ambient air temperatures

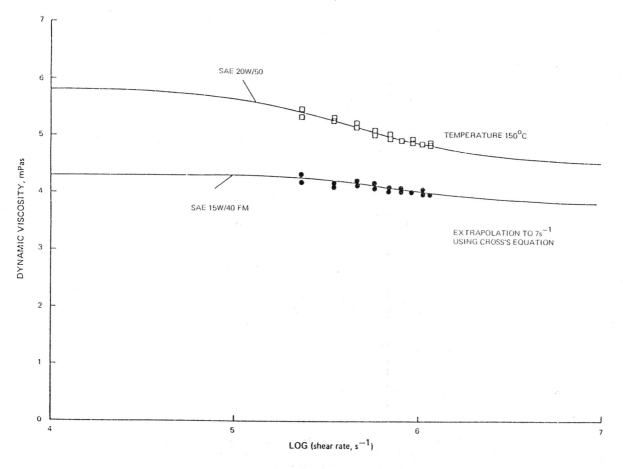

Fig 3 Viscosity at high temperature and high shear rates

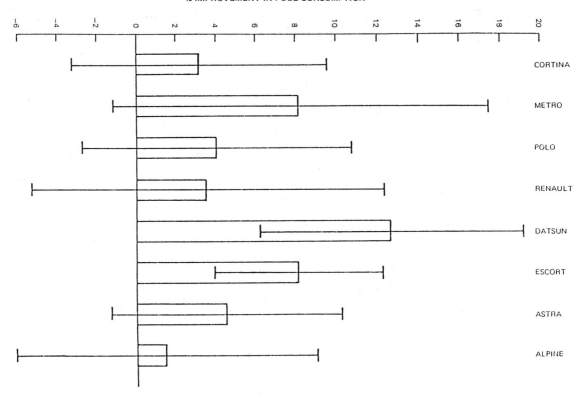

Fig 4 Percentage improvement in fuel consumption when running on test oil compared to reference oil — lap 1 from cold start rejecting 2 outlying data points

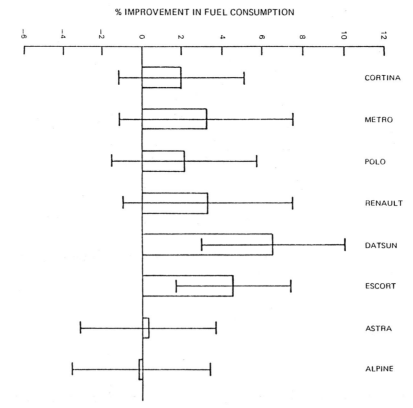

Fig 5 Percentage improvement in fuel consumption when running on test oil compared to reference oil — laps 1–5 rejecting 2 outlying data points

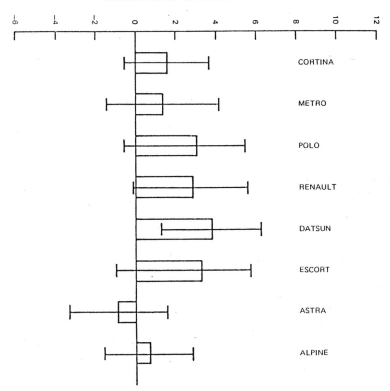

Fig 6 Percentage improvement in fuel consumption when running on test oil compared to reference oil — laps 11–15

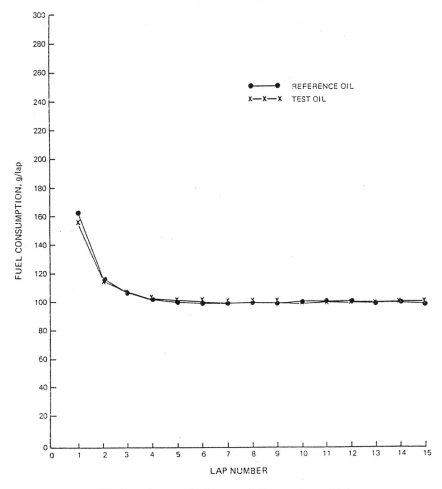

Fig 7 Average fuel consumption per lap — Alpine

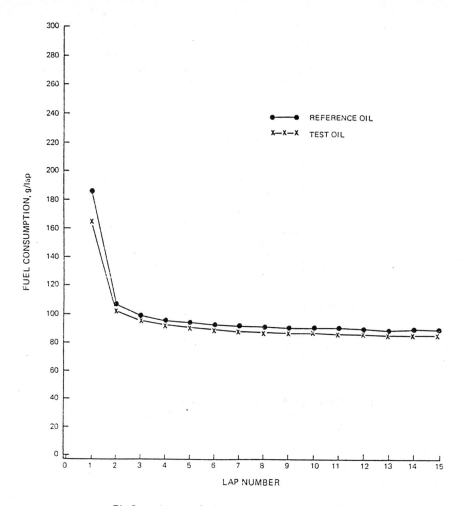

Fig 8 Average fuel consumption per lap — Datsun

C430/84

SAE 841291

Impact of fuel composition on octane requirement

F J MARSEE, J P SUNNE, BME, and W E ADAMS, BME
Ethyl Corporation, Ferndale, Michigan, USA

SYNOPSIS Fuel composition can affect the octane requirement of a spark ignition engine. Three cars of one European model were run to 80,000 km on fuels containing 0, 0.15, and 0.4 g Pb/litre. Differences in octane requirement among the various fuels were found as the distance traveled increased. This paper also describes octane requirement differences after combustion chamber deposits were removed.

1 INTRODUCTION

Fuel composition can influence the increase in the octane requirement of a car's engine with distance traveled. Accumulation of engine deposits with time leads to higher octane requirements. The engine that operates knock-free when new may develop a knock problem after running several thousand kilometers. This paper describes a test designed to determine the octane requirement increase (ORI) of three 1983 European cars of Make BA operating on fuels containing different amounts of lead antiknock compound, but nearly the same octane numbers. Interest in this type of test was stimulated by U.S. studies (1,2,3,4), which showed that unleaded fuels gave higher octane requirement increase than leaded fuels.

The car model chosen for this test was equipped with an engine having a fast-burn compact combustion chamber. Little has been published about the ORI characteristics of this type of engine. The cylinder head is a flat design, with the combustion chamber in the piston. The published compression ratio of this engine is 11:1. Ignition timing is controlled by a microprocessor that adjusts the spark advance for intake manifold pressure, coolant temperature, and engine speed. This processor is not equipped with a knock detector. The specifications for the test cars are shown in Table 1.

Table 1

Test Car Specifications
1983 Model BA

Engine Type:	4-Cylinder, In-Line, Liquid-Cooled, Single Overhead Cam, Cast-Iron Block, Aluminum Head, Compact Combustion Chamber In Piston.
Engine Displacement:	1.272 Litres
Compression Ratio:	11.0:1
Bore:	75 mm
Stroke:	72 mm
Maximum Power:	55 kW @ 5800 RPM
Maximum Torque:	104 Nm @ 3600 RPM
Transmission:	Synchronized 4-Speed Manual Gearbox
Vehicle Weight:	750 kg

2 TEST DESIGN

The test was designed to simulate a car driven most of the time in moderate urban driving conditions in Europe. High-speed driving was added to this schedule to simulate periodic trips on an autobahn, although this inclusion probably reduced the ORI (5). This type of combined schedule was chosen because it was felt that the deposits formed would be very stable (2). Octane number requirement, fuel consumption, and compression ratio were measured at intervals of 12,000 to 14,000 km.

Because of the accelerated schedule of nearly continuous driving, the octane number requirements (ONR) of these engines may have been somewhat different from those of an engine operating on an all-urban or autobahn schedule. The ORI was probably reduced by the low number of cold starts compared to vehicles driven in customer service. The small sample (three cars) meant that high precision was required to produce meaningful results. The goal of the test was to establish only the trend of the octane requirement increase caused by the three fuels. Since the absolute octane number requirement was not a major objective, it was not necessary to be concerned about small differences in engine construction or control settings among the cars. However, great effort was made to maintain constant settings during the test. Also, care was exercised in establishing operating conditions so that ratings were as accurate and repeatable as possible. However, no measure of rating repeatability is available.

Most ORI tests have been terminated after 15,000-25,000 km because the octane requirement is said to be stable at this distance. This test was extended to 80,000 km to determine if the ORI trends change after a greater distance has been traveled.

3 VEHICLE INSPECTION AND MAINTENANCE

Each car was inspected and set to factory specifications before the start of the test. Compression pressures and compression ratios were measured initially for each car. Routine maintenance was performed following the factory-recommended intervals. Additional inspections were made before each octane rating or fuel consumption test.

4 MEASUREMENT TECHNIQUES

4.1 Compression Ratio Measurements

The cars were soaked at 22°C for 12 to 16 hours before the compression ratios were measured. The compression volume indicator used for this test is a sonic device for measuring the combustion chamber volume. The instrument operates on the Helmholtz principle that a volume of gas contained in an enclosure having a small opening will resonate at a single frequency when properly excited by an external sound source. With the engine at top center on the compression stroke, a simple air whistle is placed in the spark plug opening, with low-pressure air providing the acoustical signal. By carefully matching the resonance frequency of the compression volume with that of a known volume, it is possible to accurately measure the unknown volume without disassembling the engine.

4.2 Fuel Consumption Measurements

Fuel consumption was measured on the chassis dynamometer for the hot ECE cycle and steady-state operation at 90 and 129 kph. These tests were performed at 3,000, 15,000, 28,000, 40,000, and 80,000 km. For these measurements, a dedicated set of test tires was mounted on each vehicle. Fuel consumption was measured using a strain-gauge weigh scale that was accurate to 4 grams. Dynamometer roll counts were used for the total distance traveled. Initial fuel consumption tests were run on each of the three test fuels in each car in order to determine any possible effects of fuel composition on fuel consumption. None were detected.

4.3 Measurement of Octane Number Requirement

The octane number requirements for accelerating conditions were determined using the Cooperative Octane Requirement Committee (CORC) Technique No. 1980/3. The CORC Technique No. 76/5 was used for determining the octane requirements at constant speed. The octane requirement tests were performed in a weather-controlled chassis dynamometer room. Temperature was controlled to 21°C and humidity at 46% relative. Air speed at the fan in front of the vehicle was held at a constant 50 km/hr for the ONR determination under accelerating conditions. The fan speed was matched to the vehicle speed for the constant-speed tests.

The chassis dynamometer used for this test was an electric model with a 1.2-meter roll diameter. Inertia disks were used to simulate vehicle weight.

Idle carbon monoxide concentration and basic timing were checked, and set if necessary, prior to each series of ONR determinations. In addition to the octane requirements, the following data were recorded for each data point:

- Barometric pressure
- Room temperature
- Relative humidity
- Engine speed
- Intake manifold vacuum
- Vehicle speed
- Ignition timing
- Engine coolant outlet temperature
- Air-fuel ratio

The manufacturer's instructions were to limit engine speed to 3,000 rpm when the engine was new. Therefore, the ONR tests at 0 km were limited to speeds of 3,000 rpm or below. Complete CORC tests were conducted at 3,000, 40,000, 54,000, 64,000 and 80,000 km. Additional tests were performed at 14,000 and 28,000 km, but these tests did not include the part-throttle or high-speed-acceleration tests.

5 TEST FUELS AND OIL

5.1 Rating Fuels

CORC Series 7 reference fuels were used for all octane ratings. The rating fuels were blended from 88 to 100 RON in steps of 1.0 octane number. Specifications for these fuels are shown in Table 2.

Table 2
Properties of CORC Series 7 Reference Fuels

	CORC Series 7 Fuel
RON Range	88-100.1
Sensitivity, O.N.	8-11
Rvp, bar	0.45-0.65
Olefin Content, vol %	10-20
Pb Content, g/L	0.10-0.15
ASTM Distillation, °C @	
10% Evaporated	40-60
50% Evaporated	85-115
90% Evaporated	140-180

5.2 Test Fuels

The three fuels compared were unleaded, 0.15 g Pb/litre, and 0.4 g Pb/litre. The fuels containing 0.15 and 0.4 g Pb/litre were obtained from Europe. The third fuel was an American unleaded premium. The aromatic content of this fuel was lower than that of either of the two European fuels. Because this probably would not be representative of an unleaded gasoline marketed in Europe, seven percent toluene was added to increase the aromaticity of this fuel. Table 3 shows the properties of the three fuels.

5.3 Oil

Each car was operated on the same type of engine oil -- a commercial product obtained from a European refinery. Specifications are shown in Table 4.

Table 4
Engine Oil Specifications

API Classification	SF/CC
SAE Class	10W-40
Specific Gravity @ 15°C	0.880
Viscosity	
@ 100°C, mm^2/s	15.5
@ 50°C, mm^2/s	65.3
@ 40°C, mm^2/s	96.9
@ -18°C, mPa.s	2300
Viscosity Index	170
Flame Point, °C	210
Evaporation Loss, wt %	18
Pour Point, °C	-33
Ash Oxidation, wt %	0.65
Sulfate, wt %	0.80
Additive Elements, wt %	
Calcium	0.16
Zinc	0.13
Phosphorus	0.12
Base Number, mg KOH/g	6.0
Shear Stability Viscosity Loss, %	15

Table 3
Test Fuel Properties

Test Fuel	U.S. 0 Pb Premium + 7% Toluene	European Low-Lead	European High-Lead
Pb Content, g/L	0	0.14	0.39
Research O.N.	98.9	101.1	98.8
Motor O.N.	87.6	87.8	89.9
API Gravity	52.2	51.6	56.9
Density, kg/L	0.768	0.771	0.750
Rvp, bar	0.83	0.81	0.88
ASTM Distillation, °C @			
10% Evaporated	44	43	42
50% Evaporated	106	96	93
90% Evaporated	163	147	161
Hydrocarbon Content, vol %			
Aromatics	42.0	50.8	38.6
Olefins	3.6	11.8	1.9
Saturates	54.4	37.3	59.5

6 DRIVING SCHEDULE

The three cars were driven on an EPA-approved urban durability route at approximately the same time of day. Each car accumulated 750 km in 16 hours at an average speed of 50 km/hr. Three hundred stops and starts occurred during this time. The cars were operated on this schedule for 8,000 to 12,000 km. An additional two days of high-speed driving was added to the distance to simulate periodic trips. At the test track, each car was driven at a speed of 135 km/hr for 1.5 hours. The car was then rested for 15 minutes before the high-speed run was repeated. After the high-speed driving, the cars were tested for ONR. This particular place in the driving schedule was chosen to try to obtain maximum deposit stability. The combined urban and autobahn schedule accumulated 12,000 to 14,000 km between octane ratings.

7 TEST RESULTS

7.1 Fuel Consumption

Accurate records of fuel consumption were kept for both durability route and high-speed portions of the test. Throughout the test, the fuel consumption values for the three cars were within 3% of each other. This would be expected, since the fuels were nearly equal in density and aromaticity.

Fuel consumption also was measured on the chassis dynamometer. These data were obtained for the hot ECE cycle, as well as for the steady-state operation at 90 and 120 km/hr. As shown in Table 5, these results also were very similar for the three cars throughout the test.

Additional fuel consumption data were obtained on the chassis dynamometer at 80,000 km. These tests were conducted at basic timing settings of 5° and 3° BTC. The settings represent the manufacturer's recommended timing setting and a 2° retarded condition. Multiple hot ECE tests and steady-state tests at 90 and 120 km/hr were conducted at each spark setting in order to establish a valid relationship between spark retard and fuel consumption penalty. It was found that a combined fuel consumption loss of 0.6% per degree of spark retard was a valid average for the three cars tested.

7.2 Oil Consumption

The engine lubricating oil was changed after the first 7,500 km of driving. The oil drain interval was then extended to 15,000 km as recommended by the manufacturer. There was no visible sign of any oil being used by the three engines during the entire test. The low oil consumption of these engines indicates that lubricating oil would be expected to have little effect on those combustion chamber deposits that might affect ONR.

7.3 Octane Number Requirement at Constant Speed

Wide-open-throttle ONR measurements were made at constant speeds of 2,500, 3,000, 3,500, 4,000 and 4,500 rpm at five intervals from 3,000 to 80,000 km. Measurements were made only at 2,500 and 3,000 rpm at the start of the test because the manufacturer recommended that the engines not exceed 3,000 rpm when new. These start-of-test values were not included in the data plots, since they did not constitute full data sets.

Table 5

Fuel Consumption On The Chassis Dynamometer

Fuel Consumption, L/100 km

Test Km	0.0 g Pb/L			0.15 g Pb/L			0.40 g Pb/L		
	Hot ECE	90 KPH	120 KPH	Hot ECE	90 KPH	120 KPH	Hot ECE	90 KPH	120 KPH
300	7.86	5.85	7.56	8.16	5.81	7.54	7.84	5.72	7.42
3,000	-	5.55	7.16	-	5.69	7.33	-	5.72	7.40
14,000	-	5.58	7.14	-	5.43	7.08	-	5.47	7.42
28,000	-	5.63	7.24	-	5.51	7.19	-	5.68	7.27
40,000	-	5.59	7.19	-	5.76	7.13	-	5.78	7.07
80,000	7.72	5.49	7.14	8.11	5.70	7.26	8.17	5.66	7.32

Figure 1 shows the observed constant-speed data for the maximum ONR (which normally occurred at 2,500 rpm for all three cars) and for the ONR at 4,500 rpm. As is usually found in ONR testing, the data vary from rating to rating. However, there is clearly an increase in the 4,500-rpm ONR of the unleaded-fuel car throughout the test.

A close inspection of the constant-speed data showed that the spark advance, which had been equal for the three cars at the start of test, had begun to show differences of 1 to 3 crankangle degrees among the three cars at a given engine speed. Since the objective of the test was to discern the effects of fuel composition on ONR and not to study the effects of other variables, it was necessary to adjust all ONR values to the same conditions.

Tests were run on all three cars at the 80,000-km test point to determine the relationship between spark advance and octane number requirement at each speed. The results showed a 1-RON decrease in ONR per degree of spark retard. This figure agrees well with published Coordinating Research Council data (6), which show a 0.8-RON decrease per degree of retard for premium-fueled U.S. cars rated on full-boiling reference fuels.

With this relationship established, all ONR data were adjusted to the same spark advance that all three vehicles had exhibited at the 3,000-km test point. Figure 2 shows the ONR at 4,500 rpm and the maximum ONR, with all values adjusted for spark advance. Although the absolute values are changed from the unadjusted data, the trends remain the same.

The adjusted data for maximum ONR in Figure 2 show relatively small trends with distance traveled. The unleaded-fuel ONR line is nearly flat, whereas the lines for 0.15 g Pb and 0.4 g Pb tend to go up slightly and then fall near the end of the test. This may seem unusual, but the relatively severe driving schedule may account for the fact that there were small increases in the maximum ONR during the test. As noted earlier, maximum ONR usually resulted from the ratings performed at 2,500 rpm, the lowest speed measured.

The ONR at 4,500 rpm show different relationships with distance traveled. The cars using 0.4 g and 0.15 g Pb gave increasing ONR to 54,000 and 69,000 km, respectively, and then decreased somewhat. Compared with the car run on 0.15 g Pb, the 0.4 g Pb car gave a lower ONR by 1 to 4 O.N. after the initial rating.

The ONR for the unleaded-fuel car increased steadily throughout the test to 69,000 km and then held constant at the 80,000 km rating. Although the last ONR rating was the same as the one preceding it, the one rating is not sufficient evidence that the ONR has stabilized. Therefore, there is a risk of a continuing increase in ONR and high-speed problems after 100,000 km or so of service when using unleaded fuel.

When all constant-speed and accelerating ONR ratings were completed at 80,000 km, the engine blocks were flushed to remove any scale or deposits that might have accumulated in the cooling system and affected the ONR. The constant-speed requirements were measured again before the fuel-related deposits were removed. There were no changes in any of the ONR data that could be attributed to cooling system deposits.

7.4 Octane Number Requirement Under Accelerating Conditions

Octane number requirement under accelerating conditions was measured using CORC Method 1980/3. Wide-open-throttle ONR's during low-speed and high-speed conditions (Parts I and II of the CORC Procedure) were determined at each test interval. Part-throttle ONR data (Part III of the CORC Procedure) were obtained at each test interval except 14,000 and 28,000 km.

Results of ONR determinations under accelerating conditions are shown in Figure 3. The upper plot gives the maximum ONR at wide open throttle, and the lower plot represents maximum ONR under part-throttle conditions. It should be noted that these data are of less importance than the constant-speed ONR data for two reasons. First, engine damage from knock is primarily associated with prolonged operation at high engine speeds. Second, it was not possible to correct for the spark advance changes that occurred during the test as discussed in the previous section.

Under WOT accelerating conditions, the deposit removal procedure at the end of the test indicated a maximum octane requirement increase of 2 O.N. for cars run on unleaded fuel and 0.4 g Pb fuel and 3 O.N. for the car operated on 0.15 g Pb.

For the part-throttle data, the plotted values are the maximum of the ONR obtained at four manifold vacuums -- 60, 120, 180, and 240 m bar. Deposit removal at the end of test showed a maximum ORI of 2.5 O.N. for the unleaded-fuel car, 1.5 O.N. for the car using 0.15 g Pb and 1 O.N. for the car using 0.4 g Pb.

7.5 Combustion Chamber Deposits

The three engines were disassembled at the end of the test, and the combustion chamber deposits were inspected and removed. Deposits also were removed from the underheads of the intake valves and the intake ports. The amount of deposits on the underheads of the valves and in the intake ports was very light for all three engines. Although intake port deposits have been shown to affect ORI (7), the small amount of these deposits found in this test should have a minimal impact.

In the 0.4 g Pb engine, the deposits on the cylinder head and piston tops appeared dark in color, granular, and medium hard. Some of the areas showed a small amount of flaking. The exhaust valves had a tan-color deposit that was granular and medium to heavy in thickness. The deposits in the 0.15 g Pb engine were irregular and lighter in color. These deposits were granular and hard, and many areas showed signs of flaking. The exhaust valve had tan-color deposits of average thickness. The pistons and combustion chambers were covered with similar deposits. Deposits in the unleaded-fuel engine appeared uniform and smooth and had a varnish-like appearance on the cylinder head. Figure 4 shows the head and valve deposits for one cylinder of each engine.

Deposits affect knock in three ways -- change in combustion chamber volume, catalytic effect, and thermal insulation (8). Table 6 shows the changes in combustion chamber volume during the test. These data show that the volume change of the unleaded-fuel car was about half that of the leaded-fuel cars.

The volume measurements were used to calculate the changes in compression ratio during the test. Table 6 also shows the changes in compression ratio during the test. Since the volume changes were small for all three cars, most of the ORI could be attributed to other influences.

Table 6

Effect of Deposits on Combustion Chamber Volume

Lead Content, g/L	0.0	0.15	0.40
	Combustion Chamber Volume, cm^3		
Start of Test	33.58	33.79	34.08
End of Test	32.92	32.68	32.92
Deposit Volume	0.66	1.11	1.16
After Deposit Removal	33.43	33.43	34.04
	Compression Ratio		
Start of Test	10.47	10.41	10.33
End of Test	10.64	10.72	10.67
After Deposit Removal	10.51	10.51	10.34

7.6 ONR Determinations After Deposit Removal

Following removal of deposits, the engines were reassembled and inspected. The full CORC Procedure was then run on each car. Results of the ratings obtained before and after deposit removal are shown in Figure 5. These data were adjusted to a constant spark advance condition as previously discussed.

Octane number requirement data obtained under accelerating conditions are shown in Figure 3. The ORI data for these conditions can be interpreted from the end-of-test data after deposit removal that are also shown in Figure 3. As noted earlier, there was not much difference in ORI among the three fuels under accelerating conditions.

8 DISCUSSION OF DATA

In 1972, it was reported (9) that the limited number of European cars tested up to that time did not show any difference in octane requirement increase between cars run on leaded fuels and those run on unleaded fuels. Also, it was concluded that most but not all cars had reached a stabilized level of ONR by 16,000 km.

As noted previously, several studies have shown that the ORI of U.S. cars run on unleaded fuel is higher than that of cars run on leaded fuel. Data from this test indicate that the same results may occur for some European cars operating under more severe conditions of engine speed and load than those typical in the U.S. Although the small sample size tends to lessen the degree of statistical significance of absolute ONR values, there is a consistent indication of higher ORI for the unleaded-fuel car at full-throttle, constant-speed conditions.

Our experience in octane number requirement studies over the years has led us to believe that the best measure of ORI is obtained from ONR ratings before and after combustion chamber deposit removal (5). That technique minimizes the effects of uncontrolled atmospheric variables and unavoidable mechanical or electrical changes that can occur over a period of time. By this technique, the ORI of the unleaded-fuel car exceeded that of the 0.4 g Pb car by 4.5 O.N. and that of the 0.15 g Pb car by 1.5 O.N. at 4,500 rpm, as shown in Figure 6. The ORI of the unleaded-fuel car was 2.5 O.N. higher than that of the 0.4 g Pb car and 1.0 O.N. higher than that of the 0.15 g Pb car at maximum ONR. Finally, the ORI of the unleaded-fuel car was 2.2 O.N. higher than that of the 0.4 g Pb car and 0.2 O.N. higher than that of the 0.15 g Pb car when compared on the basis of five constant speeds.

Based on results of previous work, the maximum ORI for these three vehicles appears quite low regardless of the fuel composition. However, the ORI was definitely affected by lead antiknock level and fuel composition. As noted in Table 6, the change in cylinder volume due to fuel deposits was inverse to the ORI for the three engines (i.e., the engine with the least deposit volume had the highest ORI). The intake charge heating and thermal insulation effects obviously had more influence on ORI than did the uneven and porous lead deposits.

These findings are not surprising, because it was discovered many years ago that the thermal effect of deposits far outweighed their volumetric effect. Dumont demonstrated that 60% to 80% of ORI can be attributed to the catalytic or thermal insulation effect (8). Warren showed that 70% to 100% of the ORI could be attributed to the thermal effects of deposits (10). Two possible types of thermal effects are (1) insulation and (2) heat capacity. The insulation effect can be thought of as restricting the heat flow to the coolant and thereby raising the surface temperature of the combustion chamber. Heat capacity is the product of specific heat and deposit weight and is a measure of the capability of deposits to absorb heat from one combustion cycle and transfer it to the following air-fuel charge during the intake and compression strokes. Additional work (11) showed that higher heat capacity of deposits gave higher ORI and that heat capacity is greater when deposits contain more carbon. Deposits from operation on leaded fuel normally contain less carbon because of the catalytic action of lead on carbon combustion. Thus, the above mechanism is proposed as one that would explain the results from the current test.

Of concern is the fact that the high-speed ONR of the unleaded-fuel car displays an increasing trend that might continue. The lower ORI exhibited by the 0.40 g Pb car at the end of this test could be used as a safety margin to provide greater knock resistance for this type of engine while maintaining the same fuel consumption as that obtained using the other fuels. Alternatively, the ORI difference of 2.5 O.N. at maximum ONR could be translated to a fuel consumption benefit for the 0.4 g Pb car with the same degree of built-in knock protection as the unleaded-fuel car. Using data discussed earlier under Fuel Consumption Test Results, a fuel consumption advantage of 1.5% would be calculated for the leaded-fuel car.

9 CONCLUSIONS

1. This study of one European engine indicates that, compared to a fully leaded gasoline, there is a trend to higher ORI at constant-speed, wide-open-throttle conditions with unleaded and low-lead gasolines. This higher ORI could cause an increase in fuel consumption if the same margin of knock protection is provided.

2. At the high-speed conditions, the ONR of the unleaded-fuel engine generally continued to increase with distance traveled.

3. Under accelerating conditions, there was little difference in ORI among the fuels.

REFERENCES

1. "Influence of Leaded and Unleaded Fuels on Octane Requirement Increase in 1971 Model Cars--Phase I: 1970-71 CRC Road Rating Program", CRC Report No. 451, Coordinating Research Council, September 1972.

2. Saillant, R. B., Pedrys, F. J., and Kidder, H. E., "More Data on ORI Variables", SAE Paper 760196.

3. Niles, H. T., McConnell, R. J., Roberts, M. A., and Saillant, R. B., "Establishment of ORI Characteristics as a Function of Selected Fuels and Engine Families", SAE Paper 750451.

4. Ernest, R. P., "A Unique Cooling Approach Makes Aluminum Alloy Cylinder Heads Cost Effective", SAE Paper 770832.

5. Gibson, H. J., Hall, C. A., and Hirschler, D. A., "Combustion Chamber Deposition and Knock", SAE National Fuels and Lubricants Meeting, Tulsa, Oklahoma, November 6-7, 1952.

6. "Octane Number Requirement Survey--1970", CRC Report No. 446, Coordinating Research Council.

7. Graiff, L. B., "Some New Aspects of Deposit Effects on Engine Octane Requirement Increase and Fuel Economy", SAE Paper 790938.

8. Dumont, L. F., "Possible Mechanisms by Which Combustion Chamber Deposits Accumulate and Influence Knock", SAE Quarterly Transactions, Vol. 5, No. 4 (1951).

9. "The Effect of Removing Lead Alkyl Antiknock From European Motor Gasolines", Octane Requirement Increase Report by the Co-operative Octane Requirement Committee, October 1972.

10. Warren, J. A., "Combustion Chamber Deposits and Octane Number Requirement", SAE Transactions (1954).

11. Forster, E. J. and Stinson, L. E. "Effects of Leaded Vs. Unleaded Gasolines on Stabilized Octane Requirements", NPRA Paper F&L 70-46, National Petroleum Refiners Association Fuels and Lubricants Meeting, New York City, September 10-11, 1970.

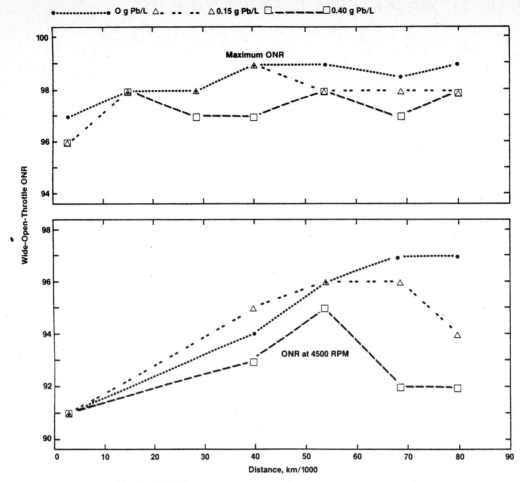

Fig 1   WOT octane number requirement at constant speed. Data not adjusted for spark advance

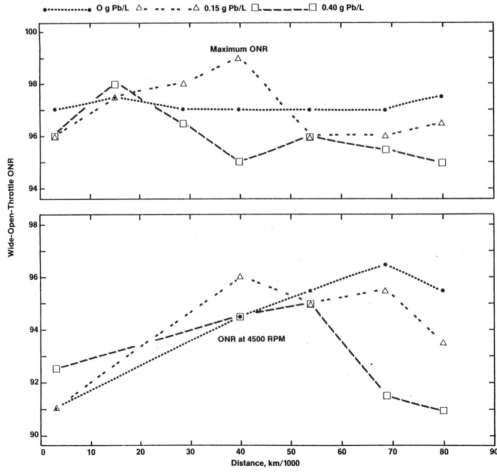

Fig 2   WOT octane number requirement at constant speed. Data adjusted for spark advance

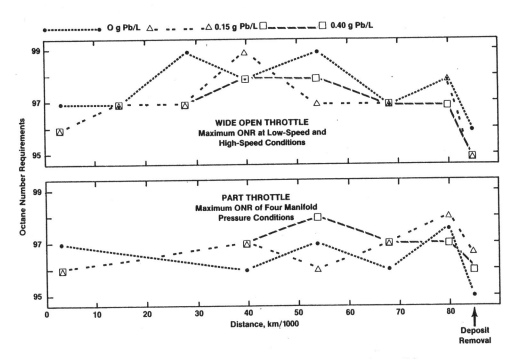

Fig 3   Octane number requirement under accelerating conditions

Fig 4   Cylinder head deposits

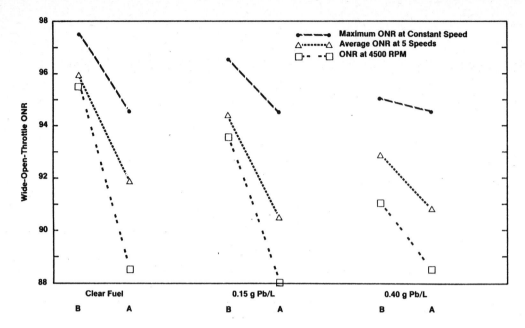

Fig 5 WOT octane number requirement at constant speed. Before (B) and after (A) removal of deposits from piston tops, combustion chambers and intake ports. Data adjusted for spark advance

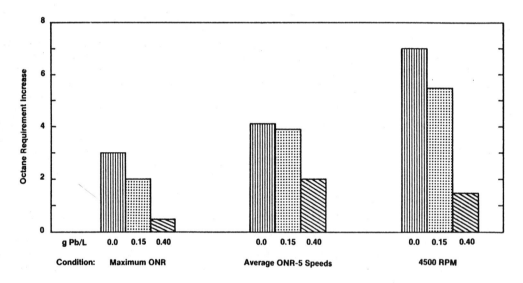

Fig 6 Octane requirement increase at constant speed using deposit removal technique

C433/84

SAE 841292

New temperature control criteria for more efficient gasoline engines

H-P WILLUMEIT, Dr-Ing and P STEINBERG, Dipl-Ing
Technische Universität, Berlin
H ÖTTING, Dr-Ing, B SCHEIBNER and W LEE, Dr-Ing
Volkswagenwerk AG, Wolfsburg, West Germany

SYNOPSIS Calculating the heat balance of Otto engines one can recognize that the ratio between heat transfer and power increases in the part load range, i.e. city driving. To improve the efficiency of such engines it is necessary to reduce the quantity of heat transfer and also of friction losses. Both effects can be achieved by increasing the temperatures of the coolant and the lubricant. This gives a reduction of fuel consumption as well as of exhaust emissions. Both effects are valid even in the warm-up phase as well as under steady state operating temperature as are confirmed by the results of our experiments.

INTRODUCTION

High fuel consumption and exhaust emission is a property of engines which are running in the warm-up phase, that is the time interval between cold start and steady state operating temperatures. In the USA the most frequent trip length is 8 km (1), whereas in Europe the most frequent trip length for instance in Vienna (2) is 4 km. Both of those trips are conducted in the warm-up phase. The potential improvement in fuel economy for this warm-up phase is estimated to be between 2.5 % and 4.7 % (3). It is well known that the exhaust emissions for CO and HC are also significant during the warm-up phase.

There is additional potential fuel consumption reduction during the so-called warm-up phase of engine operation. In this phase the coolant temperature has reached its steady state value, whereas the temperature of the individual engine components and the lubricant depends upon load. At part load the lubricant and engine component temperatures are lower than desirable for fuel economy.

HEAT BALANCE IN STEADY STATE OPERATION WITH STANDARD COOLING

The heat balance investigation on a four cylinder spark ignited engine is shown in Fig. 1 a, under full load conditions. It can be seen that the heat loss by convection to the coolant has a nearly constant ratio to the power output. Under part throttle road load conditions, Fig. 1 b, the proportion of heat to coolant is much greater than power output but falls steadily with increasing speed. After separating load and engine speed influences on the heat transfer it can be seen that the heat transfer is nearly independent of load but increases with engine speed, Figures 1 c, 1 d.

The mechanical connection of cooling water pump with the engine shaft is an important reason for this behaviour. High engine speed leads to high flow speed of cooling water which improves the convection boundary conditions. Under constant engine speed, the difference of temperature between cooling medium and engine components increases with the load. The heat transfer coefficient will not change until the onset of nucleate boiling because of the constant flow speed of the cooling medium.

The partial heat transfer fluxes into the engine block and into the cylinder head are approximately equal, Figures 1 b, 1 c, 1 d.

TRANSIENT WARM-UP PHASE (COLD START)

In the transient warm-up phase of the engine the influence of the load on the heat transfer is negligible. Coolant temperature, Fig. 2 b, at the end of a trip is essentially dependent upon length of the trip and is virtually independent of the speed profile. The same property is valid for the lubricant temperatures. Fig. 2 a emphasizes how the cooling water temperature approaches the operating values much earlier than the oil which means high mechanical losses over a relatively long period.

ENGINE COMPONENT TEMPERATURES UNDER CONVENTIONAL COOLING IN STEADY STATE OPERATION

Methods of improving the thermal behaviour of an engine can be developed and judged only if the engine component temperatures in critical ranges are known. The critical temperature of the cylinder block occurs in the cylinder wall opposite to the top ring when the piston is at top dead center. Fig. 3 shows this critical cylinder wall temperature as a function of engine speed for full load and no load conditions.

Although the coolant temperature is constant the cylinder wall temperature varies over a wide range. This is due to the typical cooling concept which maintains a constant coolant temperature not directly related to any engine component temperature. In practice, the goal of any cooling concept is to protect the engine components from reaching critical temperatures against overheating und undercooling without unnecessary high differences between operating temperatures and critical temperatures.

INFLUENCE OF COOLANT AND LUBRICANT TEMPERATURES ON THE HEAT BALANCE IN STEADY STATE OPERATION

If the common cooling concept of maintaining a constant coolant temperature is abandoned then coolant becomes a system variable. In Fig. 4 a the heat balance is shown as a function of coolant temperature for constant lubricant temperature. The heat, carried off by coolant, decreases with increasing coolant temperature. On the other hand, the heat transfer by convection to the lubricant and the heat transfer by external radiation and convection (i.e. from engine to environment) both increase with increased coolant temperature. Fuel consumption is lower for engines which operate at higher temperatures than the typical production engine temperatures of the present.

This result is achieved because there is an improvement in the combustion process and reduction of friction in the crankshaft/piston group. In Fig. 4 b the changes in the heat balance under constant cooling water temperatures but under variable lubricant temperatures are shown. Also, mutual change of the heat transfer of lubricant and coolant occurs here. The influence of those conditions on the external radiation and convection is negligible.

FUEL CONSUMPTION AND EXHAUST EMISSIONS UNDER CONTROLLED ENGINE COMPONENT TEMPERATURES AT STEADY STATE CONDITIONS

A cooling concept which controls constant engine component temperatures instead of constant water temperatures reduces the heat transfer into the coolant and the lubricant. Thus, conditions can be achieved where both parts, coolant as well as components, have high temperatures and the fuel consumption is consequently reduced.

Fig. 5 a shows the fuel consumption reduction, achieved by application of the above described means over the entire operating range with $\lambda = 1.0$ (CO = 1 %). Fuel savings up to 20 % can be achieved in the lower part load range.

An example of the HC-emission by controlling constant component temperature is shown in Fig. 5 b. The reduction is between 10 % and 55 %.

The described effects are achieved in practice in a car if the carburettor is not changed corresponding to the new cooling concept.

FUEL CONSUMPTION AND EXHAUST EMISSIONS UNDER CONTROLLED ENGINE COMPONENT TEMPERATURES AT TRANSIENT CONDITIONS

In Fig. 6 a controlled component temperature concept is shown. An adjusting valve regulates the carried off heat transferred by the coolant. The control of the adjusting valve takes the temperature of the cylinder wall in height of the top dead center as input value. The results of exhaust emission and fuel consumption are shown in Fig. 6 for an ECE-cycle (warm start).

SHIFT OF THE OPERATION LIMITS TO THE LEAN BY CONTROLLED COMPONENT TEMPERATURES

In Fig. 7 the typical emissions HC, CO, NO of a gasoline engine as a function of air-fuel ratio are shown. It is only in the lean range, close to the lean limits, that the exhaust emissions have low concentrations. To avoid ignition failures the engine usually must be operated with air-fuel ratio slightly leaner than the stoichiometric which causes drastically increased NO-emissions. It is suspected that high cylinder wall temperatures improve the combustion process in the lean range which would extend the operating limits to higher air-fuel ratios.

The experiments with a controlled engine component temperature (shown in Fig. 8) confirm this shift of $\Delta\lambda = +\ 0.25$ in the lower part load range.

INFLUENCE OF THE CONTROLLED COMPONENT TEMPERATURE COOLING CONCEPT ON THE TRANSIENT WARM-UP PHASE

In the cooling concept of controlled engine component temperature the coolant does not move until certain component temperatures are reached. Once the threshold temperature is reached the coolant is allowed to flow by means of a control valve. A reduced warm-up time interval is obtained by that concept, Fig. 9. As a consequence of the reduced warm-up time the fuel consumption and exhaust emissions are reduced too.

In Fig. 10 is shown how that concept can be put into practice. The coolant flow is not stopped in the transient warm-up phase for the entire engine but only for the engine block. Inside the cylinder head the coolant flow is maintained to use the generated heat of the cylinder head for heating the passenger compartment. The friction losses inside the engine block are reduced and the behaviour of the passenger compartment heating is improved, for the heat quantities generated inside engine block and cylinder head are almost equal (Fig. 1) but the heat capacity of the cylinder head is much smaller than of the engine block.

Results of the experiments concerning this aspect are shown in Fig. 10.

LEGEND

n	engine speed
t	time
v	vehicle speed
bmep	brake mean effective pressure
M	torque
P_E	power output
T	temperature
Q	heat quantity
Q_B	heat quantity conducted by the cooling water from the engine block
Q_H	heat quantity conducted by the cooling water from the cylinder head
Q_{EX}	heat quantity conducted by the exhaust gas
Q_R	residual heat quantity
Q_D	delivered heat quantity by fuel
	air-fuel ratio
Q_W	heat quantity conducted by the cooling water
Q_{OIL}	heat quantity conducted by the lubricant oil

REFERENCES

(1) AUSTIN, Th.C.; HELLMANN, K.H. Passenger Car Fuel Economy as Influenced by Trip Length. SAE-Paper 750004

(2) DORFWIRTH, J.R.; HERRY, M.; PASSEK, G. Verkehrsmodell für Wien, 1976. Institut für Verkehrstechnik der TU Wien, entstanden im Auftrag der Gemeinde Wien, MA-18, 1976

(3) MOSER, F. Über das Verhalten von Fahrzeug-Ottomotoren, insbesondere Vergasermotoren, im nicht betriebswarmen Zustand und Wege zur Verbesserung von Abgasemission und Kraftstoffverbrauch in der Warmlaufphase. VDI-Fortschritt-Berichte, Reihe 6, Nr. 54, Juni 1978

(4) ZECHNALL, R.; BAUMANN, G.; Reines Abgas bei Ottomotoren durch geschlossenen Regelkreis. MTZ 34, 1973, D. 7 - 11

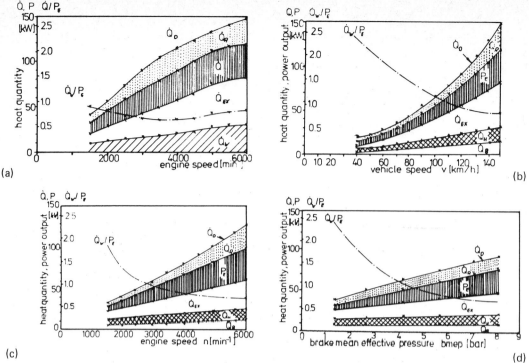

Fig 1 Heat balance of a four cylinder spark ignited engine with standard cooling
(a) under full load
(b) under road load
(c) versus engine speed with constant brake mean effective pressure (5 bar)
(d) versus brake mean effective pressure with constant engine speed (3 500 min^{-1})

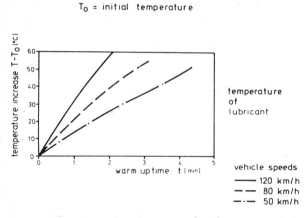

Fig 2a Temperature of lubricant versus warm-up time at different vehicle velocities

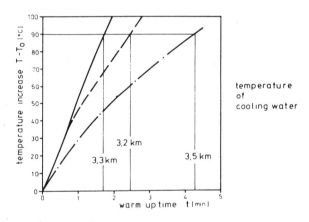

Fig 2b Temperature of cooling water versus warm-up time at different vehicle velocities, measured in fourth gear with standard cooling circuit

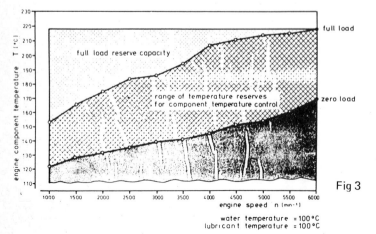

Fig 3 Cylinder temperature versus engine speed for full load and zero load conditions at a position corresponding to the top ring position with the piston at top dead centre

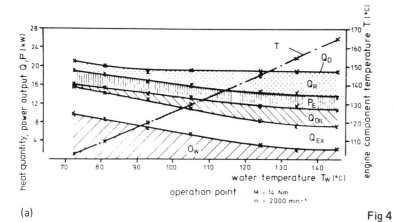

(a)

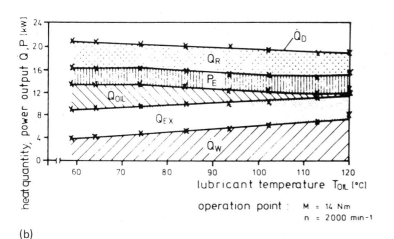

(b)

Fig 4 Heat balance and cylinder temperature at a position corresponding to the top ring location with the piston at top dead centre
(a) versus variable temperature of the coolant with constant lubricant temperature
(b) versus variable lubricant temperature with constant temperature of the coolant

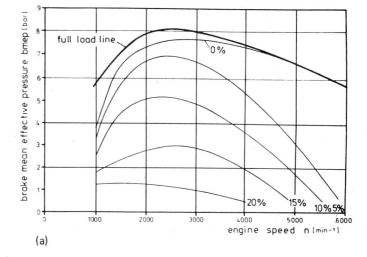

(a)

Fig 5 Brake mean effective pressure versus engine speed
(a) percentage in improvement of fuel consumption
(b) percentage in improvement in HC-reduction

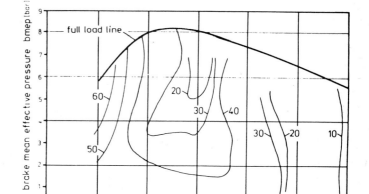

(b)

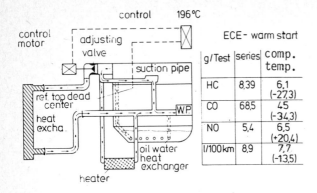

Fig 6   Controlled component temperature concept exhaust emissions and fuel consumption compared to conventional cooling

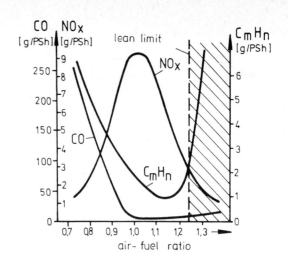

Fig 7   CO, C_mH_n and NO_x emissions versus air–fuel ratio of a spark ignited engine (4)

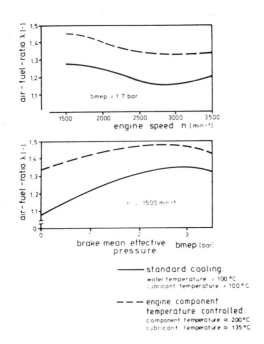

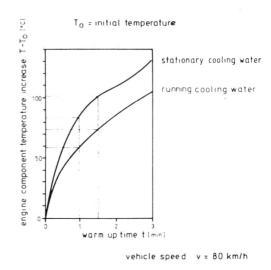

Fig 9   Increase of the cylinder wall temperature versus warm-up time

Fig 8   Air-fuel ratio versus engine speed and air–fuel ratio versus brake mean effective pressure comparing the performance of the standard cooling circuit and the concept of a controlled component temperature

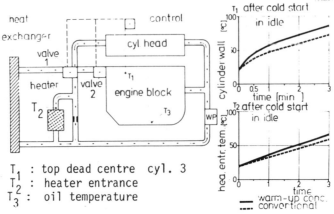

T_1 : top dead centre cyl. 3
T_2 : heater entrance
T_3 : oil temperature

Fig 10   Warm up concept
resting coolant in the engine block
heater supply from cylinder head

C446/84

SAE 841293

Austin-Rover Montego programmed ignition system

E W MEYER, BSEE, MSAE and R GREEN, BSc
Austin Rover Group Limited, Birmingham
M H COPS, MIERE
Lucas Electrical Limited, Birmingham

I. INTRODUCTION

For any engineer serious about consistently achieving maximum power and fuel economy from a reciprocating, internal combustion engine, a complete knowledge of how the ignition system should function as opposed to the way it has been functioning is absolutely necessary. Most engineers today recognize that the mechanical distributor using contacts is the most unreliable part of a motor car. Not all engineers however recognize that the distributor provides only a first approximation of the spark timing that the engine requires for optimum achievement and even that with large errors. Beginning in the early 1970's, breakerless ignition, using variable reluctance transducers and transistors in place of cams, rubbing blocks and contacts, reduced failures by 99.8%. This change has been accomplished almost universally throughout the world in the succeeding twelve years. Since 1976 the use of solid-state electronics in an ignition system to provide optimum spark timing exactly as the engine requires has been occurring. The engineers of Austin Rover and Lucas have created a whole new design of a programmed ignition system with detonation control that we believe is a significant technical step forward by itself and is one of a series of steps we are taking to achieve superior fuel economy on our automobiles. In this paper we hope to show the benefits of our solution, describe the resulting hardware and show how we achieved outstanding reliability.

II. THE LIMITATIONS OF A DISTRIBUTOR

When operating an internal combustion engine at full load, spark advance should be adjusted against engine speed to follow detonation borderline or maximum power curve whichever is the lower. Any single centrifugal spark advance mechanism of a distributor can be adjusted reasonably well to match these curves. This is specifically illustrated in Figure 1a which displays the maximum power spark advance on an A-R 1.6L production engine, its detonation borderline spark advance curve, and the nominal speed related distributor spark advance actually used.

The large tolerances typical in distributor use, $\pm 1°$ in setting and $\pm 2°$ in centrifugal mechanisms, means in practice that the speed responsive spark advance has to be set retarded from detonation to avoid engine damage. This is illustrated in Figure 1b, which is the nominal curve in Figure 1a, expanded to its tolerance limits. (Note that most of the error band lies below detonation). A total error band of $\pm 5°$ is possible and can be considerably retarded from detonation. If detonation occurs in one cylinder and not in the others, as it most commonly does, a distributor cannot provide the correct spark advance in the other 3, 5 or 7 cylinders, since it must follow the lowest detonation curve produced by this one cylinder.

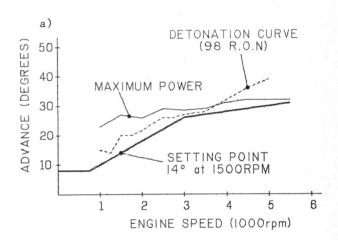

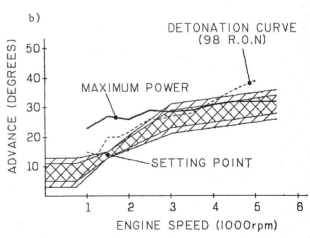

Setting point chosen to keep most distributors below detonation.

▨ Error band 50% of distributor production

▧ Error band 100% of distributor production

Fig 1 Distributor curve — engine full load

At part throttle the spark timing problem is normally different. Detonation is not a problem, or at least it is not with high octane leaded fuel, but the limitations in response of the load-related spark advance device, the vacuum capsule, and its additional timing tolerances make it very difficult to achieve optimum spark timing. This is illustrated in Figure 2a which displays nominal speed-related and load-related spark advance curves of a distributor, and the composite of them, plus the spark advance curve required by an A-R 1.6L engine for optimum fuel economy. Because of the crudity of the vacuum device and because the speed-related spark advance is already established based on other criteria, the resulting total spark advance range achieves a very poor fit with the required spark advance curve for best road load power and fuel economy except in midload. Figure 2b shows the worse situation where the range of spark timing variation of a good, commercial distributor properly applied on an actual A-R 1.6L engine is compared to optimum at road load.

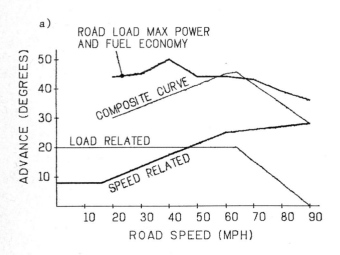

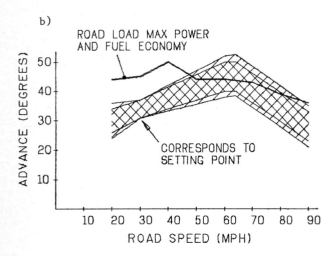

Fig 2 Distributor curve — engine part load

However, all of this spark timing error does not produce direct losses in full load power and road load fuel economy because reciprocating internal combustion engines have rather shallow curves relating timing to power or consumption. Also in the past when power and fuel consumption were not as important, the mismatches in spark timing produced by a distributor were accepted because nothing better was available and distributors were cheap. Today, however, achieving best power and fuel economy is competitively very important and better solutions have been created. The A-R solution in concept and the benefits derived are expressed in Section III.

III. PROGRAMMED IGNITION, CONCEPT AND BENEFITS

The solid-state programmed ignition system introduced on the new A-R Montego has four specific functions designed to improve fuel economy and/or power as compared to the conventional distributor. These are:

- Spark timing accuracy.

    The spark timing error of this system is $\pm 0.9°$ engine.

- Capability to produce any spark timing needed.

    There are 256 speed-load combinations to produce any spark advance desired plus no rate or slope or hysteresis limitations.

- Adjustable spark timing during warm-up.

    A separate coolant temperature - load matrix of 64 sites co-ordinated with electronic fuel control carburetter matrix for both stronger driveability and improved fuel economy during warm-up.

- Cylinder by cylinder detonation control.

    This system can sense detonation, cylinder by cylinder, and adaptively adjust individual cylinder spark timing to prevent both engine damage at speeds up to 6,000 RPM and improve fuel economy and power at lower speeds all without any audible signals reaching the driver or passengers of the car.

To illustrate the benefits of these features we applied them to the A-R 1.6L engine described in Figures 1a and 1b and 2a and 2b and, through use of the engine's map data, analytically determined the actual benefits that could be achieved compared to a distributor. The results are illustrated in Figures 3 and 4. Because distributors have such wide tolerances, the power and fuel economy benefits are plotted against a 50 percentile unit (mean) and a 100 percentile unit (worst case). Distributor tolerances are too large to ignore in any serious study but equally attempting to deal with them in a more sophisticated method would make this paper into a statistical treatise which would simply be a distraction in our judgement.

The programmed ignition system in this study is calibrated for the spark advance which will produce full load maximum power at all speeds. Below 3,750 RPM the cylinder by cylinder detonation control system is used to control spark advance and audible detonation on the one cylinder where detonation is occurring. We believe this is a 'worst case' viable strategy. Even if detonation occurs at lower engine speeds it would not damage the engine or proceed into pre-ignition or engine overheating and thus the risk is low.

Figure 3 shows full load engine power increases possible using this strategy with the A-R programmed ignition as compared to the above described 50 and 100 percentile distributors. Because the gains are largest at lower speeds the driver would sense the benefit in better driveability and acceleration.

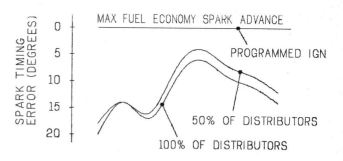

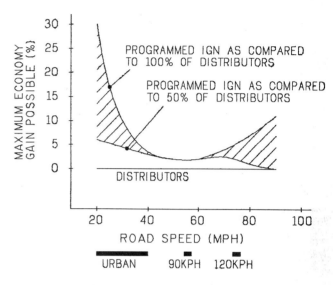

Fig 4   Road load — spark timing error and economy gain

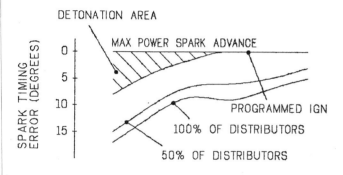

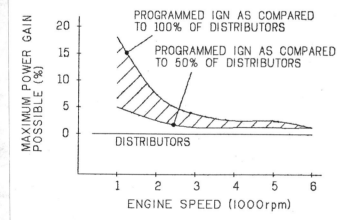

Fig 3   Full load — spark timing error and power gain

Figure 4 similarly shows the fuel economy gains possible by just applying optimum spark timing at road load as compared to the 50 and 100 percentile distributors. If engine mounts are inadequate or the clearances in the drive train are too large it may be difficult to achieve this fuel economy benefit without producing the unpleasant transient reaction colloquially known as 'shunt'. However, these problems can be straightforwardly either improved or eliminated if better driveability and fuel economy are paramount goals.

Because most engines have characteristics not unlike this one we believe this analysis is typical. Of course if an engine is not detonation limited then programmed ignition will allow compression ratio to be increased until it is. This will increase both power at full load and fuel economy at road load. While any single engine can be 'tuned' to optimum spark advance at some conditions with a distributor, we believe programmed ignition with detonation control comprehensively meets the requirements of the engine at all conditions.

IV. SENSORS AND ELECTRONICS (Figure 5)

To achieve the results envisioned in the preceding sections, both sensors and electronic circuitry had to be designed following preceding practice but embodying features that provided the rigorous reliability and functional advantages desired. Clearly the speed sensing must never be asynchronous with the engine nor inaccurate through mechanical tolerances. The load sensor must meet rigorous accuracy requirements. The detonation sensor must be able to clearly control even subtle detonation but must equally distinctly discriminate between engine generated noise and detonation. The high voltage system must be as simple as possible and yet last the lifetime of the car. The following descriptions of the hardware designed or selected will clarify how this was achieved.

Sensors

The single crankshaft position sensor (illustrated in Figure 6) generates a voltage by sensing increasing and decreasing magnetic flux as it interacts with the rotating 34 tooth reluctor. In this way it provides both crank angle information in $10°$ intervals and a specific indication of TDC for each cylinder. Typically this type of sensor requires very small air gaps to generate the voltage amplitudes necessary at low rotating speeds and is thus affected by reluctor eccentricity or axial movement through end float. These conditions were significantly improved in this application by the contour of the reluctor and the resulting axial position of the sensor. This design allows air gaps of at least 1.00 mm in all directions around the pole piece without losing either accuracy or signal strength. The sensor consists of the conventional magnet and coil wound on an armature with the pole piece extended to interact with the reluctor. The larger diameter of the reluctor is the feature which allows both great accuracy even with ordinary mechanical tolerances and adequate voltage, even at instantaneous crankshaft speeds of 30 RPM during cranking.

Absolute manifold pressure sensing is achieved using a silicon bridge strain gauge with onboard interface and temperature compensation circuitry. This transducer, see Figure 7, has the advantage of small physical size allowing the sensor to be incorporated within the module. As there are no moving parts a high degree of reliability can be achieved. The output from the transducer is read by the controlling microprocessor through one of its four 8 bit A-D channels.

The detonation sensor illustrated in Figure 8 is screwed into a boss cast on the engine block at a precise position identified by a holographic modal analysis whilst the engine was excited by vibrations at its characteristic knock frequencies. In this position the sensor exhibits both good discrimination and a high sensitivity to knock in all the cylinders.

Many competitive systems use a tuned knock sensor which gives a high output over a narrow frequency band. This can require a sensor with a different tuned frequency centred on the knock frequency for each engine variant. Also, and perhaps more important, the centre frequency of the detonation vibrations moves with engine load and speed by a greater amount than the bandwidth of a tuned sensor.

The system described here uses a flat response piezoelectric accelerometer as a detonation sensor and the tuning required is performed electronically within the programmed ignition unit resulting in greater flexibility in tailoring the system to a specific engine variant and the necessity for only one type of sensor for all engines. The sensor contains a disc of piezoelectric material clamped between the base of the device and a seismic mass. Thus axial vibrations on the sensor change the compression force on the disc and cause a large relative voltage to be developed between its faces even feeding a capacitive input circuit.

Although it is possible to eliminate high voltage distribution by the use of a multiplicity of ignition coils, the cost of these, the increased complexity of the electronics to drive them and the resulting

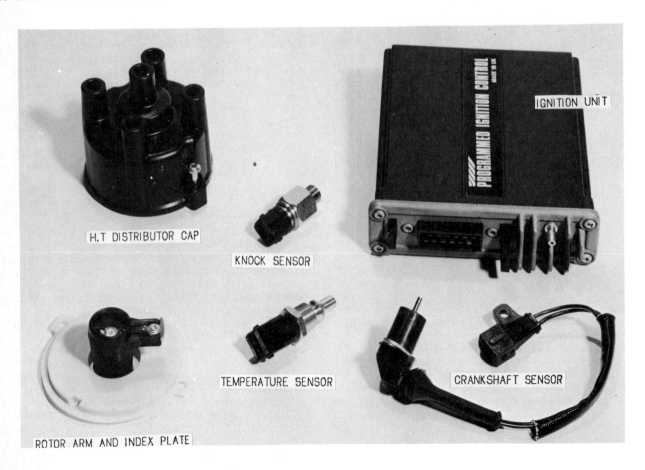

Fig 5 Sensors and electronics

reliability reduction cannot be justified against the lower cost and proven reliability of a rotor arm and distributor cap. In this application no distributor body is used. Instead the high dielectric strength thermal polyester rotor arm incorporating a die-cast insert is screw mounted to a stub shaft, which is fitted into the non-driven end of the camshaft via a resilient bush to reduce torsional vibrations. A resistor is incorporated to reduce electromagnetic interference. The large diameter thermal polyester cap ensures more than adequate high voltage flashover capability. The material was chosen for its extremely low susceptibility to insulation deterioration under humid and contaminated conditions. It is located on the cylinder head as illustrated in Figure 8 at the correct angular position with respect to the rotor by an indexing mounting plate which also acts both as a flash-shield and to encourage a high level of scavenging of ozone and oxides of nitrogen by the pumping action of the rotor arm.

The coolant temperature sensor is a negative temperature coefficient thermistor mounted in a drawn brass housing with integral connector and is retained in the housing on the engine side of the water thermostat by a nut and olive ring. Its high accuracy ($\pm 1\%$) output is shared between the programmed ignition system, the electronic fuel controlled carburetter and driver information systems.

Electronic Circuits (Figure 7)

The solid-state circuits of the A-R programmed ignition system are based around two 6805 microprocessors. Other circuit components include transient suppression of all inputs, a voltage regulator protected from vehicle supply transients, output drive stage for switching the coil, watchdog circuit, pressure to voltage conversion, instrument gauge drive and knock processing circuitry.

Fig 6 Crankshaft sensor and reluctor ring

The main function of the microprocessors is to produce a correctly timed spark by controlling the angular point when the coil current is switched off. This off point is influenced by several different factors. These are engine speed and load (manifold vacuum) but secondary factors include coolant temperature and detonation control.

The primary factors, speed and load, are contained within a 16 x 16 software matrix in the controlling microprocessor's memory. The speed break points are fully programmable with linear interpolation between these points. Load break points are linearly traced but also employ linear interpolation. The interpolated advance produced from the speed and load matrix is then supplemented by the secondary factors. Temperature compensation is contained within a separate 8 x 8 software matrix with load. The correction is read directly from this matrix and then imposed on the speed and load interpolated advance along with any knock correction.

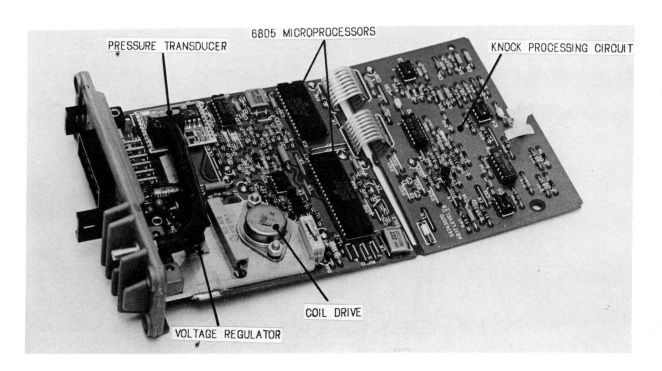

Fig 7 Electronic control unit

As a reliability feature each microprocessor generates a signal to drive a hardware watchdog as well as performing an internal software watchdog function. An additional element to the hardware watchdog is the voltage regulator which incorporates a reset function. A pulse is available, after a programmable delay, to reset the microprocessors at the power-on phase. Also a reset pulse is generated after supply under or overvoltage fault conditions. Another feature of this regulator is its very low dropout voltage which allows a regulated voltage to be maintained even during the cranking phase when the battery voltage could drop as low as 6V.

An important consideration in the design of this system was maintaining as low a dissipation as possible in the coil drive stage. It can be seen from Figure 9 that an angle based system, such as this, has a much lower power dissipation than a time based system. This lower power dissipation not only permits a compact heat sinking arrangement to be used but also improves the reliability of the output stage. Furthermore, accurate control of ignition timing at low engine speeds is assured, unlike the time-based systems whose response to engine speed changes cause large errors.

In addition to this true constant energy coil current control, the unit also incorporates a software controlled stall turn-off.

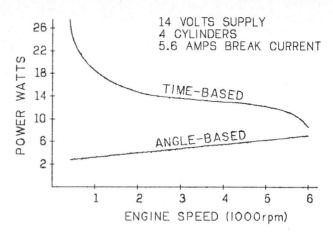

Fig 9 Coil drive power dissipation

Speed and position information is gained from a single variable reluctance sensor detecting teeth every ten degrees except at top dead centre (TDC) where there is a missing tooth. The output from this sensor has to be converted from analogue to digital so that it can be processed by a microprocessor. This conversion is achieved using an adaptive sense amplifier, which provides a single-shot pulse output triggered by the negative going zero crossing of the ground referenced input signal. From this information a reference signal can be produced.

Fig 8 Engine layout - HT components and detonation sensor

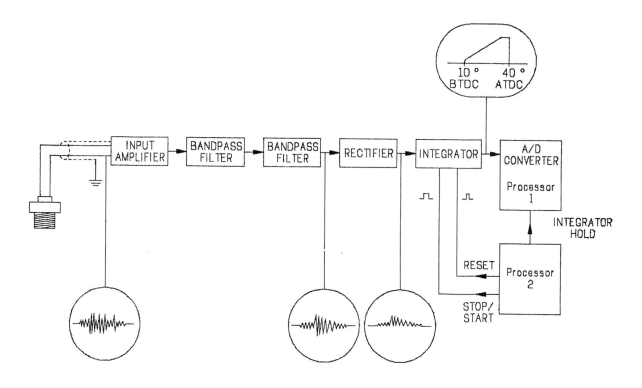

Fig 10 Detonation control circuit (block diagram)

The single piezoelectric accelerometer described earlier is fixed to the engine block providing detonation information from all four cylinders. As is seen in Figure 10 the output from the accelerometer is fed, firstly, into an input amplifier: this amplifier allows the gain of the input circuitry to be set for a particular application. From the input amplifier the signal is fed through a fourth-order band pass filter. Again, this part of the circuit can be set for a particular application by varying the frequency pass band. After filtering the signal is precision rectified and fed to an integrator whose period of integration is controlled by one of the microprocessors, such that integration takes place over a predetermined angle. This angle (the detection window) is chosen to exclude extraneous mechanical noises, such as those produced by valve closures, whose amplitude can be larger than the signal, due to detonation alone.

The voltage on the integrator is read through one of the A-D ports of the controlling microprocessor and a weighted average of that cylinder's past events is stored in a background average table. During the A-D conversion the background average from the relevant entry in the background average table is multiplied by the detection factor (which is programmable), to give the detection threshold. If the reading is less than the threshold then a new background average is computed and placed in the table. If the reading is greater than or equal to the threshold, detonation is detected. In this case no new background average is computed, thus the background average is not corrupted by the high levels produced by detonation. A programmable correction of $5°$ is then added to the location for that cylinder, in the detonation correction table, unless a maximum correction of $20°$ is reached, in which case the correction is held at maximum. This correction is the angle that the spark for that particular cylinder is retarded. In addition to the retard steps being programmable the advance steps back towards mapped timing and their rate of advance are also programmable. The actual rate of return to normal spark timing as programmed in this design is $0.625°/16$ sparks.

The need to treat each cylinder individually for both detonation of knock and subsequent retardation of the ignition arises, firstly, because the sensitivity of the accelerometer to each cylinder varies owing to the differing cylinder to accelerometer transfer functions. Secondly, the background noise level of each cylinder is different. Thirdly, there can be a large difference in the knock borderline timings from cylinder to cylinder. Retarding all cylinders on the basis that one is knocking can seriously reduce engine efficiency and torque. This is illustrated by the following test data. In this test a 1.6L Montego was driven following 8 different driving cycles including full throttle using fuels with research octane numbers between 98 and 89. As Figure 11 shows, the detonation control system of the A-R programmable ignition system is so sensitive and discriminating that even though detonation measurably increased as octane levels were reduced, average fuel economy was unaffected. Only at full throttle did power and fuel economy decrease but, even here, for less than most engineers would have expected. We believe this is a striking example of the benefits of this type of detonation control.

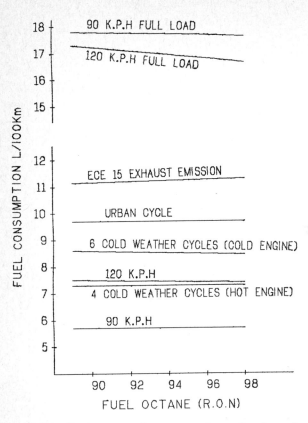

Fig 11 Fuel consumption versus octane rating for various drive conditions

V. SYSTEM DEVELOPMENT

There are three specific operating modes that cause trouble for any programmed ignition system and must be carefully addressed during the development period. These are:

- synchronisation of the ignition system and the engine during the transient speed variations of cold starting and rapid accelerations at higher speeds;

- noise immunity from both airborne input and lineborne sources on the car itself and

- electromagnetic susceptibility from sources external to the car.

Beyond these conditions, development testing must be applied to the detonation control system to ensure the signal to noise threshold is set correctly for all engines in all conditions, that the window where detonation signals are accepted is properly selected and, lastly, that the detonation system is consistently controlling all meaningful detonation.

Low Speed Synchronisation

Very early in the development work it was recognised that under cold cranking conditions ($0°C$ to $-30°C$) the system did not always recognise the missing reluctor tooth at TDC and synchronise with the engine.

The original strategy chosen to do this, (Figure 12), relied upon measuring the time between tooth intervals and comparing the last measurement taken with 1.5 times the previous measurement. When this factor was exceeded the system recognised it as a missing tooth interval (TDC). This worked fine at constant engine speed, but under low temperature cranking conditions when the cyclic speed variations were large, the tooth intervals prior to TDC were being increased and falsely recognised as missing teeth. The 2.0L A-R engine with its greater acceleration potential suffered from this more than the less powerful 1.6L A-R engine.

The strategy was revised (Figure 12) to look for a short tooth period (less than 0.6 times previous measurement) and thus recognise the tooth interval after TDC because cyclic speed variations guaranteed that this was always much shorter than the missing tooth interval. Trials with a hot engine under rapid acceleration/deceleration (simulated by quick clutch engagement on a rolling road) confirmed that this technique was a sound solution.

High Speed Synchronisation

Some prototype engines were fitted with damaged reluctor rings, which gave larger signals as distorted teeth passed the sensor. The input circuit which processed this signal had been compromised to provide best noise immunity (from coil turn off) at low speed. This resulted in a high speed misfire after processing the distorted teeth. The problem was solved by simply clamping the signal with zener diodes to remove large peaks.

The results of this synchronisation work at high and low speed resulted in a system which:-

- synchronised effectively at all engine speeds

- synchronised effectively under engine acceleration/deceleration

- had good noise immunity to ignition noise

- achieved this with a single variable reluctance sensor.

Electromagnetic Compatibility

A programme of EMC testing was conducted. The objective was to test the system up to 50 V/M radiation over a frequency range of 1 MHz to 1 GHz. By a specification previously agreed within Austin Rover it was permissible for this level of radiation to cause the electronics to shift timing slightly providing they recovered after the field was removed and the driveability of the engine was not impaired.

Initial trials were on two vehicles having 1.6L and 2.0L engines. This work was intended to test various harness layouts and configurations together with the programmed ignition design level which was the intended production release.

The harness work indicated which route should be used for minimum susceptibility and that screened leads on both the crank sensor and knock sensor were essential.

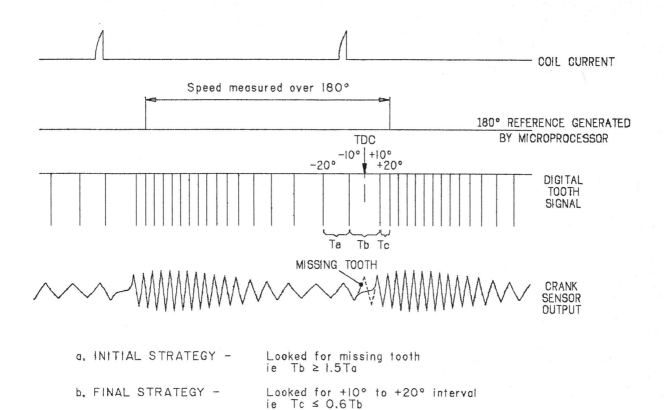

Fig 12 Missing tooth recognition strategy

Susceptibility work on the module itself showed that it suffered from a variety of problems below 10 V/M radiation - misfire, severe timing variations, knock retard initiation - all generally caused the engine to stop. Following these unacceptable results the circuitry was modified to include a ground plane on the printed circuit-board and RF filter components on the input leads. This was followed by further EMC trials. These showed a vast improvement over earlier tests. At 40 V/M only slight timing variations were observed and at 50 V/M slight engine faltering occurred but not severe enough to cause the engine to stall. This level of electromagnetic radiation is quite severe and it appears that the A-R programmed ignition system has as little susceptibility to this level of radiation as any vehicle electronic system known.

During cold cranking work, below about $-5°C$ both the 1.6L and 2.0L engines frequently failed to run after apparently cranking and firing correctly. The engine always stalled on release of the ignition switch. Signal traces identified that there was a fast voltage dip on the supply to the ECU. This was causing the internal regulator to reset the microprocessors (the regulator used provides a power up reset pulse). Under ambient temperature conditions the inertia of the engine carried it through the reset period to start normally. But when cold, the increased drag caused it to stop within this period. Laboratory tests showed that the problem was caused mainly by the grease used in the ignition switch. At low temperatures it became very viscose and induced transient contact separation. A simple solution, modification to the lubricant specification, solved the problem although much testing was done thereafter to confirm this.

VI. RELIABILITY

After the completion of the design and the anticipated development, laboratory and vehicle reliability began. To assess the reliability of the system, almost seventy programme ignition systems were subjected to 400 hour engine tests. An additional 25 vehicles with programme ignition accumulated over 500,000 miles in durability, hot and cold testing and pavé tests. A further quantity of programme ignition units were subjected to electromagnetic susceptibility, high temperature operation, salt spray, vibration, high and low temperature soak and thermal cycles in laboratory tests. Following our standard technique as failures occurred in the above tests each failure had to be analysed, a solution determined and the solution tested in the environment in which the original failure occurred. Through this intensive process all failure modes were forced out and corrections found and proven so that confidence developed that this programme ignition system was an intrinsically reliable design.

For each electronic product we establish an analytically determined reliability and keep improving the design until the analytical reliability exceeds the reliability goal established at the start of the programme. These goals are always 99% or better. To prove to ourselves that the final design met our reliability goals statistical verification tests were then begun. Over 161 electronic modules and harness connectors were electrically exercised at the design temperature extremes for 600 hours. Actually because we had some early failures a total of 322 units were subjected to this test to prove that we had met our reliability goal. In addition, to even more rigorous reliability requirements Austin Rover had tested 803 detonation sensors and 803 crankshaft variable reluctance sensors to prove that they were 99.8% reliable. Many engineers find this level of testing astonishing and even perhaps unnecessary. Even some of our engineers thought this initially. None does today. The value of this final rigorous validation test has been proven completely and is necessary if electronics are to demonstrate the high level of reliability that is required in modern automobile usage.

VII. CONCLUSIONS

The Austin Rover programmed ignition system has been shown to exhibit high reliability - proven by extensive testing. Its static timing reference is fixed on engine build by the tolerance of assembled parts and requires no setting up either then, or throughout the life of the vehicle. This gives significant benefits to both the manufacturer and the customer, by reducing his service costs and providing consistent economy/performance.

The individual cylinder knock control allows the engine to run at its optimum ignition advance under all conditions, it also compensates for variability between engines of the same type and provides tolerance to fuel octane rating changes.

As such it has made a major step towards introducing electronic engine controls onto mass produced cars within the United Kingdom.

C427/84

SAE 841294

A study on dual circuit cooling for higher compression ratio

H KOBAYASHI, K YOSHIMURA and T HIRAYAMA
Toyota Motor Corporation, Shizuoka, Japan

SYNOPSIS In this paper, feasibility of reducing knock and thereby achieving higher compression ratio in a spark ignition engine has been studied by controlling cylinder head and cylinder block coolant temperatures independently with the aim of reducing the temperature of unburnt portion of mixture during flame propagation, while preventing adverse effects on the engine performance. The study was performed using a 1.3 litre OHV water cooled engine equipped with a dual circuit cooling system.

In a fundamental study at the base compression ratio of 9:1, it was found that when the cylinder head coolant temperature was lowered while maintaining the cylinder block coolant temperature at 80°C engine output at WOT was increased significantly, because of considerable knock reduction and increased charging efficiency without increase in friction torque. These results indicated a great potentiality of the dual circuit cooling system for raising the compression ratio.

Based on the results of the fundamental study, engine output at WOT and fuel economy at partial load were therefore evaluated at various compression ratios from 9:1 to 15:1 for various cylinder head coolant temperatures. It was found that the decreased engine output caused by the increased knock intensity, with raised compression ratio at the coolant temperature for both circuits of 80°C, was restored by lowering the cylinder head coolant temperature without deteriorating improved fuel economy at partial load.

Given controlling temperatures of the cylinder head and cylinder block coolants at 50°C, probably the minimum for practical use, and 80°C respectively, and allowing that WOT engine output must be at least equal to that at the base compression ratio, it was concluded from the above investigations that maximum compression ratio attainable by the dual circuit cooling system was 12:1, when using regular gasoline of 91 RON. With these specifications, WOT engine output increased by about 10% at high engine speeds without any sacrifice of power at low engine speeds and fuel economy at partial load improved by about 5%.

1 INTRODUCTION

Raising compression ratio is an efficient approach to higher fuel economy in a spark ignition engine. However, since the engine tends to knock when the compression ratio is raised, it has not been practically feasible to use high compression ratios. The phenomenon of knock has been the subject of many studies and much effort has been made in order to reduce knock. Papers have been published on combustion chamber geometry (1), squish and swirl (2), and mixture strength (3) and so on. However, very little work has been done on the possibility of reducing knock by means of coolant temperature control.

The subject of this paper is an investigation into the possibility of reducing knock, and thereby enabling compression ratio to be raised, by controlling the cylinder head and block coolant temperatures independently, aiming at reducing the temperature of unburnt portion of mixture during flame propagation while preventing adverse effects on the engine performance. For this purpose, an experimental engine equipped with a dual circuit cooling system was prepared and effects of coolant temperature of each circuit on knock and engine output at the base compression ratio were examined and then effects of compression ratio on the engine performance for various cylinder head coolant temperatures and the potentiality of this dual circuit cooling system for higher compression ratio in practical use were also examined.

2 APPARATUS AND PROCEDURE

A conventional OHV 1.3 litre, four cylinder, four stroke spark ignition engine was used. This engine has wedge type combustion chamber geometry and normally operates with a compression ratio of 9:1 and a coolant temperature of 80°C.

As shown in Fig. 1, the cooling system was separated into independent circuits for the cylinder head and cylinder block by means of a cylinder head gasket without water holes. In order to control the coolant temperatures independently each circuit was equipped with a water pump, a heat exchanger, a flow control valve, an electromagnetic flow meter and thermocouples.

Using this engine equipped with the dual circuit cooling system, this study was conducted in the following three stages.

(1) A fundamental study at base compression ratio of 9:1: This study was carried out with the objective of determining effects that controlling engine coolant at various temperatures would have on trace knock spark timing (TKST) and engine output at TKST at

wide open throttle (WOT).

(2) An investigation of engine output and fuel economy as compression ratio was raised : Characteristics of engine performance at TKST at WOT and fuel economy at partial load for various coolant temperatures were examined as the compression ratio was raised gradually from 9:1 to 15:1 by shaving the bottom surface of the cylinder head as shown in Fig. 2.

(3) A study of potentiality of the dual circuit cooling system for higher compression ratio in practical use : Maximum compression ratio that could be achieved in practical use by the dual circuit cooling system was determined from the findings of the above investigation. And whole engine performance characteristics were evaluated at the engine specifications determined above.

For an accurate assessment to be made of the effects of coolant temperature on knock, it was neccessary to analyze heat release rate and to measure combustion chamber wall temperatures and heat losses to coolants. Heat release rate was calculated from cylinder pressure data using Sanda's method (4). Combustion chamber wall temperatures were measured by sheath thermocouples of 0.5mm diameter buried in the wall as close to the surface as possible. A mixing box was installed in each cooling circuit to restrain fluctuations in coolant temperature at the measuring points within 0.1° C in order to determine the heat loss to coolant accurately.

Tests were conducted using regular gasoline of 91 RON and multigrade oil of SAE 10W-30.

3 FUNDAMENTAL STUDY AT BASE COMPRESSION RATIO OF 9:1

3.1 Effects of coolant temperature on knock and WOT output

The temperature of coolant in each circuit was lowered in turn while maintaining the coolant temperature of the other circuit at 80°C, and WOT engine output was measured at TKST. TKST was defined as a spark timing where sound of knock began to be audible when advancing spark timing gradually from retarded timing. The tests were carried out at low engine speeds, where engine output at WOT was most restricted by knock. As shown in Fig. 3, the output increased significantly as the cylinder head coolant temperature was lowered but only very marginally when the cylinder block coolant temperature was reduced.

For better understanding of these results, the effects on trace knock spark timing(TKST), charging efficiency and friction torque produced by the reduction in coolant temperature in each circuit are shown in Fig. 4 and also the effects of each coolant temperature on heat release rates at fixed spark timing and TKST are shown in Fig. 5.

(a) Trace knock spark timing :
The advanced value of trace knock spark timing shown in Fig. 4(a) was regarded as an index of knock reduction since changes in coolant temperature for either circuit had negligible effect on MBT, which was evidenced also by the fact that the heat release rate at a fixed spark timing did not change as shown in Fig. 5(a). It was found that the degree of knock reduction obtained by lowering the cylinder head coolant temperature was about twice as much as that when the cylinder block coolant temperature was lowered.

This greater knock reduction obtained by lowering the cylinder head coolant temperature contributed to the larger gain in engine output as evidenced also by the increased combustion velocity with the advanced TKST as shown in Fig. 5(b). The cause of knock reduction is discussed further in paragraph 3.2.

(b) Charging efficiency :
Fig. 4(b) shows the effects on charging efficiency. As shown, lowering the coolant temperature in either circuit similarly resulted in a significant increase in charging efficiency.

(c) Friction torque :
As shown in Fig. 4(c), friction torque remained constant when the cylinder head coolant temperature was reduced but it increased considerably as the cylinder block coolant temperature was lowered, because the major sliding parts in an OHV engine are contained in the cylinder block. This increase in friction torque largely caused the much smaller gain in output obtained by the reduction of cylinder block coolant temperature.

In addition, the heat loss to coolant in each circuit was measured for various cylinder head coolant temperatures. As shown in Fig. 6, it was found that the heat loss to the cylinder head circuit increased as the cylinder head coolant temperature was lowered whilst that to the cylinder block circuit decreased. Conversely similar results were obtained when the cylinder block coolant temperature was lowered. In either case total heat losses to coolants remained unchanged. This indicates that lowering either coolant temperature does not have meaningful influence on the cooling heat loss related to thermal efficiency, although it affects the distribution of heat loss to each circuit. It has been confirmed by experiment that this change in the distribution of heat loss to each circuit was mainly caused by a change of heat flow through contact surfaces between the cylinder head and the cylinder block due to the temperature difference between both coolants.

From these findings, it was evident that lowering the coolant temperature in the cylinder head circuit not only reduced knock but also increased charging efficiency without any increase in friction torque. This resulted in significant gains in WOT output. The results furthermore indicate a great potential of the dual circuit cooling system for raising the compression ratio.

3.2 Knock reduction resulting from lowered cylinder head coolant temperature

Knock is generally regarded as being generated by the self-ignition of the unburnt portion of mixture during flame propagation. Given a change in coolant temperature, two factors most affecting the temperature of the unburnt portion of the mixture seem to be the combustion chamber wall temperature and the temperature of mixture immediately prior to ignition. These factors were therefore examined to find out why the greater reduction in knock could be obtained by increasing the cylinder head cooling rather than the cylinder block cooling.

(1) Combustion chamber wall temperature
Temperature measurement of the upper wall of the combustion chamber, composed of the cylinder head wall, was taken at TKST at seven points for various coolant temperatures. As shown in Fig. 7, the change in the wall temperature was 90% of the change in coolant temperature when the cylinder head coolant temperature was lowered; increasing the cylinder block cooling did not affect the upper wall temperature.

The temperature of the lower part of the combustion chamber, composed of the upper surface of the piston, was calculated according to the method presented by Hasebe, et al (5). The change in the wall temperature was found to be only 30% of the change in cylinder block coolant temperature; there was no change when the cylinder head coolant temperature was reduced.

(2) Mixture temperature
With the condition of the spark timing at TDC, the mixture temperature immediately prior to ignition, that is at the end of compression stroke, was estimated for various coolant temperatures.

This was done as follows : Firstly, the mixture temperature at the end of intake stroke was approximately determined based on the state equation of gas, assuming that the cylinder pressure was equal to the pressure measured in the intake manifold, and using the mixture quantity obtained by measurement. Next, using the mixture temperature calculated above as the initial condition, the mixture temperature at the end of compression stroke was obtained based on the first law of thermodynamics by splitting the compression stroke into small elements and calculating mixture temperature step by step through the compression stroke. Heat transfer coefficient values necessary in this calculation were obtained by Woschni's equation (6) ; temperature measurements were taken of the combustion chamber upper wall and cylinder liner ; and the temperature of the piston upper surface was calculated.

The results thus obtained showed that when the coolant temperature in either circuit was lowered from 80°C to 30°C, the mixture temperature at the end of compression stroke fell by about 20°C.

From these findings it may be evident that lowering the cylinder head coolant temperature is more effective in reducing knock than lowering the cylinder block coolant temperature. That is because the former case seems to result in a greater cooling of the unburnt portion of mixture during flame propagation due to the larger decrease in combustion chamber wall temperature, though the both cases similarly result in a decrease in mixture temperature immediately prior to ignition.

4 INVESTIGATION OF ENGINE OUTPUT AND FUEL ECONOMY AS COMPRESSION RATIO IS RAISED

It was indicated in the previous chapter that there was the great possibility of raising compression ratio by lowering the cylinder head coolant temperature while maintaining the cylinder block coolant temperature unchanged. Therefore, the characteristics of engine performance for the various cylinder head coolant temperatures were evaluated as the compression ratio was gradually raised from 9:1 to 15:1.

The results are given in Fig. 8, which shows WOT engine output at TKST and fuel economy at partial load at various compression ratios for various cylinder head coolant temperatures.

(1) WOT engine output at TKST
With conventional coolant temperature, i.e. both circuits at 80°C, engine output deteriorated as the compression ratio was raised since TKST had to be retarded due to knock. However, as the cylinder head coolant temperature was lowered TKST was advanced and the output was restored. For example, engine output at the compression ratio of 13.5:1 with the cylinder head coolant temperature at 30°C was equal to the output at the base compression ratio of 9:1 with coolant temperature at 80°C.

(2) Fuel economy at partial load and idling
Fuel economy at partial load was not adversely affected by lowering the cylinder head coolant temperature since spark timing could be set at MBT and friction torque did not change, and it was improved as the compression ratio was raised.

When the cylinder head coolant temperature was lowered, for example, by 30°C at the base compression ratio, fuel economy at idling deteriorated by a few percent, probably due to a poor vaporization of the fuel in the intake port and cylinder. However, the fuel economy was improved beyond that of base engine as the compression was raised, due to the improvement in thermal efficiency.

5 POTENTIALITY OF DUAL CIRCUIT COOLING SYSTEM FOR HIGHER COMPRESSION RATIO IN PRACTICAL USE

In view of the cooling capacity of a conventional radiator, it would not be feasible to maintain cylinder head coolant at very low temperatures. However, with an exception of a continuous driving condition with high load at extremely low vehicle speed, most conventional radiators are capable of controlling the cylinder head coolant temperature at 50°C. Therefore, the controlling temperature of 50°C would be realized by practically acceptable modifications in cooling capacity and controlling system of radiator.

Given this figure of 50°C and allowing for the condition that WOT engine output at low engine speeds, where the output is most restricted by knock, must be at least equal to that at the base compression ratio, the maximum compression ratio that could be used would be 12:1, as indicated in Fig. 8.

The performance characteristics of the engine were therefore evaluated with the cylinder head coolant temperature at 50°C, the cylinder block coolant temperature at 80°C and the compression ratio at 12:1.

(1) WOT engine output
Output over the entire speed range is given in Fig. 9. As shown, the output at low speeds was the same as that of the base engine and as engine speed increased the output of the

engine became greatly superior to that of the base engine, mainly because in high speed range the spark timing can be set at MBT or near MBT even at the compression ratio of 12:1 and therefore the potential improvement in thermal efficiency due to the increased compression ratio is derived to almost full extent — the difference was more than 10% at high engine speeds. With the coolant temperature in both circuits at 80°C, the engine output was inferior to that of the base engine at low speeds.

(2) Fuel economy at partial load and idling
As shown in Fig. 8, fuel economy at partial load showed a 5% improvement over the base engine.

Fuel economy at idling was improved by about 7% over the base engine due to the improvement in thermal efficiency, although, as mentioned in the previous chapter, it deteriorated when the cylinder head coolant temperature was reduced at the base compression ratio of 9:1.

(3) Exhaust emission
NOx emission at constant engine torque did not increase significantly in mass compared to that of the base engine although the concentration increased slightly, because the charging gas quantity decreased slightly due to the improved thermal efficiency.

HC emission increased in mass by about 30%. This was caused by the reduction in exhaust gas temperature due to the improvement in thermal efficiency and probably by the increase in quenching layer thickness owing to the lower temperatures of the combustion chamber walls. The effect of raising compression ratio on the increase in HC emission was much greater than that of lowering the cylinder head coolant temperature, as shown in Fig. 10.

The above results show that, by adopting the dual circuit cooling system which controls cylinder head coolant temperature at 50°C, probably the minimum for practical radiator, and cylinder block coolant temperature at 80°C, it is possible to increase the compression ratio from 9:1 to 12:1 without any sacrifice in WOT output. Therefore, WOT engine output was improved by about 10% at high engine speeds and fuel economy was also improved about 5% at partial load and about 7% at idling.

6 CONCLUSION

The following results were obtained by using a 1.3 litre OHV engine which was equipped with a dual circuit cooling system.

(1) Major findings obtained in a fundamental study at base compression ratio of 9:1 are:
(a) The degree of knock reduction by lowering the cylinder head coolant temperature was twice as much as that by lowering the cylinder block coolant temperature. This is because lowering the cylinder head coolant temperature probably caused a greater cooling of unburnt portion of mixture during flame propagation due to the larger decrease in the combustion chamber wall temperature.
(b) Charging efficiency was almost equally increased with lowering the coolant temperature of the cylinder head and with lowering that of the cylinder block.
(c) Friction torque was not increased when the cylinder head coolant temperature was lowered, since major sliding parts are contained in the cylinder block.

From these findings it is evident that lowering the cylinder head coolant temperature results in significant gain in WOT output, since it not only reduces knock but also increases charging efficiency without any increase in friction torque. This indicates a great potentiality of the dual circuit cooling system for higher compression ratio.

(2) An investigation at higher compression ratios indicated that it was possible to raise compression ratio, without sacrificing WOT output, by lowering the cylinder head coolant temperature while maintaining the cylinder block coolant temperature unchanged.

(3) In view of practical use, compression ratio of 12:1 was achievable by adopting the dual circuit cooling system which controlled the cylinder head coolant temperature at 50°C, probably the minimum for practical radiator, and the cylinder block coolant temperature at 80°C. This resulted in a better fuel economy of about 5% at partial load and improved WOT output by more than 10% at high engine speeds without any sacrifice of output at low engine speeds.

REFERENCES

(1) Benson,G., Fletcher,E.A., Murphy,T.E. and Schever,H.C. Knock (Detonation) Control by Engine Combustion Chamber Shape. SAE Paper No.830509, 1983.

(2) Thring,R.H. and Overington,M.T. Gasoline Engine Combustion —The High Ratio Compact Chamber. SAE Paper No. 820166, 1982.

(3) May,M.G. Lower Specific Fuel Consumption with High Compression Lean Burn Spark Ignited 4 Stroke Engine. SAE Paper No.790386, 1976.

(4) Sanda,S., Toda,T., Nohira,H. and Konomi,T. Statistical Analysis of Pressure Indication Data of an Internal Combustion Engine. SAE Paper No.770882, 1977.

(5) Hasebe,T., Kaminishizono,T., Arai,F., Machida,S., Fujikake,K., Takahashi,R. A Thermal Analysis of a Spark Ignition Engine Piston. FISITA No.82019, 1982.

(6) Woschni,G. A Universally Applicable Equation for the Instantaneous Heat Transfer Coefficient in the Internal Combustion Engine. SAE Paper No.670931, 1967.

Table 1 Specifications of base engine

Number of cylinder	4
Displacement (cm^3)	1290
Bore × Stroke (mm)	75 × 73
Compression ratio	9
Combustion chamber	Wedge type
Valve train	O.H.V.
Coolant temperature	80°C

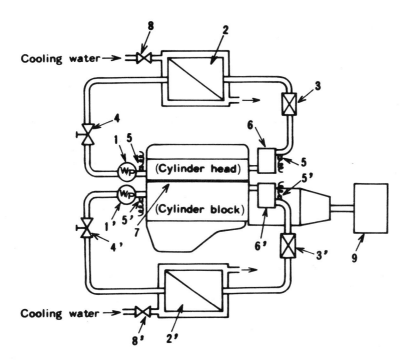

1,1'. Water pump
2,2'. Heat exchanger
3,3'. Electro-magnetic flow meter
4,4'. Flow control valve for coolant
5,5'. Thermocouple
6,6'. Mixing box
7.   Cyl. head gasket without water holes
8,8'. Flow control valve for heat exchanger
9.   DC-DY Dynamometer

Fig 1 Experimental apparatus

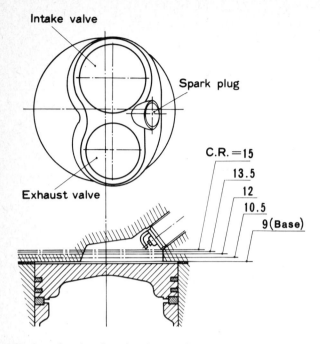

Fig 2 Combustion chamber configuration and contact surface between cylinder head and block for each compression ratio

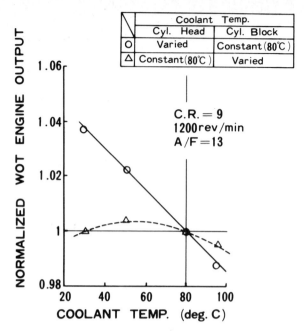

Fig 3 Effect of coolant temperature on WOT engine output at trace knock spark timing

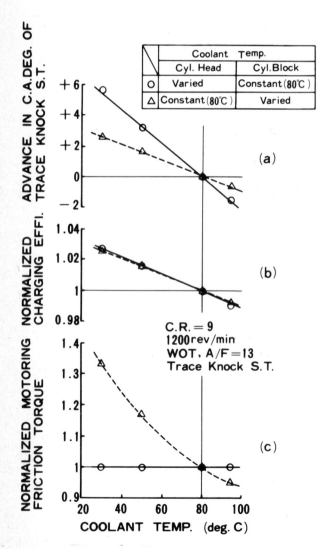

Fig 4 Effect of coolant temperature on (a) trace knock spark timing, (b) charging efficiency and (c) friction torque

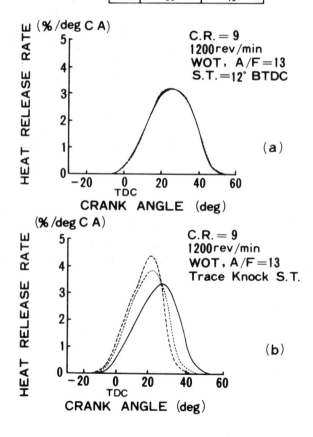

Fig 5 Effect of coolant temperature on heat release rate (a) at a fixed spark timing and (b) at trace knock spark timing

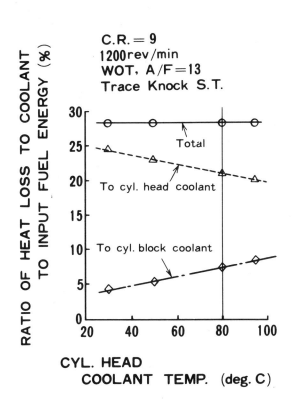

Fig 6 Effect of coolant temperature on heat loss to each coolant

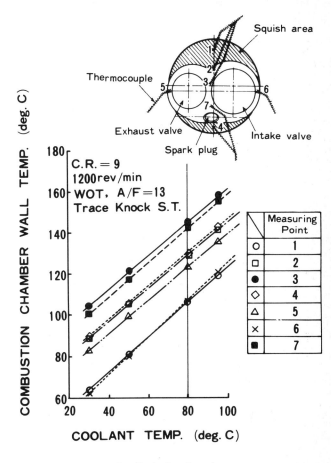

Fig 7 Effect of cylinder head coolant temperature on combustion chamber wall temperature

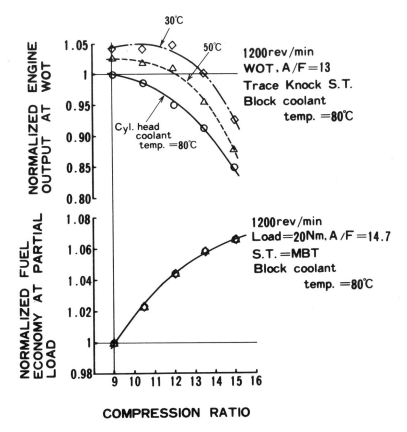

Fig 8 Effect of compression ratio on WOT output and fuel economy for various cylinder head coolant temperatures

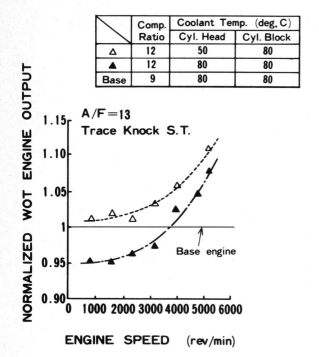

Fig 9 WOT output characteristics of engine equipped with dual circuit cooling system

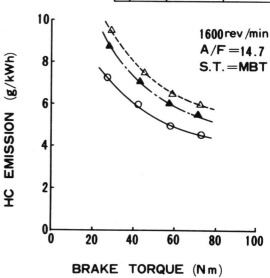

Fig 10 HC emission characteristics of engine equipped with dual circuit cooling system

A single cylinder engine for crankshaft bearings and piston friction losses measurement

R CERRATO, R GOZZELINO and R RICCI
Fiat Research Centre, Turin, Italy

1 INTRODUCTION

Reduction of engine friction is one of the most important actions to be taken in order to improve fuel economy both in S.I. and Diesel engines.

Conventional data on friction losses have normally been based either on measurements of indicator method or on motoring.

In the first method, the total loss is derived by direct measurement of the indicated power from a cylinder pressure diagram from which the brake power is subtracted. Ideally, this is the only procedure to obtain the absolute value of losses but the problems of achieving a satisfactory indicator diagram, even when using the most sophisticated electronic equipment are very heavy.

Moreover, this procedure normally results in mean values that don't take into account friction trends versus crankangle.

No satisfactory alternative procedure exists for establishing engine losses.

Only "half truth" can be derived by methods such as motoring which make it possible to determine the total loss or part of it by stripping the engine: in fact, the motoring loss so obtained is not the same as the loss under firing mode because of the different engine condition in terms of pressure on the piston, temperatures, clearances, etc.

Moreover, when these methods are used, results are still in the form of mean values, without any possibility to observe friction in its components in real operation. This is particularly difficult to be determined for piston and ring friction, which, according to the technical literature, may be responsible for as much as 60% of the friction of the whole engine, but is the most difficult to be assessed because both temperature and pressure play such a large part in determining the friction.

The primary purpose of this paper is to provide a better understanding of piston and ring friction. To obtain such information a single cylinder engine has been developed in order to measure separately friction losses in cylinder/piston and in crankshaft/bearing couplings.

In this engine, derived from a 1.3 litre Fiat commercial engine, the cylinder liner and the crankshaft bearings are mechanically disconnected from the crankcase and equipped with force sensors.

This paper refers about the setting-up of the equipment, the calibration of piston frictional signals and the best methodology.

Employing the described methodology, tests were carried out to determine the sensitivity response of the equipment in terms of friction to main engine parameters variations.

The measurements of friction forces versus crankangle were made at different speeds, loads, temperatures of cooling fluid, type of lubricant.

From the experimental data it was possible to identify the type of friction (hydrodynamic, mixed and boundary lubrication) and to obtain Sommerfeld diagrams.

Different parameters, such as piston, ring and bearing geometry, surface roughness, clearances and different materials have been finally investigated and results are reported both for motoring and firing conditions.

2 EXPERIMENTAL SET-UP

Friction losses in a reciprocating engine depend on many parameters:

- type and viscosity of the lubricant;
- linear velocity of the piston;
- quality and roughness of the sliding surfaces;
- pre-loading of piston rings;

- bore-to-stroke ratio;
- temperature of the lubricant and of the coolant;
- forces due to alternative and centrifugal masses.

A single cylinder was designed not only to measure such forces, but to make it possible to vary parameters on which friction forces depend.

The experimental set-up has been designed to allow these measurements. In particular, it has been planned to measure the friction loss in the piston and in a journal bearing using standard components of a Fiat engine.

The engine block in which the crankshaft is assembled has been designed in order to obtain an extremely rigid structure to avoid influencing signals of the measuring sensors (1). To such purpose a total balancing of alternative forces both of the first and second order has been obtained via two shafts with contro-rotating interchangeable masses.

One of the two journal bearings has been installed in the block by means of a composite ball and needle bearing which allows, through a load cell, actuated by a reaction lever, the measure of friction losses in the bearing as a function of the angular velocity.

At one end of the crankshaft an optical 360 marks encoder is mounted, phased with the top dead centre, for measuring the angular velocity and the instantaneous crankangle.

The crankcase is rigidly connected to the test bench. The cylinder block, on the contrary, may slide through recirculating ball bearings on four columns (Figs. 1a and 1b).

The intermediate section has to transfer the friction forces of the piston to two piezo-electric cells, one at each side. The assembling of these load cells to the block and to the intermediate section is obtained through bars, designed in order to have the highest possible resonance frequency (700 Hz).

Particular care has been taken in reducing the effect of side on load cells coming from the piston, in order to obtain the highest accuracy in measuring only the friction forces of the piston (Fig. 2b).

In the intermediate section there are the housing of the interchangeable cylinder, in order to vary both material (cast iron, aluminum, etc.) and type of surface (roughness, coating); and the flexible pipe for the coolant circulation whose flow rate and temperature are controlled.

The seal between the engine block and the cylinder is obtained simply by a Kalrez o-ring (Fig. 1c) resisting at high temperature (up to 300°C). No mechanical link exists between the cylinder block and the intermediate section with the obvious purpose of avoiding the negative effect of pressure on friction forces. In fact, the engine head, which has also been obtained from a standard component, is mechanically supported by a structure directly connected with the ground.

The driving of the camshaft is obtained by means of a timing belt, and the engine head is instrumented with a transducer to measure the pressure in the firing chamber.

This single-cylinder engine is connected to a brake, and has a cooling apparatus which controls independently both the lubricant and the coolant.

Measurements are performed both in "firing" and in "motoring" conditions, recording at the same time on a magnetic tape, the signals from the two load cells in the piston, from the load cells in the bearing, and from the pressure transducer, as a function of the signal coming from the encoder, to obtain the behaviours of these parameters and the maps of FMEP (Friction Mean Effective Pressure) versus speed and load.

An electronic device allows to obtain in real time the FMEP values during the tests, both for the piston and the bearing.

In fact, signals from the load cells are periodically sampled as a function of the T.D.C. signal and of the actual crankangle and fed to a multiplier which, using the value of piston velocity pre-determined in an eprom, gives the instantaneous power, which integrated for 720° shows the FMEP on a display in order to give to the experimenter the exact knowledge of what is going on in the engine in real time.

Figs. 1a, 1b show the experimental set-up obtained with standard components for a 1,300 c.c. engine with a bore of 86.4 mm and a stroke of 55.5 mm.

Only by replacing the crankshaft, can the influence of stroke to bore ratio from 0.64 to 1 be studied.

3 SIGNALS ANALYSIS

Signals coming from the load cells located on the cylinder liner and on the bearing have been critically analyzed.

The sensitivity of the sensor has been investigated by varying such parameters as speed, load, lubricant and coolant temperatures.

All these variables have been found to influence substantially friction forces.

Fig. 2 shows the signal of the piston friction forces as it comes from the load cells.

This signal and the signal coming from the load cell mounted on the bearing have been analyzed in frequency to determine the source of the noise superimposed to the base frequency.

Results have shown that the base frequency is the same as the engine frequency and may be interpreted as the real friction effect, whereas the superimposed "noise" has a frequency of 700 Hz, independent from angular velocity.

This frequency has been found to be inborn in the system, as determined through an impulse generator.

It has been decided to eliminate this frequency by a low-pass filter of 300 Hz to experiment with a "cleaner" signal.

It has been found that the "noise" amplitude is influenced by the lubricant viscosity and by the roughness of the mating surfaces, even if a quantitative effect is not so easy to be measured.

Comparing results obtained in "motoring" with different throttle openings and in "firing" conditions, it has been shown that friction signals coming from a full engine cycle in both conditions differ substantially only in the compression-expansion phase, due to different pressure forces (Fig. 3).

Effects of angular velocity on friction forces have been found to be remarkable as shown in Fig. 4 where results in "motoring" at 1,000 and 2,000 rpm, with the same throttle opening and oil and water temperature are shown.

It should be noticed that the signal corresponding to the friction between piston and cylinder was not going down to zero at the T.D.C. of the compression-expansion phase. Through a series of experimental runs, the reason for such a phenomenon has been identified in the traction effect by the o-ring on the movable cylinder setup when pressure exists in the combustion chamber.

After determining the transfer function between this pressure and the traction force of the o-ring, an analytical procedure has been developed in order to subtract from the total force signal the contribution due to the o-ring traction.

In Fig. 5a the two signals, the real and the correct one, are shown in motoring conditions; in Fig. 5b they are shown in firing conditions.

In this latter case, the adjustment is much more noticeable due to very high pressure forces.

The treatment of digital data consists in obtaining, on a statistical basis, an average cycle for each running condition from which the corresponding FMEP's are calculated.

This parameter is visualized as a map in the examples shown in Fig. 6.

4 EXAMPLES OF PARAMETRIC ANALYSIS

Employing the experimental methodologies presented before, a few examples of analysis of parameters influencing the tribologic behaviour of the mechanical couplings are given.

In particular, the variation of the friction force as a function of lubricant and coolant temperature, of piston clearance and liner surface roughness is analyzed (2), (3).

As far as the piston is concerned, the percentage of the friction force due to each ring and to the skirt surface is given (2), (3),(4).

4.1 Effect of lubricant and coolant temperatures

Several measurements have been made in "motoring" and in "firing" conditions at constant load and speed, but with different temperatures of the lubricant and of the coolant.

In Fig. 7 the friction values of the two couplings: piston-cylinder and bearing-crankshaft in "motoring" condition at 1,000 rpm with the throttle closed, are shown with inlet temperatures of the lubricant and the water of 45-45°C and 90-60°C respectively (5).

The effect of thermal conditions on the friction forces in the mentioned couplings is evident.

Quantitatively through a complete series of experimental runs, it has been found that the force and torque of friction of the two couplings are proportional to the lubricant viscosity.

As regards the piston it is interesting to note how, at constant inlet temperature of the lubricant, friction is substantially dependent on the coolant temperature.

This is explained by the effect of the coolant on the temperature of the cylinder liner and, consequently, on the viscosity of the lubricant on the cylinder boundary layer.

4.2 Mounting clearance

The influence of this parameter has been investigated only for the piston-cylinder coupling.

In Fig. 8 results between a standard set-up with the standard 0.06 mm clearance and a set-up where the clearance is 0.11 mm are compared in "motoring" condition at 2,000 rpm without internal pressure and with temperatures of the lubricant and of the coolant of 90°C and 60°C respectively.

If the clearance is wider (0.11 mm), friction decreases by about 20% compared to the standard layout (0.06 mm).

It has been found that such difference is not entirely due to the piston skirt, as a change has also been noticed in the tribological behaviour of the rings.

4.3 Surface roughness

The influence of this parameter has been investigated by comparing two cast iron cylinders having an average roughness of 0.2 and 1.4 μm.

The mounting clearance, the temperatures of lubricant and coolant were the same.

In Fig. 9 results of a test in "motoring" condition at 1,000 rpm without any pressure force are shown. It can be seen that friction near the dead centers, being of the composite-dry type, is largely dependent on surface roughness.

In the portions of the stroke where friction is mainly hydrodynamic, the effect of surface roughness is practically nil (6).

At rates exceeding 2,000 rpm the tribological behaviour of the piston-cylinder coupling is no longer affected to a remarkable extent by surface roughness on the cylinder liner.

4.4 Friction forces due to the skirt and to the rings of the piston

Results of tests in "motoring" conditions without pressure forces on the complete piston, on the skirt only and on the skirt with the various rings: 1st, 2nd and scraper are shown.

Percentages of friction forces have been attributed to the skirt and to the rings (7), (8).

In Fig. 10 the friction forces are shown for the various configurations at the same rpm and thermal conditions.

Under such experimental conditions, the skirt is responsible for 25% of the friction of the complete piston.

Among the rings, the scraper is responsible for 55% of the friction of the rings, due to the tangential load which is definitely greater than in the other rings (see Table 1).

It must be noted that in "firing" conditions and when pressure forces are present, the percentages of friction attributable to the various piston components are different.

5 CONCLUSIONS

A single cylinder engine has been developed in order to measure separately friction losses in cylinder/piston and in crankshaft/bearing couplings.

This experimental set-up allows firing and motoring measurements up to about half the maximum speed of the engine.

Further restrictions are due to the range of cooling temperatures.

Nevertheless, the described test method has been proved effective in measuring FMEP and its changes as a function of engine parameters and couplings geometrical features.

The first measurements of friction forces in piston/cylinder versus crankangle have shown what follows:

- both in motoring and firing conditions friction forces increase with increasing throttle opening near compression TDC;

- increasing engine speed results in on increase of friction forces for crankangles corresponding to hydrodynamic regions;

- increasing oil temperature friction forces have been observed to decrease in hydrodynamic regions and to increase in mixed regions (near T.D.C. and B.D.C.);

- friction forces are influenced by cylinder surface roughness only in mixed regions;

- friction forces are influenced by piston/cylinder clearance only in hydrodynamic regions.

As to measurements of crankshaft bearing friction torque versus crankangle, the following trends have been observed:

- increasing throttle opening results in a torque increase near compression TDC;

- increasing engine speed results in an increase of mean friction torque; however friction torque peak near compression TDC is lower;

- increasing oil temperature results in an almost linear decrease of mean friction torque.

FMEP values for each compling have confirmed the reported trends.

Future work will be aimed at quantifying FMEP variations and the influence of these variations on engine efficiency.

6 REFERENCES

(1) FURUHAMA-TAKIGUCHI. Measurement of Piston Frictional Force in Actual Operating Diesel Engine. SAE Paper 790855.

(2) FURUHAMA-TAKIGUCHI. Effect of Piston and Ring Designs on the Piston Friction Forces in Diesel Engines. SAE Paper 810977.

(3) MILLINGTON-HARTLES. Frictional Losses in Diesel Engines. SAE Paper 680590.

(4) HAMILTON-MOORE. Comparison between Measured and Calculated Thickness of the Oil Film Lubricating Piston Rings. Proc. Instn. Mech. Engrs., Vol. 188, 20/74.

(5) DOWSON. Piston Ring Lubrication. Energy Lubrication Technology, Winter Annual Meeting of the American Society of Mechanical Engineers, December 1979.

(6) HATA-NAKAHARA-AOKI. Measurement of Friction in Lightly Load Hydrodynamic Sliders with Striated Roughness. The Winter Annual Meeting of the American Society of Mechanical Engineers, Chicago, November 1980.

(7) RHODES-PARKER. The AE Coneguide Low Friction Piston Skirt Design. AE Symposium, April 1982.

(8) BRUNI-RAGGI. An Experimental Lightweight Piston. AE Symposium, 1978.

Table 1 Tangential load of rings

1st ring	$T = 14.4$ N
2nd ring	$T = 12.6$ N
oil-ring	$T = 49.4$ N

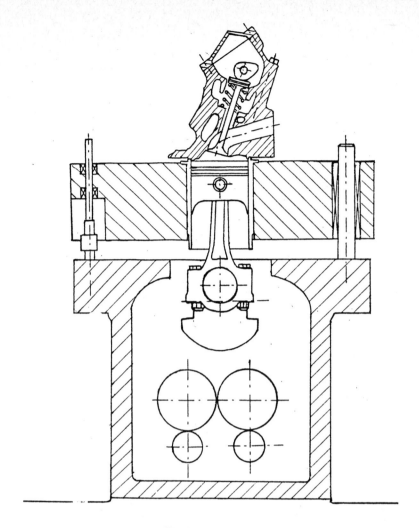

Fig 1a Cross-section

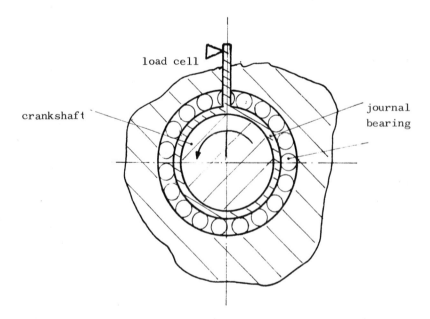

Fig 1b Detail of the instrumented bearing

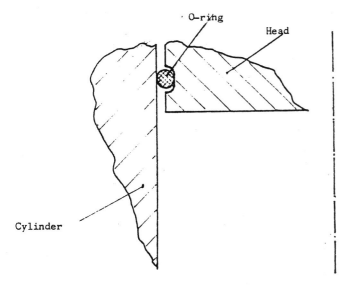

Fig 1c   Detail of the O-ring

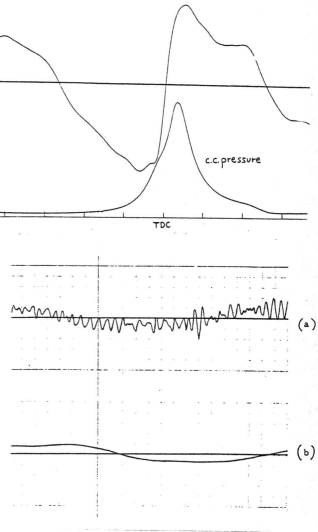

Fig 2   Frictional force signal
(a) Original electric signal
(b) Filtered electric signal

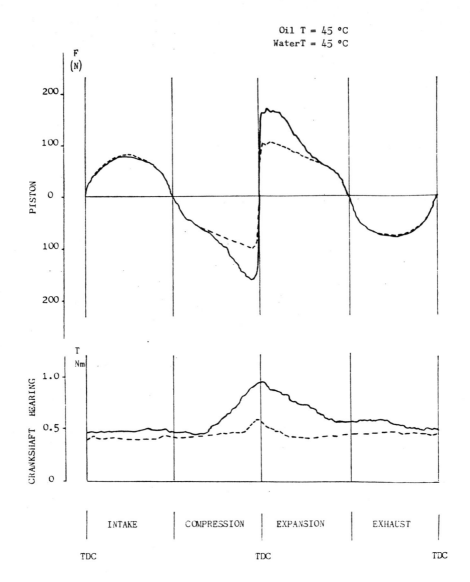

Fig 3 Piston frictional force and crankshaft bearing frictional torque
Motoring 1000 r/min
——— WOT
- - - - idle throttle position

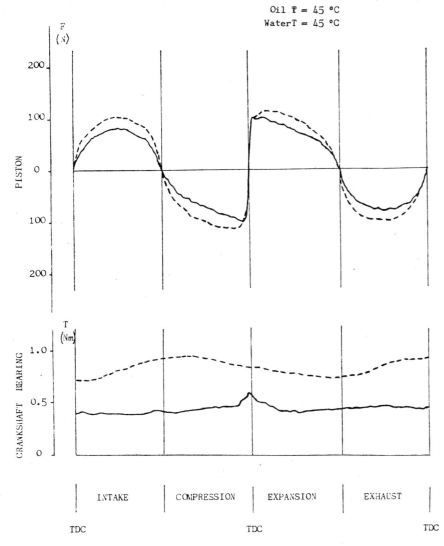

Fig 4 Piston frictional force and crankshaft bearing frictional torque
Motoring idle throttle position
——— 1000 r/min
– – – – 2000 r/min

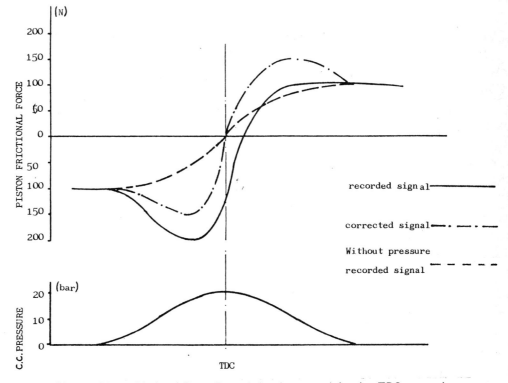

Fig 5a Piston frictional force. Recorded and corrected signal at TDC comparison

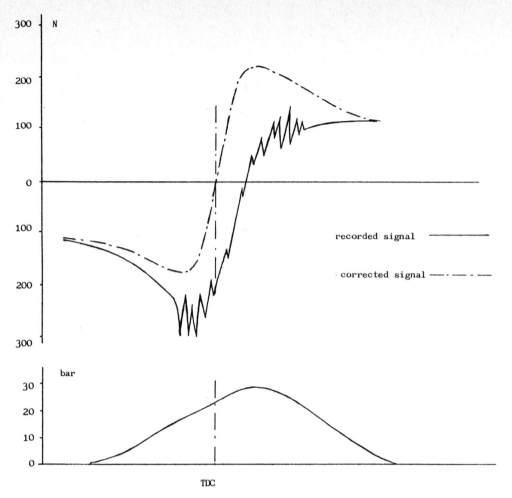

Fig 5b Piston frictional force firing condition. Recorded and corrected signal at TDC comparison

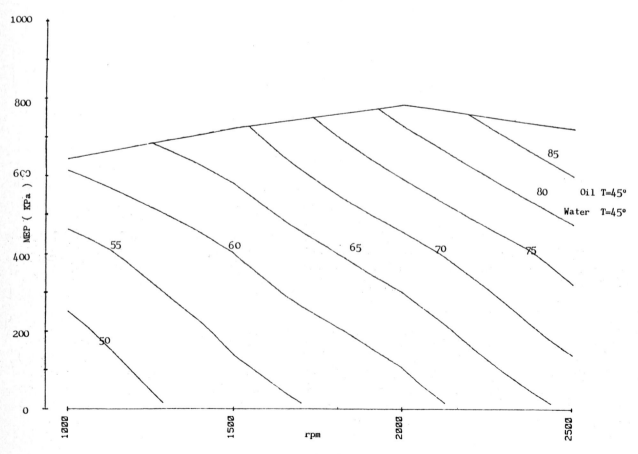

Fig 6a Typical piston FMEP (KPa) map — firing condition

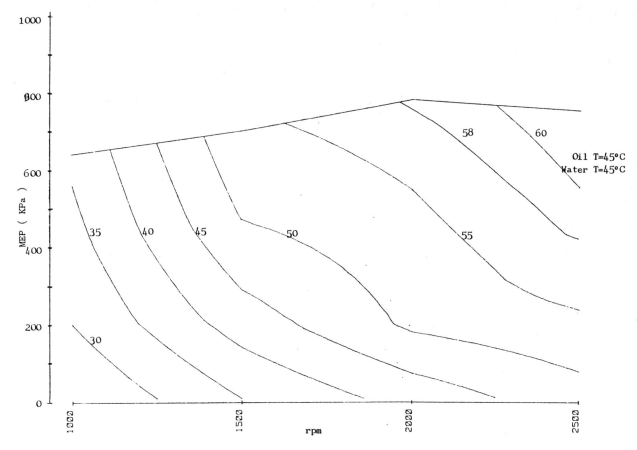

Fig 6b   Typical crankshaft bearing FMEP (KPa) map – firing condition

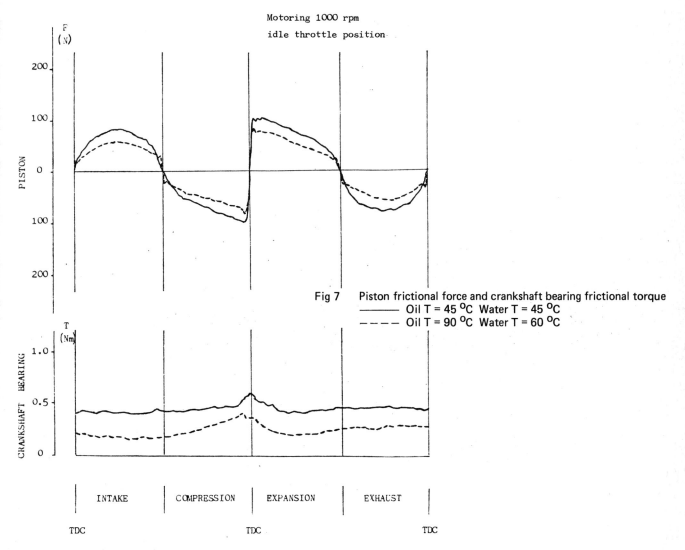

Fig 7   Piston frictional force and crankshaft bearing frictional torque
——— Oil T = 45 °C Water T = 45 °C
- - - - Oil T = 90 °C Water T = 60 °C

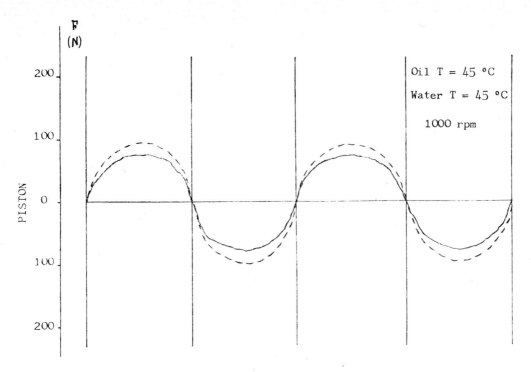

Fig 8 Piston frictional force — motoring without pressure
————— 0.11 piston clearance
- - - - 0.06 (mm)

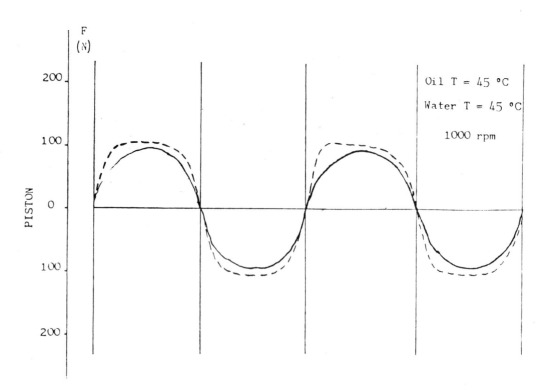

Fig 9 Piston frictional force — motoring without pressure
————— 0.2 surface roughness
- - - - 1.4 (μm)

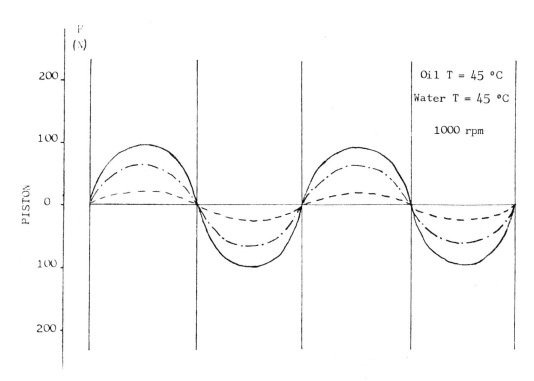

Fig 10 Piston frictional force — motoring without pressure
——— piston with all rings
– – – – skirt
– . – . skirt + oil-ring

C449/84

SAE 841296

Piston and ring mechanical losses

L FEUGA
Régie Nationale des Usines Renault (RNUR), Rueil-Malmaison, France
C BURY
École Nationale Supérieure d'Arts et Métiers (ENSAM), Paris, France

SYNOPSIS An engine was equipped with a view to measuring piston - ring-liner friction. Problems involving leakage, vibration and elimination of spurious forces were apprehended and resolved. The effect of some parameters was studied. Substantial gains seem possible which would improve engine fuel consumption.

1 INTRODUCTION

The piston - ring-liner assembly has already been subjected to numerous studies. The related problems involve fabrication, the operation and finally eventual degradation of the various parts composing the assembly. A trade-off between noise, oil consumption, blow-by and friction losses is necessary to assure a good operation of this assembly. In an internal combustion engine, such losses represent 20 % to 50 % of the total mechanical losses, and correspond to 2 to 40% fuel comsumption, depending on load conditions (1). Experimental study is indispensable because of the difficulties of describing all the phenomena by theoretical study using mathematical models. An engine was equipped to measure piston - ring-liner friction. After resolving setup and measurement problems, the influence of some parameters was studied.

2 THEORETICAL ANALYSIS

2.1 Friction equations

In order to take the various types of friction (hydrodynamic or mixed) encountered into account, it is necessary to resolve a system of equations:

- The basic hydrodynamic friction formula is the Reynolds equation. To allow for surface roughness, flow factors (2) must be introduced. The tangential speed variations can be negligible.

- Film shear over asperities can be calculated using the mean local pressure field, obtained from the nominal thickness of the film, considering deformations to be elastic (Hertz model).

2.2 Supplementary hypotheses and equations

It is necessary to impose fluid film pressure limit conditions in the divergent zones (Sommerfeld conditions).

It is possible to simplify the ring friction model by ignoring its gap or neglecting its torsion.

It is impossible to resolve friction equations without knowing the clearances between the moving parts. The position of the piston in the liner (3) and the piston rings in their grooves (4) under thermal and mechanical stress must be known. To obtain an accurate idea of piston lateral movement, it is necessary to consider the friction due to the piston pin and its ring in their grooves. Taking only one of these reasons for friction into consideration is not recommended (5).

A system of equations allowing for gas pressure, inertial forces and piston and ring reaction with the piston rod in the liner must therefore be concurrently resolved using a system of friction equations.

The actual viscosity of the lubricant depends on its temperature and on shear conditions. Throughout the cycle, these parameters vary constantly, so it is therefore necessary to add a new equation to the above systems.

2.3 Conclusion

The list of problems to be resolved described herein is not exhaustive. This simply gives an idea of the difficulties encountered in perfecting mathematical models. A parallel experimental study on simulators provides data which are difficult to apprehend by calculation, and enables the validity of the results obtained to be checked.

3 EXPERIMENTAL SETUP

3.1 Existing simulators

Much research has been conducted concerning piston-liner and piston - ring-liner friction using complex experimental machines or production engines.

Uras and Paterson of the University of Michigan (6) used the IMEP method on a production motor. The friction force is obtained from the difference between the longitudinal force of the connecting rod and the inertial and pressure forces applied to the piston. An accurate estimate of this difference can be obtained only on low load, and at very low speed.

Furuhama, Takiguchi and Sasaki (7) (8) of the University of Tokyo have measured piston and ring friction on the liner directly. They developed original technologies to eliminate the effect of gas pressure on the test. However, important modifications are required in the case of a single cylinder engine.

3.2 Experimental setup used

In the study described below, friction was measured on the liner of a production engine, with minimum transformation work. The actual operational conditions are therfore conserved. The use of production parts facilitates study of the effect of geometrical differences.

The spark ignited engine used was obtained from the Renault - 2 litre range.

- Engine characteristics :

  bore : 88 mm
  stroke : 82 mm
  compression ratio : 9.5 : 1
  connecting rod length : 137 mm

- Piston characteristics :

  pin offset : 1 mm
  2 compression rings
  1 scraper ring

The test system on which this engine was equipped is shown in figure 1.

The liner under test was isolated from the cylinder head and crankcase. Eight horizontal blades and three vertical blades were attached to the band arranged around the liner. The horizontal blades take up radial forces on the liner due to the piston. The vertical blades, located at 120°, transmit the friction force to three piezo-electric sensors.

This blade arrangement provides high sensitivity in measuring friction. The three vertical blades enable mechanical decoupling of piston radial forces. This arrangement considerably reduces the transverse component of the force supported by the sensors. However, accurate setting of the horizontality of the eight blades is indispensable. This is performed with the cylinder head removed, using a lever device which enables the radial force diagram to be applied to the liner.

The measurement system was calibrated with the engine hot by suspending weights from the conrod, which was disconnected from the crankpin.

As far as gas leakage is concerned, the equipressure device shown in figure 2 and used in (7) was inadequate. A level difference δ caused by crushing of the cylinder head gasket on tightening, differential expansion between the liner and ring and movement of the cylinder head due to the pressure forces would have rendered this completely inefficient.

Figure 2b shows the solution used. Sealing is provided by a viton ring seal, which offers a minimum surface with respect to hot gases ; this offered longer service life.

3.3 Acquisition and Shaping of Signals

The sealing device used did not enable balancing of the pressures applied to the liner. This resulted in an error around the explosion point, which was corrected by means of pressure measured in the chamber. The correction force results from the pressure applied to the non-balanced area Δs (fig 2b).

The engine was equipped with its original mounts system. Due to non-balanced inertial forces, as a consequence of its design, the engine was subjected to vertical acceleration. Measurement of this acceleration at the sensors enabled this to be allowed for.

After correction, the friction force is written :

$$F(t) = f(t) - p\Delta S - m\gamma$$

$f(t)$ = friction force measured by load cells
p = pressure measured in chamber
Δs = annular surface of thickness 0.24 mm and diameter (determined by experiment)
γ = acceleration measured on crankcase
m = weight of setup supported by the force sensors = 2.13 kg.

The signals obtained from the pressure, acceleration and force sensors are simultaneously averaged over several cycles, and stored in a microcomputer. The sampling frequency is that of the optical encoder located on the engine crankshaft (1024 points/cycle). The microcomputer calculates the piston instantaneous speed, the corrected friction force and losses due to friction over one cycle. This work is expressed as a friction mean effective pressure, according to the formula :

$$FMEP = \frac{1}{V} \int_0^{4\pi} F(\alpha) V(\alpha) d\alpha$$

V = swept volume
$V(\alpha)$ = piston velocity
α = crank angle

3.4 Friction measurement

The piston, rings and liner used in the setup were taken from a run-in engine having operated for 20 hours on test set. Temperatures Tw, for the water around the liner, and To for the crankcase oil were held constant throughout the tests.

Figure (3) shows the friction force obtained from the acceleration, pressure and friction signals measured, for two different engine rates (1000 and 2000 rpm). The respective FMEP's of each of the components are given.

The different shapes of the corrected curves obtained show the importance of acceleration. This effect is all the more noticeable as engine speed increases. Rigid attachment of the motor to the seismic slab of the test bench would be one method of attenuating this correction. Conversely, the FMEP of acceleration quantities ($m\gamma$) is very low. This was due to the form of the acceleration, which basically consisted of a $\cos 2\theta$ term ($\theta = 0$ at TDC), which is not balanced in a four-cylinder in-line engine ; the cos 2θ FMEP is null due to piston speed symmetry.

The force correction due to pressure is important around the explosion point. This is more or less proportional to the engine load, and was determined by experience. Effect on FMEP remains low, due to the relatively low piston speed in this area.

Supplementary correction is performed on the zero level of the corrected force curve, the latter not being known due to the use of piezoelectric sensors. The zero line is determined artificially by adding a constant, such that the maximum force in the exhaust phase is equal to the maximum force in the induction phase. The work in the respective induction, compression and exhaust phases becomes more or less identical after correction.

4 FRICTION ANALYSIS

Friction forces are consistent with mixed lubrication behavior around top and bottom dead centers, and hydrodynamic when piston speed is sufficient. Therefore, a force peak was obtained near top and bottom dead center, and a bump was obtained between two dead centers. The force curves (figure 3 and figure 4) are non-symmetrical for a piston ascending or descending phase ; this is due to profile assymetry (piston skirt rings). The effect of pressure on the piston affects friction : increase in piston/liner lateral force resulting from obliqueness of the connecting rod and an increase in the force seating the segments on the liner.

The main parameters affecting the force curve are therefore related to engine operation (speed and load), to contacting surface temperature (expansions and lubricant viscosity) and the geometry (architecture, dimensions and shape of parts, surface conditions, etc...).

4.1 Parameters related to engine operation

An increase in engine speed causes an increase in hydrodynamic friction and a decrease in mixed friction. In terms of dissipated power, the mixed friction located near top and bottom dead centers have little effect due to the low speed of the piston. The friction level has a tendency to increase globally with engine speed. This increase is relatively higher between idle speed and 1500 rpm, than beyond this point (fig. 5).

As on a complete engine, load variations had little effect on losses in the piston-ring/liner assembly, except for very low loads (fig. 4). Conversely, the shape of the force curve may be noticeably affected. The gas pressure behind the top ring increases the seating force against the liner. Moreover, lateral movement of the piston is affected by pressure variations. This shape change is essentially noticeable during the combustion stroke, and sometimes during the compression stroke (9).

4.2 Parameters related to temperature

Installation of thermocouples or fusable plugs in a piston provides an indication of the temperature reached under various rate and load conditions.

It is then possible to apply these data to an expansion model, after meshing of the piston (10).

The Stribeck curve fig8 shows that an optimum viscosity, and therefore temperature, exists for the lubricant. This temperature should be reached rapidly, then regulated. Knowledge of the piston and liner temperatures provides an estimation of lubricant temperature. Lubricant viscosity depends on temperature, but also on the shear forces to which it is submitted. The local temperature depends only slightly on the oil intake temperature, but, above all, depends on the cooling fluid temperature. This explains why the oil intake temperature has little effect on piston friction, conversely to the water intake temperature (fig. 6) : approximately 7 KPa/10°C between 40°C and 80°C.

Around the dead centers, the low rate of movement of the parts noticeable decreases carrying capacity. Oil is maintained between the parts due to squeeze film phenomena.

4.3 Parameters related to geometry

Due to its inclination, the connecting rod applies part of the force due to pressure on the piston, to the liner. Therefore, it is important to decrease this inclination as much as possible by increasing the length of the connecting rod or decreasing stroke. Maximum bulk and combustion imperatives (bore x stroke) allow little latitude for variation in parameter λ (= rod length / half-stroke), which is between 2.5 and 4.5 (fig. 7). This problem can be limited or avoided by the use of a different engine design : swivel, barrel or rotary engine.

The high shear factor which may exist between two parallel moving surfaces separated by a viscous liquid film must be reduced to a minimum. The piston skirt profile is therefore designed to create lifting capacity with minimum shear. Optimization of these surfaces, both in area and shape, enables E.C.E. cycle fuel consumption gains of up to 2 % for a 2-litre engine.

Similar reasoning may be applied to the piston rings, the shape of which is basically related to wear. The materials at present used do not enable conservation of the initial profile. Moreover, the dimensions and calibration pressure are often imposed by blow-by and minimum permissible oil consumption levels (11). A decrease in height enables a friction drop but increases wear, fuel consumption and blow-by.

Due to its mechanical characteristics (deformation, wear, friction coefficient) and thermal characteristics (espansion, thermal conductivity), the material affects friction.

New materials (plastics and composites, ceramics and sintered materials) must be envisaged to provide lightening and improvement in friction, or suppression of the lubricant. However, the laws governing friction in these materials vary widely. For plastic materials and composites, the friction coefficient is no longer proportional to applied load, but to slippage rate, and increases with temperature. As far as ceramics are concerned, liquid lubrication gives rise to heat dissipation problems (low thermal conductivity), and therefore lubricant behavior. "Life" lubrication can be envisaged for sintered materials.

Liner machining is one of the important parameters, in particular due to its effect on roughness anisotropy. It is necessary to maintain a balance between machining quality, oil consumption and wetting of surfaces to provide a fluid film in the hydrodynamic or elastohydrodynamic modes.

The impact of these friction reductions is important on the fuel economy at low load and weaker at full load.

At idle, friction losses represent the major part of the total losses. At full load, this part is reduced down to 10 %.

So, a 50 % reduction of the piston-ring-liner friction will reduce the fuel consumption of 40 % at idle and only 2 % at full load.

5 CONCLUSION

A study of friction losses on piston-ring-liner assemblies was conducted. Mathematic modelization was rendered difficult due to the large number of parameters to be taken into account. Therefore, it is indispensable to conduct a study on a special machine enabling the theoretical results to be validated.

A 2-litre engine was equipped. Cylinder leakage, vibration and the problems of eliminating spurious force were resolved. The interest of working with a real engine and not with a simulator is that the actual thermal and mechanical deformations are conserved, together with a temperature range and correct lubrication modes.

An increase in engine speed increases friction losses. The load (pressure in the cylinder) has a major effect on the shape of the force applied to the liner by the piston and the rings. However, the effect of this on F.M.E.P. is low. Temperature, due to its effect on lubricant viscosity, is the most important friction parameter (approximately 7 KPa/10°C).

To improve comprehension of the phenomena, it will be necessary to install sensors enabling measurement of operating clearances and local temperatures.

BIBLIOGRAPHY

1. FEUGA L. Modification of Engine internal friction due to architecture, design and materials. VII European Automotive Symposium, Brussels, 20-21 oct. 83

2. PATIR N. and CHENG H.S.
An average flow model for determining the effect of three dimensional roughness on partial hydrodynamic lubrication, Trans. A.S.M.E., J. Lub. Tech., 1978, 100 (N°1), 12-17

3. LI D.F., ROHDE S.M. and EZZAT H.A. :
An automotive piston lubrication model
ASLE Preprint N° 82-AM-2E-3, 37th Annual Meeting in Cincinnati, Ohio, May 10-13, 1982.

4. BEAUBIEN S.J. and CATTANEO A.G. :
Piston lubrication
SAE journal, V 54, Oct. 1946

5. TSCHOKE H. and ESSERS U. :
The effect of the friction produced by the piston and connecting rod on transverse piston movement, MTZ 44 (1983) 3,89-93.

6. URAS H.M. and PATTERSON D.J. :
Measurement of piston and ring assembly friction Instantaneous IMEP method
SAE paper 830416

7. FURUHAMA S. and TAKIGUSHI :
Measurement of piston frictional force in actual operating Diesel engine.
SAE paper 79 0855, off-highway vehicle meeting and EXPOSITION MECCA, Milwaukee, Sept. 10-13, 1979.

8. FURUHAMA S. and SASAKI S. :
New device for the measurement of piston frictional forces in small engines
SAE paper 83 12 84, International off-highway meeting and exposition, Milwaukee, Wisconsin, Sept. 12-15, 1983

9. MOLLENHAUER K. and BRUCHNER K. :
Contribution to measuring the influence of cylinder pressure and engine speed on internal combustion engine friction losses.
M.T.Z. 41 (1980) 6, 265-268

10. CHIN-HSIU U.
Piston thermal deformation and friction considerations SAE paper 82 00 86, International Congress and Exposition Detroit, Michigan, Feb. 22-26, 1982

11. MORSBACH M.
The influence of the axial width of piston rings on their performance.
MTZ 718 1982.

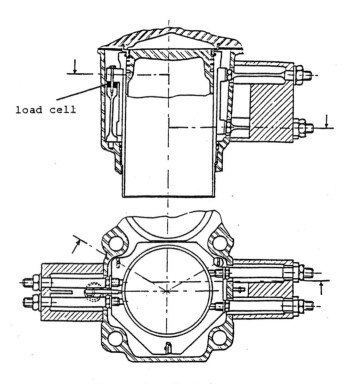

Fig 1   Measuring device

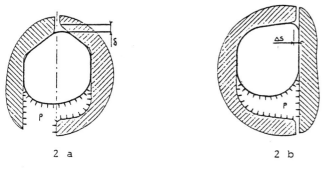

Fig 2 Gas seal O-ring device

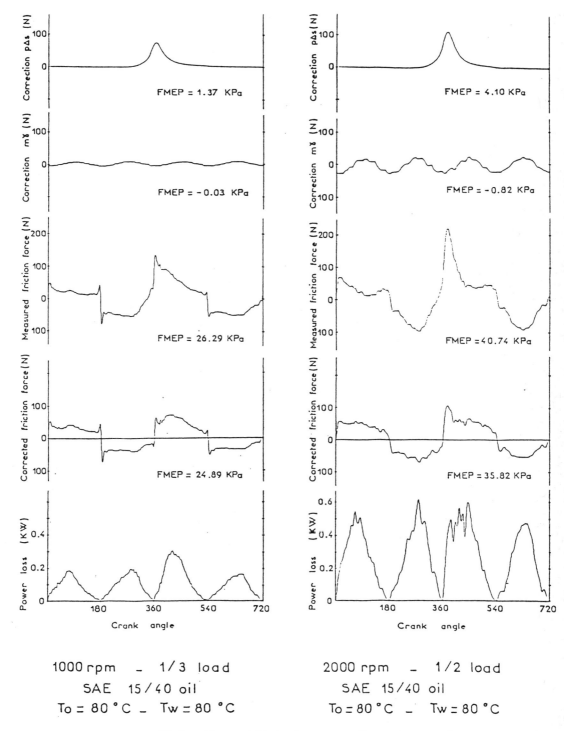

1000 rpm — 1/3 load
SAE 15/40 oil
To = 80 °C — Tw = 80 °C

2000 rpm — 1/2 load
SAE 15/40 oil
To = 80 °C — Tw = 80 °C

Fig 3 Measured friction force, corrections and corrected friction force. Power loss

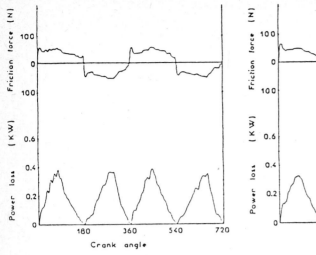
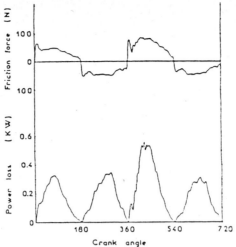

No – load , 1500 rpm , SAE 15/40 oil , To = 80 °C
Tw = 80 °C , Friction MEP = 27.8 KPa

Half – load , 1500 rpm , SAE 15/40 oil , To = 50 °C
Tw = 80 °C , Friction MEP = 33.5 KPa

Fig 4　Effect of load on the frictional force and the power loss at 1500 r/min

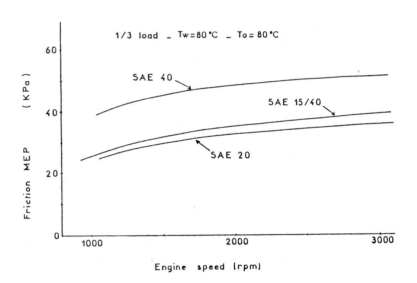

Fig 5　Effect of rotational speed on the FMEP measured on the experimental set up

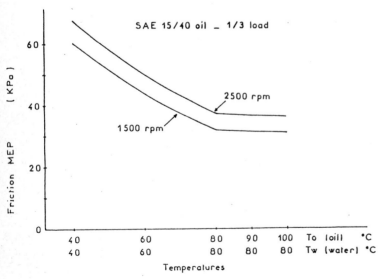

Fig 6　Effect of the temperature on the FMEP measured on the experimental set up

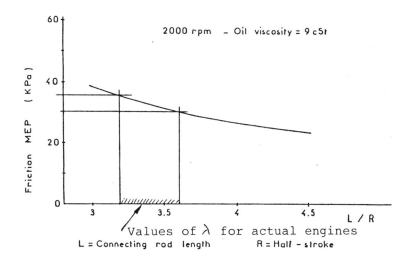

Fig 7 Calculated effect of
λ = (connecting rod length/half stroke) on FMEP

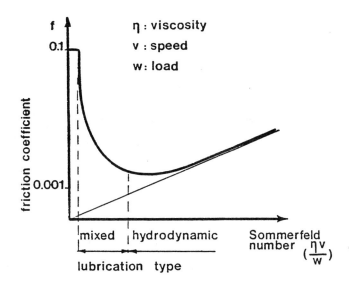

Fig 8 Stribeck curve

Adaptive air/fuel control applied to a single point injection system for SI engines

E M GAMBERG, BS and T A HULS, PhD
Ford Motor Company, Dearborn, Michigan, USA

The paper describes a method of improving the air/fuel (A/F) accuracy of a single point fuel injection system that is controlled by a microprocessor. The method develops correction factors as a function of engine speed and manifold absolute pressure. The correction factors are stored in keep alive memory (KAM) for future use. The correction factors reduce A/F deviations during both open and closed loop operation thereby improving three way catalyst (TWC) efficiency and eliminating the need for a secondary air system.

1 INTRODUCTION

The control system discussed in this paper consists of a digital electronic engine controller, appropriate software, and a typical set of production sensors and actuators used to control the A/F of a spark ignition engine. In particular, the adaptive algorithm which is used to determine and then store a table of factors that correct for inaccuracies in the open loop transfer functions of the various system components is described. The purpose of the adaptive method is: to reduce A/F control sensitivity to component variability resulting in cost and complexity reductions; obtain better emission control through improved closed loop A/F controllability at stoichiometry; and improved emissions, startup and driveaway during open loop operation.

2 BACKGROUND

The combined general objectives of simultaneously meeting government mandated emission (1) and fuel economy requirements while achieving good driveability and reasonable costs have brought about dramatic changes to engine control technology over the past several years. An important goal of this technology is to provide the precise A/F control required to concurrently maximize the reduction of hydrocarbon (HC), carbon monoxide (CO) and oxides of nitrogen (NOx) emissions using a TWC. A method of providing good A/F control is to employ an exhaust gas oxygen (EGO) sensor in a feedback control loop to correct any A/F errors. The time period to correct an error by this integral control method is proportional to the magnitude of the error. In general, the magnitude of errors that require correction are not constant over the operating range of the engine. Therefore, the feedback loop is seeking the set point, in this case stoichiometry, in proportion to the extent that the operating point is changing as well as to the magnitude of the errors. One method of reducing these deviations from the set point is to construct an algorithm which recognizes the magnitude of the errors, develops correction factors as a function of engine operating condition, and stores them for future use. Successful application of this methodology should improve closed loop A/F control and thereby improve HC, CO and NOx conversion efficiencies (2).

A second and related A/F control phenomenon which is mitigated to a large degree by adaptive fuel control occurs during vehicle startup and run. During this time it is necessary to operate open loop because the EGO sensor is not hot enough to produce a meaningful signal. Further it is desirable to eliminate the secondary air system for reasons of cost, reliability and fuel economy. To meet emission requirements without secondary air, it is necessary to operate lean of stoichiometry while maintaining sufficient mixture strength and consistency to obtain good driveability. This A/F operation range has a width of approximately 1.5 A/F. Consequently, a means of providing relatively accurate A/F control is needed prior to closed loop operation. This may be provided by a high accuracy A/F control system with its consequent cost and reliability implications or by using an adaptive system which not only tends to correct errors but also retains the information from one vehicle trip to the next. This information retention feature is provided by the keep alive memory (KAM). The corrected closed loop A/F control information stored in KAM provides more accurate reference information on which to base open loop A/F commands.

Thus the adaptive algorithm used for A/F control has two main features: 1) during closed loop operation it reduces the A/F deviations from stoichiometry thereby improving TWC conversion efficiency, and 2) when combined with KAM it provides a reasonably accurate reference from which open loop requirements may be specified (including startup and run).

These features may be utilized in different ways such as reducing costs, increasing reliability, or improving fuel economy. In fact, they may be utilized to accomplish all three simultaneously. Costs can be reduced by eliminating the secondary air system and the associated oxidation catalyst, and by using a manifold absolute pressure sensor (speed-density) instead of a mass air flow sensor. Reliability can be improved by using a less complex engine load sensor and by the elimination of the oxidation catalyst and secondary air system. Fuel economy improvements are modest but nevertheless real 1) by elimination of the secondary air system's weight and power requirements, and 2) by the improved A/F control increasing catalyst conversion efficiencies which allow closer to optimum spark and exhaust gas recirculation (EGR) schedules.

3 SYSTEM CONFIGURATION

Fig 1 is a block diagram of the engine control system. The hardware components have been described in other publications (3) (4). The KAM capability was provided through the use of a calibration console described in Reference 4. Additional circuitry can be added to production control modules to provide a similar function. The fuel control strategy includes transient fuel compensation as described in References 5 and 6. The test vehicles were 1982 Escorts (North American) each equipped with a 2.3L high swirl combustion (HSC) four cylinder engine and a centrally located single injector.

4 AIR/FUEL RATIO CONTROL

The low pressure fuel system used is designed to inject fuel in proportion to the time the injector pintle is open. This time is called the injector pulsewidth (PW). An offset value is added to compensate for the time it takes to open and close the pintle. The steady state portion of the fuel requirement is calculated using the following formula:

$$PW = \frac{Q}{S \times E \times D \times C \times B \times N} + A$$

where PW = Injector pintle open time, sec,
Q = Airflow, lbm/min,
A = Injector offset, sec,
B = Injector slope, lbm fuel/sec,
C = Injections per revolution, /rev,
D = Primary feedback loop term,
E = Adaptive multiplier,
N = Engine speed, rev/min, and
S = Stoichiometric A/F, lbm air/lbm fuel.

For a centrally located single fuel injector system on a four cylinder, four stroke cycle engine, C is set equal to two for a schedule of one injection per cylinder filling. The primary loop feedback term, D, is used in at least two ways: 1) as an integral controller to ramp the A/F in the opposite direction from the EGO sensor state, and 2) as an open loop A/F command. This mode is used during startup and run before the EGO sensor has reached operational temperature and may be used at other conditions, such as wide open throttle, to specify a maximum power equivalence ratio.

The use of an EGO sensor and an integral limit cycle resulted in a major improvement in A/F control. However, significant A/F errors can still occur because of inaccurate air flow measurements, inaccurate fuel system transfer function constants and assumptions made in deriving the relatively simple pulsewidth equation. Obviously, the open loop portion of start up and driveaway does not benefit from the EGO sensor information. Also, significant A/F errors can occur during the time immediately subsequent to a rapid change in operating point.

(The authors recognize that fueling characteristics of engines during transients are very complex phenomena and that the adaptive algorithm only superficially treats any errors associated with them. In this strategy, adaptive updates are not allowed when transient fuel compensation exceeds a calibratable amount. This adaptive algorithm is designed only to address the steady state errors of fuel control.) An example of errors subsequent to a rapid change in operating point is illustrated in Fig 2. In this example, a reduction in airflow results in a rich condition until the feedback term, D, is incrementally increased to compensate for the errors in the A/F control system. Similarly, a lean condition is compensated by incrementally reducing D.

One goal of the adaptive strategy is to significantly reduce the time it takes for the feedback loop to find the set point subsequent to a change in operating point or upon going closed loop. To do this an adaptive term, E, is used as a multiplier in the denominator of the first term of the PW equation. The adaptive term is an interpolated table lookup value as a function of engine speed and load (manifold pressure). Initially, all the table values are set equal to one. At calculated intervals, the cell value for a particular operating point is increased or decreased a small amount depending on whether D is greater or less than one. The primary and adaptive control loops are illustrated in Fig 3. By this process, the inaccuracies initially corrected by the global D term are gradually transferred to the appropriate adaptive table E terms for a speed and manifold pressure region. This maturation process is illustrated in Fig 4. The conventional feedback value for D is compared with the adaptive value for D. The difference between the two D values is accounted for by changes to one or more adaptive cell values. In theory, once the adaptive table has matured, the D term would only have to deviate from one by enough to ramp through the limit cycle to cause EGO sensor switching. In practice, the adaptive table is continuously being updated to account for changes in ambient conditions, changes in fuel, operational changes, etc. However, these changes usually will be minor or vary slowly compared with changes during the initial maturation period. In any event, the total error will tend to be reduced.

To take full advantage of this process, a means is required to retain the accumulated information from one vehicle trip to the next. This is provided by storing the adaptive table in some type of KAM.

5 ADAPTIVE TABLE

The adaptive table used in this program was a 7 by 11 matrix of adaptive multipliers as a function of manifold absolute pressure (MAP) and engine speed. The cell sizes were made identical to the lookup table used for airflow. The adaptive multiplier used in a specific PW calculation is determined by two dimensional interpolation between the four adaptive table values which surround the current operating point. The adaptive table update method used in this project was to update one cell in the vicinity of the operating point by a calibratable amount per a calabratable unit of time. This procedure allowed for a wide range in update rates. One of the four cells surrounding the operating point is selected to receive the update on a probabilistic basis. The probability of selecting any one of the four cells is inversely proportional to its distance from the operating point.

6 ADAPTIVE TABLE MATURATION RATE

One of the major considerations in this development program was to mature the adaptive table at as high a rate as practical without overshoot or instability.

Several methods were used to insure a high rate of maturation. The first was to make the adaptive table update rate a direct function of the primary feedback loop slew rate rather than a function of time only. The maximum stable update rate is related to and limited by the primary loop slew rate. The primary loop slew rate is a function of engine speed, engine and control system characteristics and other calibratable control parameters. This method increases the update rate as the engine speed increases and also maintains the maximum update rate as changes are made to the underlying software and hardware.

The second method of improving the maturation process was to add a deadband region covering the expected ideal range of the D term limit cycle. If D minus one is less than one-half of the ideal range of the limit cycle, the adaptive algorithm does not change the table values. This improvement increased stability, thereby allowing higher update rates outside the deadband.

The third strategy element was to compensate D for any change in E from the previous pulsewidth calculation so that the actual primary loop slew rate does not change. This strategy element is only necessary if the adaptive loop update rate is a significant fraction of the primary loop slew rate. Without this compensation, the adaptive loop update rate, in effect, either adds to or subtracts from the primary loop slew rate.

7 EXPERIMENTAL RESULTS

Experiments were run to investigate and optimize the adaptive algorithm behavior and to determine its ability to meet emission requirements without an adverse impact on fuel economy or driveability.

7.1 Maturation rate optimization tests

The response of the system to a step error was determined at various conditions for several update rates defined as a fraction, X, of the primary loop slew rate. The engine was run at a steady state condition when the adaptive learning algorithm was enabled. Fig 5 shows typical results. As the update rate is increased from X=0.25 to X=0.50, the EGO sensor begins to indicate an atypical amount of rich operation for what would ordinarily be a steady state condition. This phenomenon worsens as X increases. The adaptive term begins to exhibit overshoot at X=1.0 and clearly overshoots at X=2.0. The primary loop begins showing signs of instability at X=1.0 and worsens at X=2.0.

Examination of this type of data and the behavior of individual adaptive table cell responses led to setting the adaptive to primary loop rate ratio at X=0.3.

7.2 Adaptive algorithm function confirmation tests

A series of tests was run to demonstrate that the adaptive algorithm functioned as intended and was capable of correcting for the expected variation in production components. The method of introducing these expected variations as 'errors' is described in Appendix I. Three series of tests were run, the first with the introduction of a uniform 13 per cent rich error, the second with a uniform 13 per cent lean error and the third with a non-uniform error. In each case a baseline with the error included was established without the adaptive algorithm functioning. Subsequently, a 1372 second preparatory cycle followed by three or four constant volume sampler, cold and hot start (CVS-CH) tests were run according to the Federal test procedure (1).

Of these, the non-uniform error test series is the most interesting because it caused the primary feedback loop to seek the set point much more frequently and for longer periods of time. The behavior of the primary loop term, D, is indicative of the performance of the adaptive algorithm.

Initially, deviations of 10 to 20 per cent were expected and then a gradual settling down as the adaptive table cell values matured. This phenomenon is shown in Fig 6. The lower five traces are of the null term for the first 700 seconds or so of the Federal CVS-CH test procedure. The corresponding vehicle and engine speed traces are the upper two traces in the figure. The first trace of D for the mean calibration with the non-uniform error (without adaptive) illustrates the expected deviations. The second trace of D for the preparatory cycle with the 13 per cent non-uniform error and with the adaptive algorithm functioning shows considerable improvement in just a few minutes of operation. The subsequent three traces of D are for the second, third and fourth CVS-CH tests respectively (the D trace for the first test was lost because of an instrument problem). These traces show modest but continued improvement in the D trace.

A second indicator of the proper performance of the adaptive algorithm is A/F control. Fig 7 presents the A/F traces for the same series of tests as presented in Fig 6. The A/F trace for the mean calibration with error and without adaptive shows deviations of almost three A/F. The subsequent repetitive tests with the adaptive algorithm functioning show considerable improvement in A/F control.

7.3 Vehicle emission data

The CVS-CH emission data for the test series are presented in Fig 8. Each set of data represents the average of three tests. The 'without adaptive' data is either the mean calibration or the mean calibration with the errors introduced. The 'with adaptive' data for the mean calibration is the data for the first three tests after the adaptive algorithm was enabled.

The 'with adaptive' data for the error tests is the average of the first three CVS-CH tests after the preparatory cycle subsequent to the error being introduced to the mean calibration.

The adaptive algorithm had very little impact on the mean calibration because there was little error to correct.

The high HC and CO emissions during the fuel rich error tests 'without adaptive' occurred during the startup and run portion of the test. Once the primary loop closes, the set point seeking value, D, slews in the lean direction to the stoichiometric point one time to correct for the rich error. From that point on the fuel control is essentially the same as the non-error case. However, the rich error is still there the next time the vehicle is started and the high HC and CO emissions are still there during the open loop startup and run portion of the emission test. In contrast, in the case where the adaptive algorithm is functioning, the rich error is first corrected by the set point seeking value, D, in the feedback loop as before but then the error is transferred to the adaptive table E values stored in KAM. Thus, the E values are available and used during future startups to achieve the desired A/F. In this test series, the rich error was sufficiently corrected during the preparatory cycle so that the vehicle met the development emission objectives in all three of the subsequent CVS-CH emission tests. While the adaptive algorithm does not correct all of the inaccuracies encountered during open loop operation, it does provide a much more accurate A/F ratio estimate from which cold start and run requirements can be referenced. This improvement greatly increases the feasibility of eliminating secondary air systems thereby reducing costs and improving fuel economy slightly (secondary air system power and weight penalty).

All of the emission tests with the lean error met the development objectives including the non-adaptive tests. During the open loop startup and run phase of the test, the NOx emissions may be higher because of the lean operation, but not many times higher than the mean calibration levels as was the case of the HC and CO levels for the rich error tests. Once the primary loop closes, the set point seeking value, D, slews in the rich direction to the stoichiometric point one time to correct for the lean error. From that point on the fuel control is essentially the same as the non-error case. The problems associated with excessively lean operation fall in the driveability area rather than emissions. As in the rich error case, the adaptive algorithm provides a more accurate A/F estimate from which the startup and run requirements can be referenced. This allows lean operation during the initial open loop portion of the test without exceeding the practical lean combustion limit.

The non-uniform error tests had the poorest fuel control. The set point seeking value, D, was continuously seeking the stoichiometric point because the errors introduced were both rich and lean. The NOx catalyst conversion efficiency was poorest for the non-uniform error, non-adaptive test series. The NOx emissions exceeded the development objective.

The adaptive non-uniform test series NOx conversion efficiency was also poor but improved enough so that the NOx emission development objective was met in addition to the HC and CO objectives. The lower NOx conversion efficiency is attributed to the poorer A/F control. The A/F control should improve with additional vehicle operation as the adaptive table fully matures. (Some of the errors in sensor and actuator transfer functions that affect transient fuel compensation are not corrected by this algorithm and will contribute to degraded A/F control performance.)

8 CONCLUSIONS

The adaptive A/F algorithm with KAM storage is capable of improving the open loop A/F accuracy and closed loop fuel control of a speed density single point fuel injection strategy sufficiently well to meet emission requirements without secondary air. Further, this conclusion is valid when errors of plus or minus 13 per cent are introduced to simulate component variability.

The adaptive A/F algorithm update rate can be set high enough to achieve sufficient maturity to meet emission requirements by the end of 20 miles of typical driving when errors of plus or minus 13 per cent are introduced to simulate component variability.

9 ACKNOWLEDGMENTS

The authors express their gratitude to L. C. Christensen, R. K. Feller, P. J. Grutter, C. C. Leiby, I. A. Messih, R. L. Robinson and L. L. Russ for their contributions to the project.

REFERENCES

(1) "Emission Regulations for 1977 and Later Model Year New Light-Duty Vehicles and New Light-Duty Trucks; Test Procedures." 40 CFR Part 86, Subpart B.

(2) FALK, C. D. and MOONEY, J. J. "Three-Way Conversion Catalysts: Effect of Closed-Loop Feed-Back Control and Other Parameters on Catalyst Efficiency." SAE Paper 800462.

(3) CZADZECK, G. H. and REID, R. A. "Ford's 1980 Central Fuel Injection System." SAE Paper 790742.

(4) SULLIVAN, J. A. "Central Fuel Injection Engine Control System." International Symposium on Automotive Technology and Automation, Cologne, 19-23 September, 1983.

(5) AQUINO, C. F. "Transient A/F Control Characteristics of the 5 Liter Central Fuel Injection Engine." SAE Paper 810494.

(6) HIRES, S. D. and OVERINGTON, M. T. "Transient Mixture Strength Excursions: An Investigation of their Causes and the Development of a Constant Mixture Strength Fueling Strategy." SAE Paper 810495.

APPENDIX I

METHOD FOR INTRODUCING 'ERRORS' FOR ADAPTIVE ALGORITHM FUNCTION CONFIRMATION TESTS

A Uniform three-Sigma 'error' description

A.1 The manifold absolute pressure (MAP) sensor transfer function 'error' was set equal to four per cent of point.

A.2 The volumetric efficiency table 'error' was set equal to five per cent of point.

A.3 The air charge temperature (ACT) sensor transfer function 'error' was set equal to one per cent of point.

A.4 The fuel system transfer function 'error' was set equal to three per cent of point

A.5 The total 'error' was equal to the sum, i.e. 13 per cent of point.

B Non-uniform three-Sigma 'error' description

B.1 The MAP sensor transfer function 'error' was:

$$E_{map} = 13 - 0.6(MAP)$$

where E_{map} = error, per cent of point

MAP = manifold absolute pressure, in. Hg.

B.2 The volumetric efficiency table 'error' was:

For N less than 4000 rev/min.

$$E_{ve} = 6 - 0.2(MAP) - 0.00125(N)$$

For N equal to or greater than 4000 rev/min.

$$E_{ve} = 1 - 0.2(MAP)$$

where E_{ve} = error, per cent of point.

N = engine speed, rev/min.

B.3 The EGR transfer function 'error' was:

For EGR valve positions equal to or greater than 10 per cent.

$$E_{egr} = -11 + 0.22(EVP)$$

where E_{egr} = error, per cent of point.

EVP = EGR valve position, per cent.

B.4 The fuel system transfer function 'error' was:

$$E_{fs} = 3 - 0.6(PW)$$

where E_{fs} = error, per cent of point.

PW = injector pintle open time, ms.

B.5 The total error:

$$E_{total} = E_{map} + E_{ve} + E_{egr} + E_{fs}$$

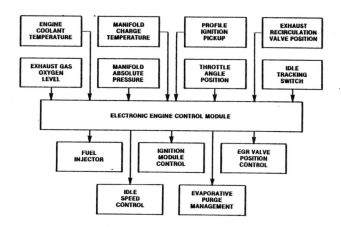

Fig 1 Single point fuel injection control system block diagram

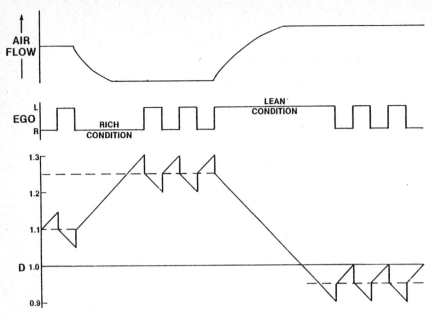

Fig 2 Conventional feedback correction term behaviour

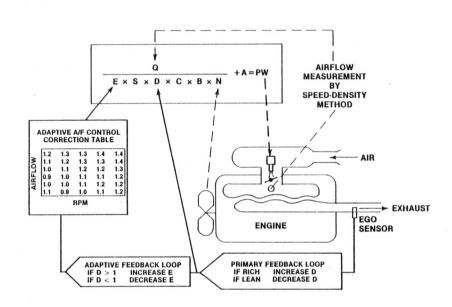

Fig 3 Adaptive A/F feedback control schematic

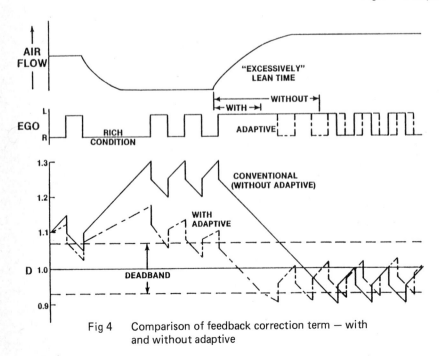

Fig 4 Comparison of feedback correction term — with and without adaptive

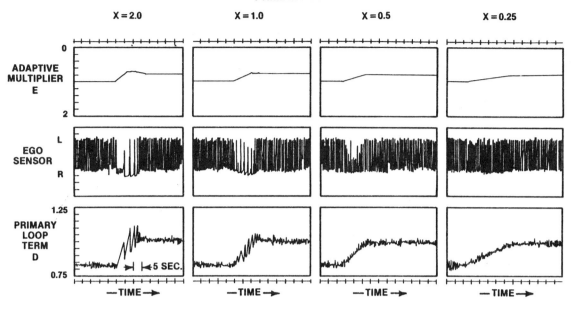

Fig 5 Typical adaptive multiplier response to a step change at selected adaptive rates

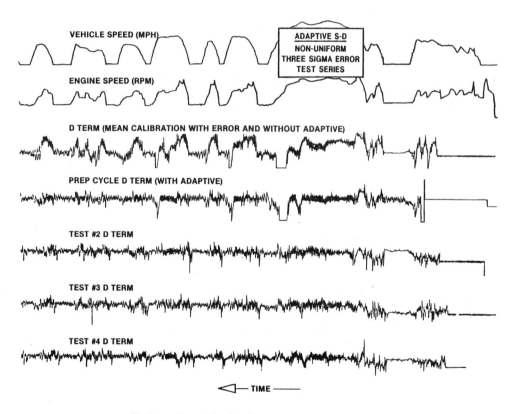

Fig 6 Actual feedback correction term traces for 'non-uniform' test series

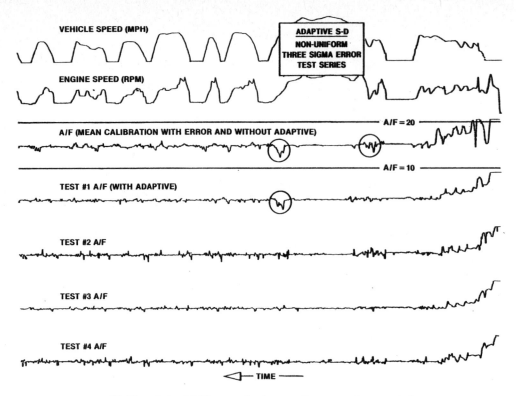

Fig 7　Actual A/F traces for 'non-uniform error' test series

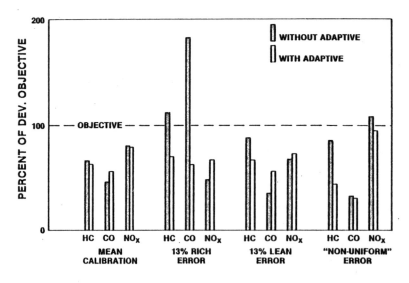

Fig 8　CVS-CH emission data without and with adaptive A/F control subsequent to 'error' introduction

C436/84

SAE 841298

Development of a high powered, economical four-valve engine, Toyota 4A-GE

Y KIMBARA, M KONISHI, N NAKAMURA and S KOBUKI
Toyota Motor Coporation, Aichi-Ken, Japan

SYNOPSIS An engine must achieve both high power to ensure good drivability and low fuel consumption. With this as the target we developed a new 4-valve engine.
This paper describes the investigation and development of the power, fuel economy, and response of the 4A-GE, a 4-valve engine installed in the new Toyota Corolla Coupe.

1. INTRODUCTION

It is an ideal of engine designers to simultaneously achieve high power, good fuel economy, and quick response. A 4-valve engine has the following features which can realize this ideal.

a) Low intake and exhaust resistance allowing an increase of aspirated air.
b) Lighter valve component mass and direct valve drive by the camshaft allowing stabilized valve movement to higher revolutions.
c) Possibility of a pent roof type combustion chamber with central spark plug, which has superior combustion characteristics, allowing higher compression ratio.
d) Two intake valves making intake air control feasible.

These features permit both high power and good fuel economy, but it is generally difficult to gain quick response owing to fuel blow back etc. at low speeds. However, we have developed a small light weight and economical 4-valve DOHC engine, the 4A-GE, which is practical and easy to handle, due to its superior performance at low-to mid-speeds as well as high speeds, in contrast with previous 4-valve engines which aimed at high power only at high speeds.

The 4A-GE engine replaces the 2-valve DOHC engine, 2T-G, installed in previous Corollas, and was developed based on a 2-valve SOHC engine 4A. Engine specifications and cross section of 4A-GE are shown in Table 1 and Fig. 1 respectively.

As shown in Table 1. the 4A-GE engine has a max. power output of 91 KW and max. torque of 142 Nm. Compared with the 2T-G engine, this is an increase of 12 KW (15 per cent) and 10 Nm (8 per cent) respectively. Furthermore engine weight is only 123 Kg, which is 29 Kg (19 per cent) lighter than the 2T-G, contributing to the reduction of vehicle weight through its light weight and compactness.

Investigation results of power, fuel economy, and response of the 4A-GE are described below by comparing 2-valve and 4-valve engines.

Table 1 : 4A-GE engine specifications

	4A-GE engine	2T-G engine
Displacement (cm³)	1587	1588
Bore × Stroke (mm)	81.0 × 77.0	85.0 × 70.0
Compression ratio	10.0	9.8
Fuel supply system	EFI	Solex PHH
Valve system	DOHC 4-valve Belt drive	DOHC 2-valve Chain drive
Combustion chamber	Pent roof	Hemisphere
Required fuel octane No.	98	98
Engine service weight (Kg)	123	152
Max power DIN	91 KW at 6600 rev/min	79 KW at 6200 rev/min
Max torque	142 Nm at 5200 rev/min	132 Nm at 5200 rev/min

2. POWER

To achieve high power from low to high speeds, the following investigations were conducted.

2-1 Comparison of 4-valve DOHC engine and 2-valve DOHC engine

In developing a DOHC engine to replace the 2T-G engine, investigations were made on prototype 2-valve and 4-valve DOHC engines based on the 4A-engine. Fig. 2 compares the power, torque, and fuel consumption of the two engines.

The max. power of the 4-valve engine is 91 KW which is an increase of 14 per cent as compared with the 2-valve engine's 80 KW. This is because 4 valves maintain high volumetric efficiency to higher speeds. The 4-valve engine has higher mid-speed torque owing to its intake air control system to be described later. Also specific fuel consumption is lower compared to the 2-valve engine.

The above results were decisive data in choosing between 2-valve and 4-valve engines.

2-2 Effects of Toyota Variable Induction System (T-VIS)

The T-VIS is an intake air control method, which increases mid-speed torques, but is also effective for improving fuel economy.

Its construction is shown in Fig. 3a. Each cylinder has two independent intake ports, one of which has an intake air control valve. The two ports are merged within the cylinder head, and fuel is injected where the ports merge.

The air control valve is controlled by engine speed, closed below, and opened above 4650 rev/min. When the valve is closed, the speed of air sucked into the cylinder is accelerated because air flows only through one of the intake manifold passages, increasing the inertia effect of the intake air column. This improves the volumetric efficiency and torque. But if the valve was always kept closed, the smaller intake port cross sectional area would cause a sharp decrease of torque at higher speeds. Thus the valve must be opened at these speeds. The two torque curves for control valve closed or opened intersect at 4650 rev/min.

As shown in Fig. 3b the addition of T-VIS improved the mid-speed torque by 5 Nm.

2-3 Intake port diameter and length

Generally speaking, the following may be said of the two intake ports, one free and the other with intake air control valve.

To improve the low- to mid-speed torque, the free port should be long and narrow. To secure high power, the other port with intake air control valve should be short and wide. If constructed in this fashion, the intake system would become bulky and complicated, so the two ports per cylinder were built to the same dimensions.

Thus, the diameter and length of the twin intake ports were investigated. Table 2 shows the increase and decrease of max. torque and power for specifications investigated, against production specification taken as 0.

Port length 310 mm showed an increase in max. torque of 0.6 Nm and 0.4 Nm when compared with port length 150 mm and 250 mm, and almost no decrease in max. power. Port length 400 mm showed a further increase of 0.4 Nm in max. torque, but a decrease of 4.5 KW in max. power. Since port length 400 mm would also entail a large increase in engine size, port length of 310 mm was adopted. To improve the max. torque, the port diameter was decreased from ⌀30 to ⌀28 but since this produced no improvement, it was further reduced to ⌀26. This increased max. torque by 0.2 Nm but decreased max. power by 1.5 KW. Furthermore, power output showed a sharp decrease after reaching peak output, so port diameter ⌀30 was adopted.

Table 2 : Performance of intake systems

Intake systems	Torque	Horse power
⌀30-150 mm	-0.6 Nm	0
⌀30-250 mm	-0.4 Nm	0
⌀30-310 mm	0	0
⌀30-400 mm	+0.4 Nm	-4.5 KW
⌀28-310 mm	0	-1.5 KW
⌀26-310 mm	+0.2 Nm	-1.5 KW

2-4 Others

The exhaust system was investigated, because lowering flow resistance of the intake system would have little effect in increasing power output if the flow resistance of the exhaust system is high. The 4A-GE adopted a dual exhaust manifold with port diameter of ⌀33, a dual front pipe with inner diameter of ⌀38.7, and a tail pipe with inner diameter of ⌀45.4, each enlarged to maximum size, limited only by problems of space and noise.

Valve timing was set with low- to mid-speed torques in mind with a small valve overlapping angle of 18 deg. The twin intake or exhaust valves for each cylinder were set at the same timing, for it was found that setting them at different timings did not improve engine performance.

High compression ratio was made possible by an electronic spark advance (ESA) system which delays spark timing under extremely high coolant temperatures.

Fuel injection position was set at the center of twin intake ports for each cylinder as max. power decreased by 1.5 KW when injection position was offset to either port. Also, an electronic fuel injection system (EFI) with D-jetronic (D-J) was adopted, thus eliminating the air flow meter, and increasing max. power by 1 KW.

The above improvements permitted the 4A-GE engine to attain top level power output for engines in the 1.6 liter class.

3. FUEL ECONOMY

It is very difficult to achieve good fuel economy while at the same time securing high power output. Realization of good fuel economy was investigated with emphasis placed on reducing friction loss, improving thermal efficiency, and minimizing fuel injection. Concerning factors which had opposing effects on power and economy, fuel economy was chosen whenever possible.

3-1 Combustion Comparison of 4-valve and 2-valve engines

It is easier to achive a compact configuration of combustion chambers of 4-valve engines, like the pent roof type, and also to place the spark plug in the center of the chamber. Moreover, since the exhaust valves face the intake valves, the former are cooled by the intake air fuel mixture. Since engine knock is alleviated by these features, it is possible to raise the compression ratio.

High compression and a combustion chamber with the above mentioned features shorten the ignition delay compared to 2-valve engines. It is further shortened when intake air control is added to 4-valve engines, owing to the turbulence generated in the air-fuel mixture.

Ignition delay is defined here as the time between sparking of the spark plug and burning of 10 per cent of the fuel, measured against the calories that complete combustion of the fuel would have generated, in terms of crank angle.

Since ignition delay is shorter, ignition timing can be delayed, improving thermal efficiency. Therefore, 4-valve engines with such features have a high potential for fuel economy.

Results of experiments are shown in Fig. 4. Brake specific fuel consumption (BSFC) with air-fuel ratio set at 15 is 4 per cent better in the 4-valve engine compared to the 2-valve engine. This is due to the factors mentioned above and because the pumping loss is smaller with the 4-valve engine. The figure clearly shows the effects of intake air control on fuel economy, in regions where the air-fuel ratio exceeds 19. Furthermore, the air-fuel ratio at which misfiring occurs is shifted towards the leaner side. These facts show the high potential of the intake air control, if EGR and/or lean burn is adopted in the future.

3-2 Friction loss comparison of 4-valve and 2-valve engines

As shown in Fig. 5a, friction loss in the 4-valve engine is smaller than the 2-valve engine it was based on. This is mainly due to the difference in pumping loss. Pumping loss of the 4-valve engine is smaller because intake and exhaust resistance is lowered by the adoption of four valves.

Fig. 5b is an indicator diagram with engine speed at 2000 rev/min. The axis of ordinate is internal pressure and the axis of abscissa is cylinder volume in natural logarithm. This figure shows that the pumping loss of the 4-valve engine is 18 per cent smaller compared to the base engine.

The above results were drawn on the premise that the same moving components may be used, but with adoption of 4 valves, reinforcement of moving components is needed to withstand the higher power output, for example, reinforcing the shafts by enlarging their diameters. However, since enlarging shaft diameters of rotating parts will increase friction loss, simple adoption of larger shafts is undesirable. To preserve the same shaft diameter as the base engine, numerous kinds of strength analyses such as FEM etc, were conducted in developing the 4A-GE engine. An example is shown in Figs. 6a and 6b. Fig. 6a shows a strength analysis of the crankshaft conducted by FEM. Computation results of FEM show good coincidence with actual measured values. As shown in the figure, applying torque to the crankshaft would cause stress concentration at the oil hole in the crankpin. Fig. 6b illustrates the relation between roughness of oil hole inner surface and fatigue strength. Reducing the surface roughness from 100μ to less than 10μ will increase the fatigue strength by approximately 50 per cent. From this experiment, oil hole inner surface roughness was set at 10μ for crankshafts of the 4A-GE engine, preserving crankpin diameter of the base engine.

For other parts also, strength analysis allowed the adoption of one of the smallest diameters for engines in the 1.6 liter class, as shown in Table 3, which contributed greatly toward reduction of friction loss.

Table 3 : Diameter comparison of rotating shaft
(1.6 liter class)

Diameter (mm)	4A-GE	2T-G	A	B
Crankjournal	⌀48	⌀58	⌀54	⌀60
Crankpin	⌀40	⌀48	⌀46	⌀50
Piston pin	⌀18	⌀22	⌀20	⌀22

3-3 Electronics and others

Apart from the features described above, the 4A-GE engine adopted, as fuel economizing methods, electronic spark advance (ESA) and electronic fuel injection (EFI) using a microcomputer, along with fuel conserving valve timing and T-VIS referred to previously.

ESA sets ignition timing to the optimized value in a wide variety of ranges. Its adoption enhanced fuel economy by 1~2 per cent. A microcomputer controls the EFI keeping fuel consumption to the minimum value by setting the air-fuel ratio to the optimized ratio in a wide variety of ranges, and by detailed matching of increments of fuel when accelerating or cold running.

By the fuel saving methods described above, the 4A-GE achieved not only high power putput but good fuel economy.

4. RESPONSE

To achieve good drivability, it is important to have not only high power output but quick response. However, 4-valve engines tend to respond more slowly than 2-valve engines, as stated previously, because the two intake valves cause fuel blow back at low speeds, and the air flow rate is slower.

To improve this response, the following investigations were conducted.

4-1 Fuel injection position

As stated previously, each cylinder of the 4A-GE has two independent intake ports which are merged within the cylinder head. The merged portion was selected as the position for a fuel injector, but there were two possibilities. The injector could either be set in the center of the two ports of offset towards one or the other, though offsetting the injector was not desirable from the view point of the engine's maximum power output and valve seat durability.

Fig. 7a illustrates torque response when the throttle valve is opened from 1000 rev/min light load to wide open throttle in 0.3 sec. Fig. 7a-Case 1 shows torque response with the injector offset towards the port without air control valve, and Fig. 7a-Case 2 shows the torque response with the injector at the center. Torque stumble is shorter for center-injected type, so the injectors of the 4A-GE were positioned at the center of the two intake ports.

4-2 Fuel injection timing

It was thought that fuel blow back occurs when exhaust pressures cause backstreaming of the intake air flow during valve overlap. Fig. 7b illustrates the relation between torque and fuel injection timing when the latter is changed from 180° BTDC to 180° ATDC. The torque shows a drop when fuel is injected between 60° BTDC and 100° ATDC, with a increase of HC and CO exhaust emissions. This is because the large fuel blow back in this region turns the air-fuel ratio lean within the cylinder, deteriorating the combustion process. In comparison, the 4A-E two-valve EFI engine which is based on the 4A exhibits little touque drop in the same region. This is because the faster air flow of the 2-valve engine accelerates fuel vaporization and suppresses backstreaming. From these results, fuel injection timing of the 4A-GE was set at 70° BTDC where this torque drop does not appear, and HC and CO exhaust emissions are reduced.

Fig. 7a-Case 2 and Fig. 7a-Case 3 illustrate the differences in torque response when fuel injection timing is changed from ignition timing to 70° BTDC.

As the figures show, this shortened torque stumble by 0.1 sec.

4-3 Extra fuel injection

Torque response was improved in the manner outlined above, but torque stumble still existed. One way to remove this torque stumble is to increase the amount of fuel injected during acceleration. There are two ways to do this. One way is to increase the amount of fuel injected at a normal single injection for each engine revolution. Another way, named "extra fuel injection", is to inject fuel apart from the normal injection timing. Improving torque stumble only by the former is not sufficient, as the necessary amount of fuel can not be injected in time. The latter method, extra fuel injection, increases the amount of injected fuel according to the throttle opening speed detected by the throttle opening sensor. With this extra fuel injection torque stumble was eliminated as shown in Fig. 7a-Case 4, and quick response was perfected, while keeping the amount of fuel injected for response improvement to a minimum value.

5. CONCLUSIONS

Through investigations and developments described above, the 4A-GE engine was completed as a high performance engine simultaneously achieving high power, good fuel economy, and quick response. As an example, relation between 0~100 Km/h acceleration time and ECE-test fuel economy is shown in Fig. 8. The Corolla Coupe powered by the 4A-GE is a well balanced car with top level acceleration and fuel economy.

Fuel economy is compared with the 2T-G engine in Table 4.

As shown in Table 4, ECE test fuel economy has been improved by 19 per cent. 4A-GE engines having these superior characteristics are now being mass produced at a rate of 10000 units per month.

Table 4 : Fuel consumption (Liter/100 Km)

	4A-GE	2T-G
90 Km/h	6.5	6.8
120 Km/h	8.1	9.4
ECE-test	8.9	11.0

REFERENCES

(1) A. Eugen, autosalon 39, Keller publishing house, 1983.

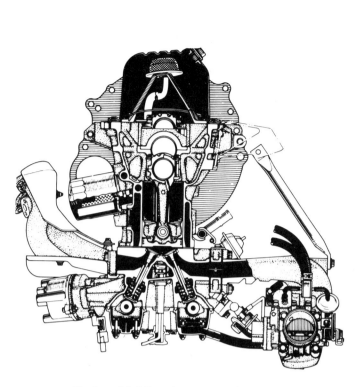

Fig 1 4A-GE engine cross-section

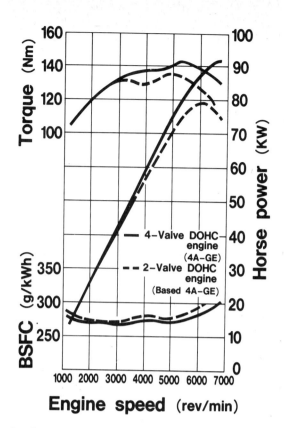

Fig 2 Comparison of performance between four-valve DOHC engine and two-valve DOHC engine

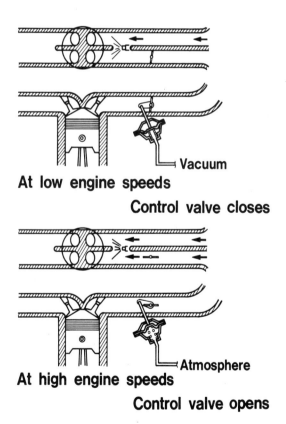

Fig 3a Toyota variable induction system (T-VIS)

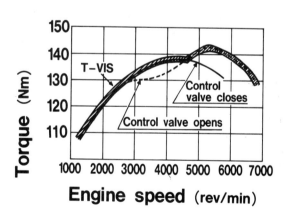

Fig 3b Effect of T-VIS

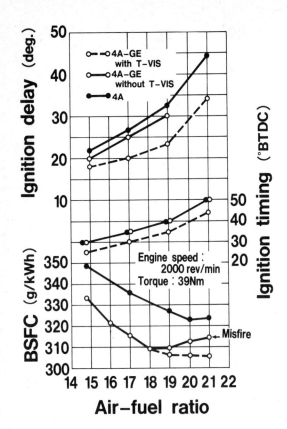

Fig 4 Comparison of fuel consumption between four-valve engine and two-valve engine

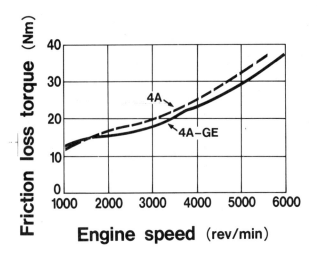

Fig 5a Friction loss

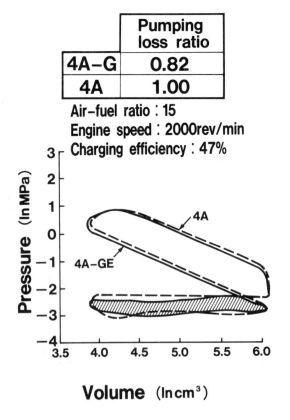

Fig 5b Logarithmic indicator diagram

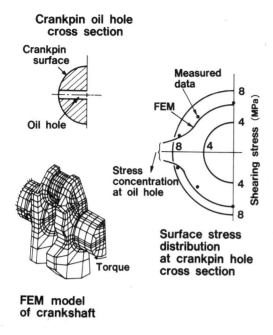

Fig 6a FEM analysis of crankshaft

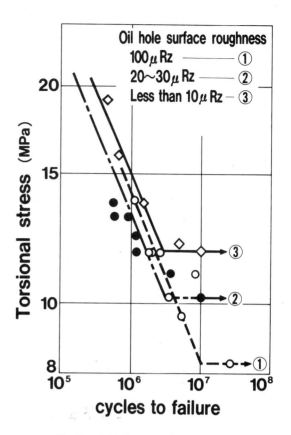

Fig 6b S-N diagram of crankshaft

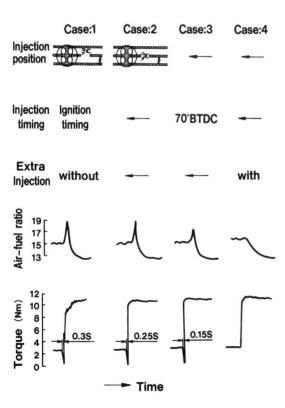

Fig 7a Comparison of response

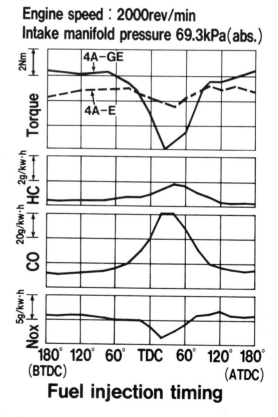

Fig 7b Effect of fuel injection timing

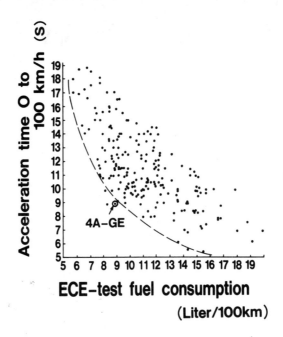

Fig 8   Performance comparison

C450/84

SAE 841299

Friction in the piston group and new ideas for piston design

H OETTING, Dr Ing, D PUNDT, Dipl Eng and W EBBINGHAUS, Dipl Eng
Volkswagenwerk AG, Wolfsburg, West Germany

This paper is dealing with the distribution of frictional losses within a petrol engine; and the majority of these losses are in the piston group. It is shown, how by innovative ideas the engine part 'piston' can be optimized in respect of friction and which effect it has on the total losses and the mechanical efficiency. It is also shown that it is difficult to improve fuel consumption in modern engines by reducing friction by single steps.

NOTATION:

A	–	Area	cm²
F_{gas}	–	Gas Force	N
F_m	–	Mass Force	N
F_n	–	Normal Force	N
F_f	–	Frictional Force	N
h	–	Height, Distance	mm
l	–	Length; here: Length of Conrod	mm
m_{osc}	–	Oscillating Masses	kg
n	–	Engine speed	RPM
MEP	–	Mean Effective Pressure	bar
IMEP	–	Indicated Mean Effective Pressure	bar
FMEP	–	Friction Mean Effective Pressure	bar
r	–	Crank Radius	mm
s	–	Stroke	mm
d	–	Piston diameter	mm
v	–	Velocity	m/sec
η_e	–	Effective Efficiency	–
η_i	–	Indicated Efficiency	–
η_m	–	Mechanical Efficiency	–
λ	–	Crank Ratio (r/l)	–
ω	–	Angle velocity	1/sec
α	–	Crank Angle	Deg
ν	–	Dynamic Viscosity	N sec/cm

1 INTRODUCTION

In addition to optimizing operation conditions, reducing the mechanical losses continues to be one of the most important tasks in engine design. Modern combustion engines have a maximum mechanical efficiency of almost 90 %, i.e. some 90 % of the work done at the piston is also available as output, but engines are seldom operated in their optimum range.

Figure 1 gives a perspective view of mechanical efficiency, plotted versus engine speed and load. The steepest gradient is found with very low mean pressures. This also results from the definition of mechanical efficiency

$$\eta_m = \frac{MEP}{MEP + FMEP}$$

when it is known that the increase in friction mean effective pressure versus load remains far below the increase in load. The decrease versus engine speed reflects the considerable influence of engine speed on friction. Optimum mechanical efficiency is thus obtained at maximum load and minimum engine speed.

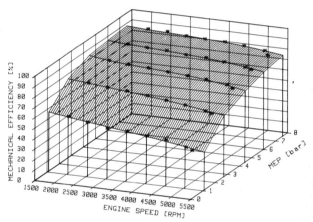

Fig 1 Map of mechanical efficiency
Engine capacity: 1272 cc
Engine performance: 60 bhp

As already mentioned above, engines are rarely operated in their optimum mechanical-efficiency range - for example full-load acceleration from low engine speeds or hill-climbing at low engine speeds - but rather in the medium engine-speed range and with relatively low load.

In this range only some 50 - 60 % of the work done by the gas is converted into mechanical output. The remainder is converted into friction heat and requires cooling. These results in the important task mentioned earlier, namely the reduction of friction losses.

2 DISTRIBUTION OF FRICTION LOSSES AND FRICTION IN THE PISTON GROUP

2.1 Relative Distribution of Friction Losses

The mechanical losses of an internal combustion engine can be split into the following categories:

- Piston-group friction
- Bearing friction
- Valve-train friction
- Auxiliary-unit friction
- Pumping losses

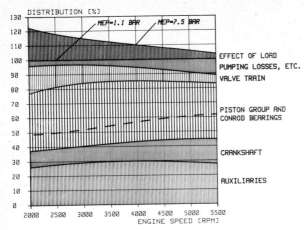

Fig 2 Relative distribution of frictional losses
Engine capacity: 1272 cc
Engine performance: 60 bhp

Figure 2 shows a relative breakdown of the individual groups related to total friction at low load in a gasoline engine. The piston group and conrod bearing are not plotted separately, since this involves considerable measurement difficulties. The dashed line results from the assumption that conrod bearing friction is of approximately the same magnitude as that of the main bearings. The influence of high and low load on friction mean effective pressure is also indicated.

The friction of the various groups were measured by motoring. The total friction of the engine under firing conditions was established by indicating the IMEP and measuring the actual output.

It is clear that, in addition to the auxiliaries, which are largely optimized in terms of cost and space required, but which are rarely designed with a view to a low friction level, the piston group has the most pronounced influence. A further factor is that the effect of load probably stems essentially from the piston group. In line with the old adage of attacking the "major evil" first, the prime task is to reduce piston friction, especially since this is where the greatest potential probably lies.

2.2 Friction in the Piston Group

Piston-group friction comprises skirt and ring friction. Ring friction is not dealt with here and the paper concentrates exluceively in the following on piston-skirt friction, or piston friction for short.

The frictional force of a body with a plane friction surface A, which moves at relative speed v over another friction surface, with an oil of dynamic viscosity ν maintaining a distance h between the two friction surfaces, can be calculated in accordance with the "Newtonian maximum shear stress theory" using the equation

$$F_f = \frac{A \cdot v \cdot \nu}{h}$$

This equation can also be applied in principle to piston friction. The lubrication gap height (h) is in turn a function of further parameters. With a piston these are, for example:

- Piston normal force on cylinder wall (resulting from gas and mass forces)
- Mass distribution at piston
- Contour, hot and cold

The greatest influence is exerted by the piston lateral force, which is automatically produced when forces act on a piston in a conventional crankshaft drive. The magnitude of the lateral force is a function of the gas and mass forces as well as of the geometry of the crank drive, which also influences the mass forces as shown by the following relationship applying to 1st and 2nd order mass forces:

$$F_m = m_{osc} \cdot r \cdot \omega^2 \cdot (\cos\alpha + \lambda \cdot \cos 2\alpha)$$

with $\lambda = r/l$

Normal force can be calculated from the equation:

$$F_n = (F_{Gas} - F_m) \frac{\lambda \cdot \sin\alpha}{1 - \lambda^2 \cdot \sin^2\alpha}$$

These equations reveal that the design parameters with the most pronounced influence are as follows:

- Crank radius r
- Length of conrod l
- Oscillating masses m_{osc}

In view of the fact that the designer can often not influence the crank radius because of its direct effect on displacement and method of operation, this paper will concentrate on conrod length and oscillating masses.

Figure 3 shows the influence of these two variables on normal forces as per equation for F_n in the high-pressure section, plotted, for the sake of clarity, versus piston stroke. The gas forces were determined from the measured pressure diagram of a 1.3 l VW spark ignition engine at n = 4500 RPM and IMEP = 9.05 bar. The associated mass forces were established in accordance with equation for F_m.

Figure 3a indicates the profile for a production engine. In the compression stroke the mass force dominates until shortly prior to TDC. This force can, however, not prevent a double change in piston contact, since the gas force is predominant shortly before TDC, thus resulting in a further crossover of the normal force.

Figure 3b shows the situation as it would be if the not unrealistic assumption were made that oscillating masses had been reduced by 40 %. As with the other two figures, the original curve for the production engine in Figure 3a is indicated with a dashed line. The most striking feature is that the piston no longer executes a side change in the compression stroke, since the gas force is already predominant in the entire second half of the compression stroke, thus even preventing the first side change. This has a positive effect on engine noise development.

As it to be expected, the reduction in oscillating masses also results in less pronounced lateral forces in the first part of the compression stroke and in the second part of the expansion stroke. The reduced mass forces unfortunately lead to increased lateral forces in the first part of the expansion stroke, since the resulting

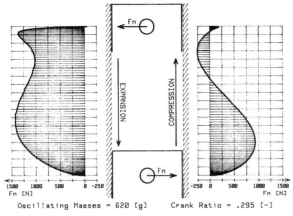

Fig. 3a PRODUCTION CONDITIONS

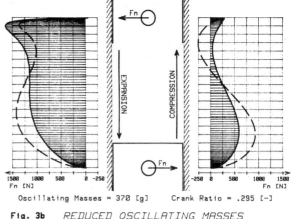

Fig. 3b REDUCED OSCILLATING MASSES

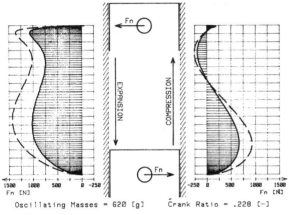

Fig. 3c IMPROVED CRANK RATIO

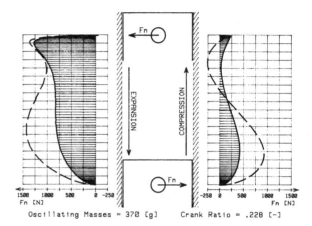

Fig. 3d REDUCED OSCILLATING MASSES AND IMPROVED CRANK RATIO

Fig 3 Plots of normal forces on the cylinder walls
Engine capacity: 1272 cc
Engine speed: 4500 r/min
Engine load: 7.5 bar

piston force is no longer reduced to the same extent by the mass force.

Whilst maintaining production engine masses, the conrod ratio was reduced by 22.5 % in Figure 3c. As illustrated, the trend exhibited by the lateral-force profile is maintained, but it is however reduced.

Figure 3d illustrates the effect of simultaneously realizing a reduction in piston masses and conrod ratio. As shown, the optimum effect is not achieved until both measures are carried out together: No side change within the compression and expansion stroke and a reduction in normal forces over the entire range.

The design goal must therefore be to make the conrod as long as possible and to reduce the oscillating masses.

Both measures are equally effective in the pumping part of the cycle strokes, since in this case it is virtually only the mass force which act.

2.3 Actual Piston Design

Figure 4 compares the new light weight piston from a 1.8 l VW engine with its predecessor. As clearly shown by the illustration, the ring zone has been reduced to a minimum and the gudgeon pin bore located directly beneath this zone, thus resulting in an optimum conrod length. It was also possible to reduce the weight of the conrod and piston by more than 10 %.

The gudgeon pin was likewise adapted to the new conditions and its diameter was also reduced by 10 %, which has advantages both in terms of weight and with regard to the conrod length.

The limits of this method are clearly indicated by the fact that the gain in terms of conrod length was only around 6 %.

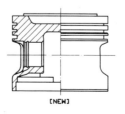
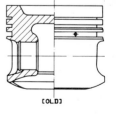

Fig 4 New and old piston
1.8 ℓ VW engine
Diameter: 81 mm

3 NEW IDEAS FOR PISTON DESIGN

As shown by the above, the most important task is to increase the conrod length in order to improve the mechanical efficiency. However, with conventional pistons, this can unfortunately scarcely be improved above and beyond the current production level.

3.1 Different Approaches and the Effect on Engine

What possibilities are there, nevertheless, for increasing conrod length with conventional pistons?

a) Transition to 2-ring piston

This has little effect on conrod length. In this case optimization is focused on the functioning and reliability of the engines.

b) Transition to short-stroke engines

Given that a specified displacement is to be maintained, this means reducing the stroke and increasing the bore. It is possible to increase the conrod length by half the stroke reduction.

The main drawback to this solution is, however, that the compression ratio - which determines consumption - is more severely restricted by the resultant combustion-chamber shape, which is less favourable in terms of combustion. The benefit gained as regards mechanical efficiency by way of the conrod length may be lost again due to reduced internal efficiency.

Moreover, it is often scarcely possible to implement this measure with given engines, since the sealing land width between the cylinders has already been minimized. With new design this leads to an increased overall engine length and thus to more weight.

The most significant drawback is, however, that the piston weight is roughly proportional to d^3. The benefit in terms of a longer conrod is thus achieved at the expense of a higher proportion of oscillating masses, i.e. this measure can also become ineffective.

c) Increasing height of block

Increasing the height of the block does not present any great design difficulties, but the disadvantages are immediately apparent:

- Increased weight
- Manufacturing problems with specific production lines
- Difficulties as regards installation with given space availability

These considerations reveal that with conventional systems improving mechanical efficiency by lengthening the conrod is offset by major drawbacks in other areas. Indeed, when reference to the effective efficiency, such a measure may even become totally ineffective.

3.2 Ideas for a "Built-Up" Piston for Passenger Car Engines

Whilst maintaining specified conditions such as

- Displacement
- Bore
- Stroke
- Block height

an attempt will now be made to show the possibilities which exist for lengthening the conrod if conventional piston design are dispensed with.

One method would be to no longer insert the gudgeon pin from outside through the piston skirt, since - as shown by <u>Figure 4</u> - there is no further potential for lengthening the conrod in this case due to the height of the ring zone. Rather, the pin would have to be attached inside the piston directly at the crown. <u>Figure 5</u> shows the prototype of such a design, still manufactured from a series piston. This piston consists of a normal piston outer section, the interior of which has been machined to accommodate the gudgeon pin. The outer and inner sections of the piston were screwed together by means of four 5 mm bolts of strength class 12.9. Weight was kept roughly constant, the length of the conrod was increased by 11.5 %. The prototype was tested in a 1.3 l spark-ignition engine and ran without problems both on the test bench and on the road.

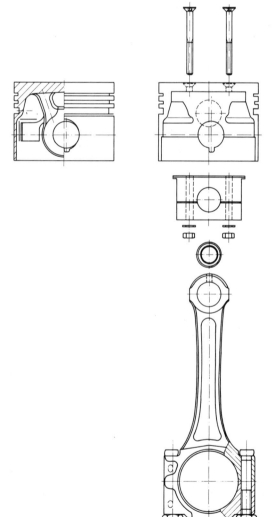

Fig 5 Prototype of the built-up piston
Diameter: 75 mm

The advantages of such a piston can already be seen from this prototype:

- The ring zone no longer restricts compression height
- The pivotal point of the piston is shifted nearer to the piston centre of gravity
- Given appropriate design such a piston may become lighter.

3.3 Results from Test Bench and Road Test

The friction losses of the piston constructed as per Figure 5 were measured in comparison with the corresponding production piston. The results are shown in Figure 6 as the improvement in friction mean effective pressure in percent as a function of engine speed and load. Clear improvements of between 4 and 5 % are also to be found in the map area most frequently encountered in operation.

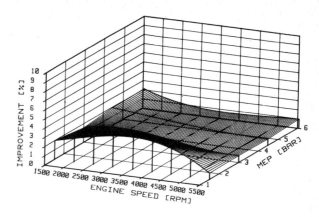

Fig 7 Improvement in mechanical efficiency
New piston / conventional piston
Engine capacity: 1272 cc
Engine performance: 60 bhp

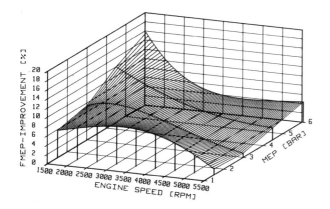

Fig 6 Improvement in FMEP
New piston / conventional piston
Engine capacity: 1272 cc
Engine performance: 60 bhp

Figure 7 illustrates the results in terms of improved mechanical efficiency. Improvements of up to 2.5 % can be achieved, which has a proportional effect on better fuel economy.

These figures are not particularly imposing at first glance. It can, however, be seen that greater improvements in consumption figures resulting from optimisation a single component are not feasible with today's engines in terms of improved mechanical efficiency.

The first design ran for approximately 40 hours on the test bench (fired and motored), before achieving a period of service of more than 15 000 km in a vehicle under road conditions and then failing due to bolt fracture.

A second piston with a similar design for a 1.8 l VW engine achieved approximately the same results in terms of friction as the first prototype, but it was subjected to far greater stress on the test bench (60 hours with high proportion of full-load operating conditions and critical conditions for the piston such as high engine speeds at low load). The piston withstood such testing without failure.

To date, the piston has run more than 20 000 km in the vehicle, without sign of failure.

These encouraging results led to the design and construction of a further type of composite piston as illustrated in Figure 8 . In this case, an inner section was dispensed with and the gudgeon pin was attached directly at the crown using two bolts. This piston has yet to run without presenting any problems.

After the first engine strip and inspection an unusual wear pattern was observed in the lower skirt area and, with one piston, signs of seizure in the lower skirt area and just below the ring zone on the major and minor thrust sides were noticed. As revealed by a '3 D analysis' performed by the piston manufacturer, Karl Schmidt in Neckarsulm, a composite piston design of this type is characterized by a differing deformation behaviour under the effect of force and temperature, with which it is possible to explain the piston damage encountered. In this particular case, this means that the piston must be modified with respect to a production piston both in terms of convexity and ovality. It also requires a greater clearance. The latter measure may under certain circumstances result in a higher noise level, but this is compensated for at least in part by the greater articulation of the conrod in the piston and the resultant improvement in secondary piston movements as well as by the reduction in lateral forces.

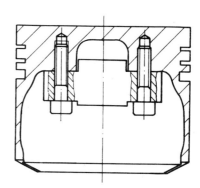

Fig 8 Built-up piston with a bolted gudgeon pin
Diameter: 81 mm

4 OUTLOOK

Even conventional piston in a new engine design require - as has been shown in practice - intensive coordination of the skirt contour. If conventional methods of piston construction are dispensed with, it cannot be assumed that these problems will disappear.

The actually expected problem, namely that of inadequate strength of the 'connection', was not encountered with these pistons. The difficulties which did arise would appear to be capable of being overcome and the advantages are clear-cut.

The pistons presented here were flat-head pistons, in which the compression space is located entirely in the cylinder head. If this compression space is shifted entirely or partially into the piston, the length of the conrod is reduced by the depth of the bowl. If, on the other hand, the piston is designed such that it does not have a flat top, the pivotal point of the conrod can be raised still further and the conrod length increases correspondingly.

Now that it has been shown that "Built-Up" pistons in passenger-vehicle engines can result in reduced friction and engine noise by increasing conrod length, future work will be directed towards establishing the limits of such a method. On the other hand, endeavours must also be made to minimize the weight of both the piston and the oscillating conrod component.

5 REFERENCES

Figures 1, 2, 3, 6 and 7 were produced on the basis of friction measurements using engine indication, which were carried out for the VW Research Department at the Instiute for Piston Engines at the University of Hanover.

C435/84

SAE 841300

Three-cylinder, 1.0 litre, diesel engine mounted on fuel-efficient vehicle of 'New Charade'

Y INO
Daihatsu Motor Co Ltd, Ikeda City, Japan

ABSTRACT This paper describes the Type CL engine, the smallest three-cylinder, one-litre, diesel engine ever produced in the world for use in passenger cars. This engine, mounted in the New Charade car, achieved a surprisingly low fuel consumption record of 4.8 litre/100km(urban driving). In this Type CL engine, efficient combustion takes place over a wide range from low-speed, low-load to high-speed, high-load, despite the fact that the small displacement diesel engine has hitherto posed various problems from the viewpoint of combustion. Hence, the engine is capable of delivering 27 KW, 60 Nm/993 cc, that is a very high specific output. Among various problems which have been encountered during the development process of this engine, this paper will explain using concrete examples specifically the matching of the injection system, the optimum setting of the combustion chamber-related components and the low-mechanical loss design, which have great effects on the achievement of a high fuel efficiency.

1 PREFACE

This paper will describe the three-cylinder, one-litre diesel engine (Type CL engine), the smallest diesel engine for passenger car use ever developed in the world. This engine has been developed by Daihatsu for an engine to be mounted in the "New Charade".

The Type CL engine is a small displacement engine with a displacement of 331 cc per cylinder.

Despite this disadvantage, the Type CL engine is capable of developing remarkable fuel consumption performance over a wide range from low-speed, low-load to high-speed, high-load. At the same time, it is capable of delivering a high output of 27 KW, 60 Nm torque. When this engine is combined with a light weight vehicle having a low running resistance, the "New Charade" diesel vehicle has made it possible to accomplish a remarkably low fuel consumption which can be rated as the top level.

In a diesel engine, low fuel consumption is not always compatible with low noise level. Therefore, one of the most difficult subjects encountered during the development work of the Type CL engine was to concurrently satisfy these two characteristics which are against each other.

This paper will explain how improvement of fuel consumption has been achieved by mentioning some concrete examples of various problems with which we were confronted in the pursuit of both low fuel consumption and low noise levels, as well as concrete examples of countermeasures which were implemented.

2 OUTLINE OF TYPE CL ENGINE

Table 1 indicates the main specifications of the Type CL engine. Furthermore, Fig. 1 gives the full-load performance, whereas Fig. 2 shows the sectional view of the engine.

The Type CL engine is a dieselized engine that has been converted from the Type CB gasoline engine. During the development, special emphasis was placed on the following three points:

(1) Satisfying both ultra-low fuel consumption and responsive running.
(2) Thorough enforcement of reduction in vibration and noise levels.
(3) Standardization of parts, machining equipment and facilities between Type CL and CB engines so that the same parts, equipment and facilities may be used in common.

Aiming at the above development objectives, a clean engine with remarkable reliability has been successfully produced. Furthermore, in order to realize an engine featuring ultra-low fuel consumption and responsive running, as outlined in the first objective, our development work was conducted with the following three items as our basic policies.

(A) Highly-efficient combustion by employing an injection pump having a high fuel injection rate and by introducing an optimum setting of the injection system incorporating a newly developed "DIA-CUT" nozzle.
(B) Attainment of high intake efficiency by combining a three-cylinder engine with a small intake interference and an inertia type supercharging method.
(c) Enforcement of extensive design modifications aiming at drastic reduction in mechanical loss, based on the Type CB engine.

As a result, as is evident from Fig. 3, the Type CL engine has made it possible to achieve an ultra-low fuel consumption, which is far lower than the fuel-efficient, gasoline-fueled "NEW Charade".

3 CONCRETE MEASURES TO ACHIEVE LOW FUEL CONSUMPTION

The two items of efficient combustion and improvement of mechanical efficiency have been accepted widely as the most effective means to achieve low fuel consumption. However, it is imperative to take into account noise level characteristics during the study of efficient combustion. Also, it is mandatory to pay thorough consideration to the durability of components and their vibrations during the study of low-mechanical loss design that is necessary for improvement of mechanical efficiency.

During the development work of the Type CL engine, various problems were solved which had been encountered within pursuit of the above-described characteristics which are in opposition to each other, without deteriorating any of these characteristics. Thus the original targets were achieved. In respect to some important and key factors which had a close bearing on determining the success or failure of the development of the Type CL engine, these will now be explained with emphasis on test results.

3.1 Difference in fuel consumption caused by size (type) of injection nozzles

Fig. 4 shows external views of two kinds of injection nozzles and their main specifications. One is the Type S nozzle which has been hitherto employed, whereas the other is the Type P nozzle whose production was started recently. The Type P nozzle features small-size and light weight, which can contribute to a compact engine. Nevertheless, the Type CL engine has employed the Type S nozzle for the following reasons:

Fig. 5 indicates the fuel consumption curves which are obtained when the nozzle type is changed. On the other hand, Fig. 6 represents the heat release rate under no-load state as well as full-load state. when the P nozzle is installed, the fuel consumption starts deteriorating at the low engine speeds of 3000 rpm or below, compared with the case where the Type S nozzle is installed. This tendency becomes more conspicuous at the high load side. Furthermore, the Type P nozzle exhibits greater output loss for the same smoke limit.

Now, studying the differences in fuel consumption and output in relation to the heat release rate, under the full-load range, compared with the Type S nozzle, the Type P nozzle exhibits a convex curve in its heat release curve at a point before TDC. This shows that the combustion takes place as negative work. Moreover, the heat release rate after TDC is gradual and its combustion period is prolonged. As a result, the combustion efficiency of the Type P nozzle is bad both before and after TDC. Therefore, if the injection timing is retarded slightly on the Type P nozzle at the low-engine-speed of 3000 rpm or below, the heat release rate before TDC can become lower. However, the heat release rate after TDC becomes lower and the combustion period becomes longer, too. Consequently, the improvement of fuel consumption is very little or zero. On the other hand, under the no-load range, both Type P and S nozzles exhibit virtually the same level of heat release rate at low engine speeds. However, as the engine speed increases, the heat release rate of Type P nozzle indicates a sharp rise. As a result, the engine noise level under the no-load state becomes higher, when the Type P nozzle is used. As is explained above, compared with Type S nozzle, the Type P nozzle demands that the injection timing be retarded at the high-speed side under the no-load state; at low speed under full load.

It is not known fully what causes the difference in fuel consumption between Type P and S nozzles. Since the D/d ratio (see Fig.4) would exert great influences on engine performance, we conducted a series of tests by changing the D/d ratio on the Type S nozzle. The test results revealed that there were minute changes in respect to the fuel consumption and engine output, but none of the major differences described above were obtained. Therefore, it is presumed that spray conditions, such as particle diameters, distribution and penetration distance of the injected fuel, vary greatly owing to the difference in the injection nozzle size.

3.2 Adoption of "DIA-CUT" nozzle and injection pump having high fuel injection rate

The difference in the fuel consumption caused by the size of injection nozzle in the preceding Paragraph 3.1 is obtained when the flow characteristics of each injection nozzle are fully exercised. If the flow characteristics vary due to carbon, etc. lodged at the orifice of the injection nozzle, then the balance between the fuel consumption and noise level will inevitably change greatly.

On the Type CL engine, the fuel injection amount during the idling operation is minute, approximately 4.5 mm^3/st. cyl. Under this condition, the lift of the injection nozzle is only within throttle range. Hence, the engine noise level will be greatly affected by minute changes in the flow characteristics of the nozzle. The prevention of these changes will constitute a key factor to achieve low-noise level and low fuel consumption.

Fig. 7 shows the change in the flow characteristics of the throttle nozzle as well as its effects on the change in noise level.
- A: Represents a new nozzle. Here, the flow characteristic is high and the noise level is slightly high.
- B: In this nozzle, flow characteristics are obtained appropriate to clogging caused by carbon, etc. Also, the noise level is good.
- C: In this case, with a fully clogged nozzle, the throttle characteristics are not exercised fully. As a result, the noise level deteriorates.

If the characteristics as indicated in B are not obtained in a stable manner, it becomes necessary to introduce injection timing retard of about 4 degrees crank or an injection pump having a lower fuel injection rate (7.9 to 6.4 mm^3/deg.) so as to contain the noise level below the target level. As a result, the deterioration in the fuel consumption rate becomes considerable, approximately 6.5 to 13.6 g/kw.h (5 to 10 g/ps.h).

With a view to preventing nozzle restrictions as well as maintaining stable flow characteristics, the "DIA-CAT" nozzle has been developed. Fig. 8 indicates the construction of this "DIA-CUT" nozzle. As is evidend from the illustration, the "DIA-CUT" nozzle has a simple construction where part of the pin section of a conventional throttle nozzle is cut off. However, its effects are very great. Furthermore, as respects the noise level, no deterioration due to lapse of time is noticed at all, as indicated in Fig. 7. Incidentally, no great differences are provided in the throttle characteristics between the "DIA-CUT" nozzle and the conventional nozzles.

Thanks to the "DIA-CUT" nozzle's guarantee for preventing the occurrence of noise, it becomes possible to employ an injection pump having a high fuel injection rate for the displacement per cylinder, as shown in Fig. 9. Consequently, we succeeded in obtaining an improvement of fuel consumption of about 6.8 g/kw.h (5 g/ps.h) and increased engine output of about 3 percent over the entire range of the engine speeds.

3.3 Improvement of fuel consumption through heat insulation around combustion chambers

As the temperature of the cooling water of the engine drops, the fuel consumption will also deteriorate. In order to reduce the engine weight drastically, the Type CL engine has employed an aluminum-alloy cast cylinder head, thereby resulting in some improvement in the fuel consumption. On the other hand, an aluminum-alloy cylinder head allows good heat transfer around the pre-combustion chambers. Hence, it was anticipated that the combustion efficiency would deteriorate because of the overcooling of the pre-combustion chambers. In fact, our experience shows that its effects are great in the case of the Type CL engine.

Fig. 10 shows how the fuel consumption and noise level alter as the wall thickness of the pre-combustion chamber and the surface roughness of the cooling water-contact surface are changed. It is apparent that the fuel consumption at low load and low engine speed side can be improved drastically by containing the heat transfer. On the other hand, slight deterioration is noticed at high load and high engine speed conditions. Besides the improvement of fuel consumption, the limit of irregular combustion (white smoke) under the no-load operation has been improved, too. However, the wall thickness around the pre-combustion chamber should be determined only from the viewpoint of the improvement of combustion, such as fuel consumption and white smoke. It must also be determined by taking into account thoroughly the high temperature strength of aluminum materials.

4 IMPROVEMENT OF MECHANICAL EFFICIENCY

With a view to attaining a thorough low-mechanical loss design, extensive improvements have been made on the basic engine of the gasoline-fueled Type CB engine. As a result, the Type CL engine indicates about 10 to 20 percent less mechanical loss, compared with the conventional diesel engine, as is evident from Fig. 11.

Among those items which made possible the reduction in mechanical loss, the adoption of the same dimensions of the crankshaft as those of the Type CB engine is notable. Although the crankshaft of the Type CB engine is rated in a smaller class among those gasoline engines of the same size, the same crankshaft dimensions have been shared in common by both Types CL and CB, thus contributing greatly to the reduction in mechanical loss.

The crankshaft for the Type CB engine is made of nodular graphite cast iron (JIS, FCD70). Furthermore, the crankpin journal diameter is 40 mm, with the main crankshaft journal diameter of 42 mm. For a crankshaft for diesel engine use, this crankshaft is rated as the lowest limit from the standpoints of materials and dimensions. As indicated in Fig. 12, the crankshaft for the Type CL engine use requires 1.5 times greater strength compared with the crankshaft for the Type CB engine use. In order to satisfy this requirement of 1.5 times greater strength, the following two methods were employed.

4.1 Improvement of stock strength

In the case of nodular graphite cast iron, an inoculant is added into the molten material in order to promote the conversion of graphite into a spheroidal form. However, the stock strength varies widely depending upon a longer or shorter period of time between the addition of an inoculant and the finish time of pouring molten iron into the cast mold.

A new inoculation method was finally developed using wire for the purpose of shortening this required time and keeping variation at a minimum level. In this wire inoculation method, the inoculant is filled into a fine pipe to form the "wire". When the molten iron starts to be poured into the cast mold, simultaneously the wire will be supplied automatically to the cast mold. As a result, the variation in stock strength is maintained at a minimum level, thus making it possible to get a repeatable high strength. Consequently, the stock strength has been improved about 40 percent even at the lower limit of variation.

4.2 Increased strength through after-treatment

Even after the stock strength has been increased by introducing the wire inoculation method, the strength of the crankshaft is still insufficient as the crankshaft for the Type CL diesel engine use. Hence, the stock strength must be improved through some after-treatments.

In our pursuit of finding effective after-treatments for nodular graphite cast iron, four methods were tried as indicated in Fig. 13. Better than expected effects were obtained, except for CO_2 laser hard quenching. However, fillet rolling was adopted after a multi-parameter study had been made from viewpoints of the productivity, strain, cost and so forth.

The improvement of strength by means of fillet rolling has been estimated to be about 30 to 40 percent at best, for steel materials. Nevertheless, in the case of nodular graphite cast iron, which comes under a lower class of material, dramatic improvements in strength ranging from 50 to 100 percent were obtained. As a result, it becomes possible to use the same geometric specifications for the crankshaft of both Type CB and CL engines. Consequently, we could reduce the mechanical loss up to 0.27 bar in terms of the mean effective pressure.

5 CONCLUSION

(1) In a small displacement diesel engine, the matching of the injection system, specifically the matching of characteristics of injection nozzles, and its maintenance in a stable manner constitute very vital keys which make it possible to obtain the improvements in engine output and fuel consumption and the reduction in noise level at the same time. Through the matching of the injection nozzles, we have succeeded in achieving the improvement in fuel consumption of about 2.5 percent and the improvement in engine output of about 3 percent, conservatively estimated, in addition to stable reduction in noise level.

(2) Deterioration in fuel consumption which is caused by the heat transfer around the pre-combustion chamber is noticeable especially at low load and low speed operating range where the heat generation per unit time is comparatively small. When we design a pre-combustion chamber and its related components, especially in the case of an aluminum-alloy cast cylinder head, thorough consideration must be given to both parts strength and fuel consumption.
(3) It has been found out that nodular graphite cast iron, which may be classified as a comparatively low grade material, can attain a strength equal to or greater than steel, if the forging method and after-treatment are carried out properly.

The foregoing explains the concrete measures taken to realize low fuel consumption during the development stage of the Type CL engine. We believe that those problems encountered have been solved as far as the item of achievement of ultra-low fuel consumption is concerned. Nevertheless, we assume that the analysis of the combustion is inadequate.

With a view to offering products which can meet the ever-increasing demand from the market and win the consumers' satisfaction, the research and development will continue so that the quality and performance of the Type CL engine may be further improved.

6 ACKNOWLEDGEMENTS

The author would like to express sincere gratitude to Mr. Tsudo, project manager of the Engine Engineering Department, Mr. Miyake and Mr. Wada who rendered to me very helpful advice and assistance while I was compiling this paper.

REFERENCE

(1) Millington, Hartles : Frictional losses in Diesel Engine, SAE Paper. 680590

Table 1 Main specifications of Type CL engine

Total Displacement	cc	993
Bore x Stroke	mm	76 × 73
Compression Ratio		21.5
Combustion System		Indirect injection Swirl chamber
Maximum Output	kw/rpm	27/4600
Maximum Torque	N·m/rpm	60/3500
Weight (with water and oil)	kg	106
Injection Order		1-2-3

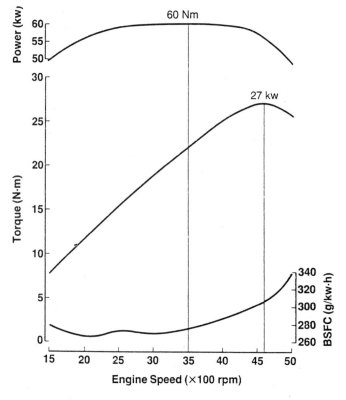

Fig 1 Performance curves of Type CL engine

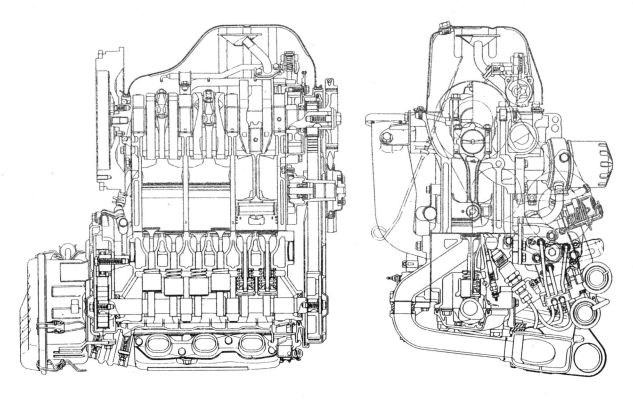

Fig 2 Sectional view of Type CL engine

Test Method : ECE (ℓ/100km)

Model	Fuel Consumption		
	Urban	90 km/h	120 km/h
CHARADE G30 (Diesel)	4.8	3.6	6.6
CHARADE G11 (Gasoline)	6.6	4.5	7.5

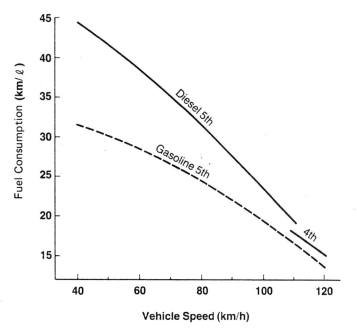

Fig 3 Comparison of diesel and gasoline engine on fuel consumption

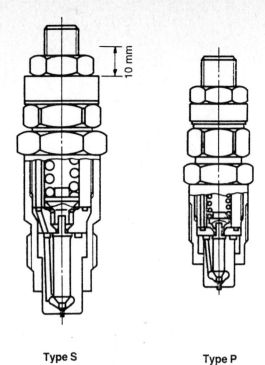

Type S   **Type P**

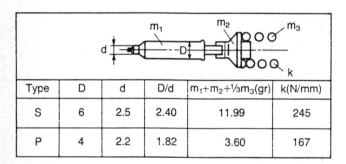

Type	D	d	D/d	$m_1+m_2+\frac{1}{3}m_3$(gr)	k(N/mm)
S	6	2.5	2.40	11.99	245
P	4	2.2	1.82	3.60	167

Fig 4   Comparison of typical injection nozzle specifications

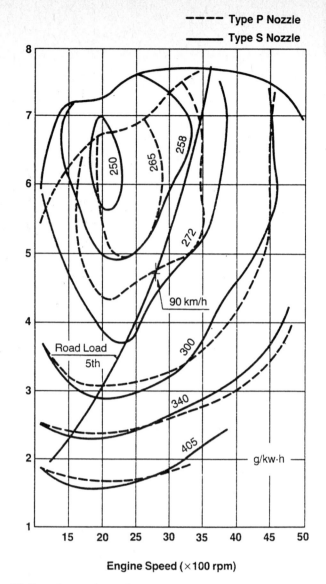

Fig 5   Comparison of fuel consumption for Type P and Type S nozzle

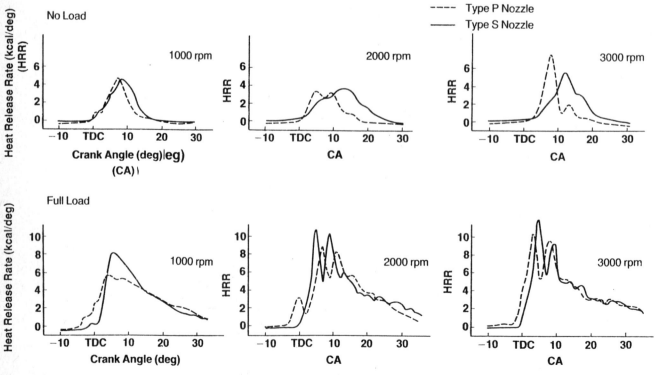

Fig 6   Comparison of heat release for Type P and Type S nozzle

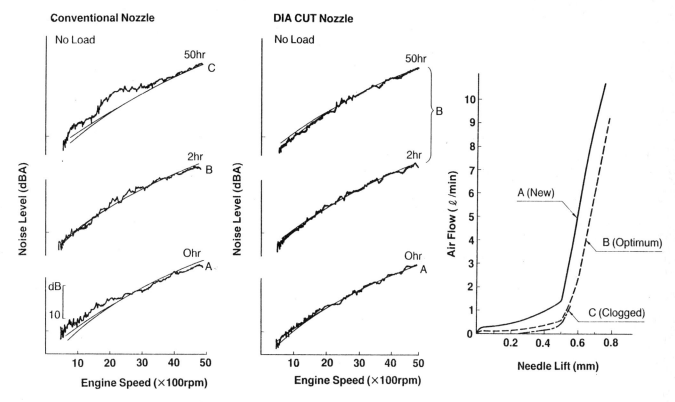

Fig 7 The change in noise of engine and flow characteristics of nozzle

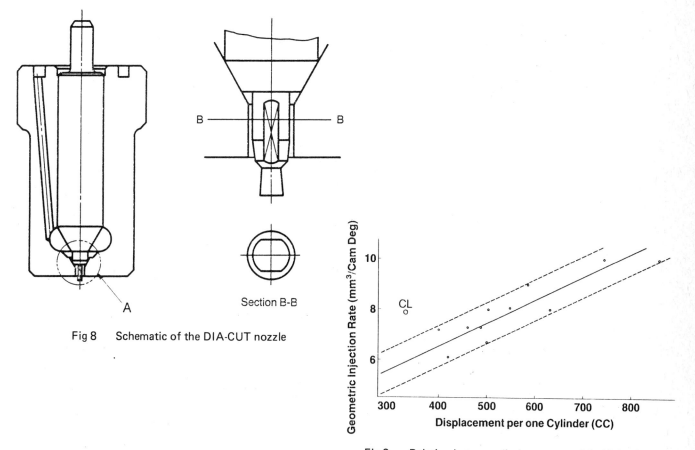

Fig 8 Schematic of the DIA-CUT nozzle

Fig 9 Relation between displacement and fuel injection rate

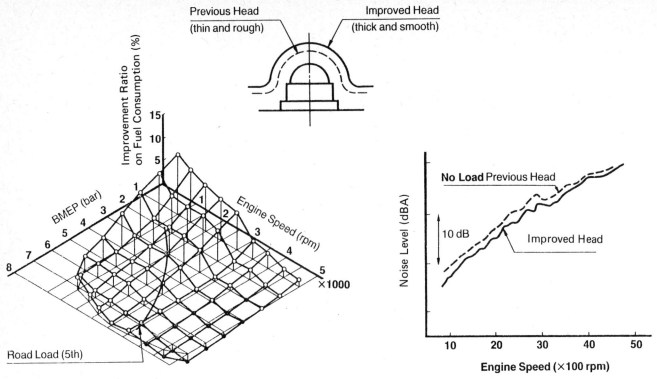

Fig 10 Effects of heat insulation around pre-combustion chamber on fuel consumption and noise

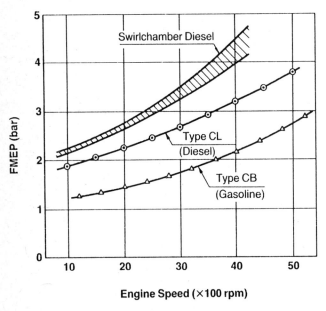

Fig 11 Comparison of FMEP at motoring

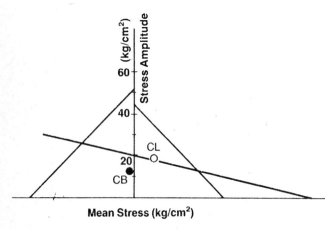

Fig 12 Life line diagram for original crankshaft of Type CL engine

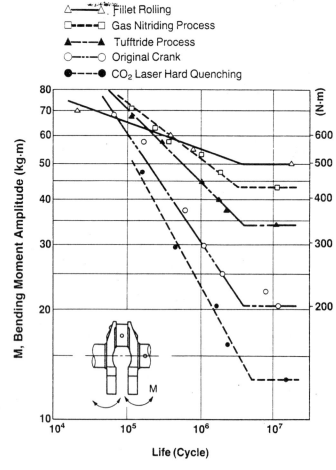

Fig 13 Fatigue test results for after-treatment

C428/84

SAE 841301

Passenger car diesel engines charged by different systems for improved fuel economy

P W MANZ, Dipl-Ing and K-D EMMENTHAL, Dr-Ing
Volkswagenwerk AG, Wolfsburg, West Germany

This report deals with different engine systems and the results of supercharged passenger car Diesel engines. Changes in the power characteristic of the engine by using a positive displacement charger, a pressure wave charger, and an exhaust gas driven turbocharger with variable area turbine nozzles are demonstrated. The different charging systems and their effect on brake specific fuel consumption are discussed. Furthermore, a comparison is made between the test results of the standard turbocharged Diesel engine of a passenger car and those of the same baseline engine equipped with a variable area turbine nozzles (VATN) turbocharger, both fitted to an engine dynamometer and to the car. A considerable gain in engine rated power, fuel economy, and torque at low engine speeds was obtained with the VATN version resulting in a much improved driveability.

INTRODUCTION

There is a growing interest in the field of supercharged diesel engines for passenger car application. The main goals are better fuel economy and higher torque and power for improved driveability.

The main advantages of a small displacement supercharged engine (1) are:

a) The small supercharged engine needs a smaller envelope for a given rated power and therefore it is suitable for installations in a passenger car where limited under bonnet space is given.

b) The supercharged engine has less friction losses. In comparison to the bigger displacement naturally aspirated engine the same torque is delivered at higher brake mean effective pressures giving better specific fuel consumption.

c) For small quantity production supercharging can offer a lower cost solution for the improved performance.

These points show the basic advantages of boosting the performance by supercharging; but nothing is said about the charging system itself. In Figure 1 different ways of solving the problem are shown: A Roots type supercharger driven, if needed, by the crankshaft via a clutch, a pressure wave charger, and an exhaust gas turbocharger with integrated wastegate. These systems were tested on a 1.6 l 4-cylinder swirlchamber diesel engine under identical conditions, the results are discussed in the following text.

Some important results in regard to efficiency of the different charging systems are shown in Figure 2. The full load performance using the 3 charging systems of Figure 1 is shown in comparison with the performance of the naturally aspirated baseline engine. Obviously the Rootscharger and the pressure wave charger lead to a considerable increase in mean pressure already at low engine speeds. The turbocharger system is characterized by less mean pressure in the range of low engine speeds. The performance characteristic of the turbocharged engine is caused by the lack of exhaust gas energy leading to less boost pressure in this region of the engine speed range. Comparing the specific fuel consumption of the 3 charging systems leads to the result that the turbocharged engine has a better fuel efficiency in a certain engine speed range. This advantage in fuel efficiency is found again at part load operation as shown in Figure 3. The results shown in Figure 2 and 3 lead to the following conclusions:

The turbocharged diesel engine is characterized by a marked lack of performance at low engine speeds that gets even bigger under dynamic operation because there is no mechanical drive from the crankshaft, just a dependance on exhaust gas energy in this operating range. On the other hand advantages in specific fuel consumption are obtained that are of importance to the fuel efficiency minded buyer of a diesel car. Because of this discrepancy there is a necessity to improve the performance of the turbocharged diesel engine both in steady state and transient modes at low engine speeds without losing the fuel efficiency benefits by using modern turbocharger technology. Therefore, a turbocharger with variable area turbine nozzles (VATN) and ball bearing system was tested on a diesel engine. The turbocharger engine system, the control of the turbocharger, and some test results of importance are dealt with in the following.

TECHNICAL DATA OF THE TEST ENGINE

For all the tests a standard 5 cylinder diesel engine of the 1983 model year Audi 100 passenger car was used, because the VATN turbocharger to be investigated was designed for an air flow bigger than that suitable for the 4 cylinder diesel engine of 1.6 l displacement mentioned before.

Engine displacement	1,986 l
Number of cylinders	5
Stroke	86.4 mm
Bore	76.5 mm
Compression ratio	23
Combustion process	swirl chamber
Injection pump	Bosch distributor type
Injection nozzle	Pintle nozzles
Exhaust and muffler system	standard
Exhaust gas turbochargers	1. standard 2. VATN TC in prototype trim

NOTATION

Power	P	(kW)
Pressure	p	(bar)
Temperature	T	(centigrade)
Mass flow	m	(kg/s)
Enthalpy drop	Δh	(kJ/kg K)
Area	A	(m²)
Pressure differential	Δp	(bar)
Radius	r	(m)
Absolute velocity	c	(m/s)
Relative velocity	w	(m/s)
Circumferential velocity	u	(m/s)
Time	t	(s)
Angle of attack	α	(degree)
Frequency	ω	(1/s)
Efficiency	η	(1)

SUBSCRIPTS

Ambient	0
Compressor exit	1
Turbine inlet	2
Close to turbine Wheel inlet	4
Turbine wheel inlet	5
Turbine wheel outlet	6
Just downstream of turbine wheel outlet	7
Turbine	T
Vacuum	V
Start	A
End	E

DESCRIPTION OF THE VATN TURBOCHARGER

In Figure 4 a VATN turbo machinery (3) is shown; the turbocharger used here was of similar design and had the following advantages:

a) The usual friction bearing system was replaced by a ball bearing system with much less friction loss. The difference in friction loss can be used to raise the boost pressure.

b) The variable area turbine nozzles can be adapted in regard to their position to the momentary gas flow. Thus the pressures in the inlet and the exhaust system can be optimized.

c) Because of the overhung compressor and turbine wheel on one side of the bearing system being positioned on the "cold" side, there is no overheating problem of the bearing system.

d) A wastegate is no longer necessary.

The main disadvantages of the VATN turbocharger used are that:

a) Because of the close neighbourhood of the compressor wheel and the turbine wheel there is a considerable heat flow towards the compressor leading to higher charge air temperatures (2).

b) Because of the given design of the turbine back wall in the turbocharger used here, there was a tendency of the back wall to buckle under high exhaust gas temperatures. At the time of the tests an exhaust gas temperature of 850 °C could not be exceeded.

c) The overhung design demands high quality balancing.

In Figure 5 the guide vanes of the turbine are shown in the extreme positions. It can be seen that even in the fully closed position of the vanes there remains a flow channel for controlled flow to the turbine wheel. Matching the point of operation (load) to the demands is done by changing the angle of attack for the impeller. There are 4 different load conditions to be distinguished: idle, low part load, high part load, full load.

During idle the vanes are brought to the fully closed position to keep the speed of the turbocharger's rotating assembly as high as possible (boost pressure 1.02-1.04 bar). Thus a high boost pressure level is reached during acceleration of the Diesel engine.

Under low part load conditions the specific fuel consumption of the engine is the important factor. Therefore the vanes are fully open to keep the back pressure on the engine at a minimum and to improve the scavenging of the engine.

For high part load and full load operation of the engine the vanes should be brought to a position where the necessary boost pressure is realized and the pressure differential in the engine

$$\Delta p = p_2 - p_3 \qquad (1)$$

is positive.

Looking at the dynamic operation of the engine just the acceleration is of importance. Pressing down the accelerator pedal in the car should be followed by the fastest possible increase in boost pressure and hence engine torque. With the aid of Fig. 6 showing an acceleration procedure of the turbine engine system the aerodynamic laws to handle the position of the vanes are explained. Two well known fundamental equations are used:

$$c_{4m} = \frac{\dot{m}_T}{A_4} \qquad (2)$$

with

$$c_{4m} = c_4 \sin \alpha_4$$

$$P_T = \dot{m}_T \cdot \omega (r_5 c_4 \cos \alpha_4 - r_6 c_7 \cos \alpha_7) \qquad (3)$$

Equation (2) shows the velocity component that is responsible for the mass flow through the turbine. Equation (3) gives the dependance of the power at the turbine shaft from the angle of attack α_4.

At the time t_o the angle of attack is α_{t_o} and the vanes are in a certain position related to the point of operation. At the start of the acceleration the position of the accelerator pedal is the input signal for the vanes to move to the fully close position. Therefore the angle of attack is changing from α_{t_o} to α_{tA}. As the mass flow remains constant for the very first time interval, the height of the velocity triangle does not change (see eq. (2)); just the velocity component $c_4 \cdot \cos \alpha_4$ is changing. Studying eq. (3) it can be seen that the power of the turbine will increase leading to more performance to drive the compressor.

At the end of the acceleration process the mass flow through the turbine has increased, leading to a higher velocity triangle; the angle of attack has changed according to the new load into α_{tE}.

Figure 7 shows a principal sketch of the control system for the VATN turbocharger. The three basic components necessary are the actuator, the rotary valve, and the vacuum pump.

The actuator consists of a system of two pistons and two coil springs. By loading the actuator with boost pressure, ambient pressure, and control pressure the rod of the actuator moves the centrally positioned ring to turn the vanes. Boost pressure and ambient pressure are taken off the inlet manifold and the airfilter respectively. The control pressure is related to the position of the accelerator pedal of the car. The control pressure is generated in the rotary valve by mixing the low pressure of the vacuum pump, the boost pressure, and the ambient pressure. According to the accelerator pedal position the actuator is pressure loaded for the necessary vane angle.

TEST CONDITIONS

To compare the standard and the VATN turbocharger engine systems the following test conditions were fixed:

1) Intake and exhaust systems were kept unchanged with the exception of the turbocharger.

2) No changes were made to the injection system.

3) The injection pump was set to a full load soot number of 3.2 units (Bosch) at 4000 rpm for both turbochargers. The amount of fuel injected under different load conditions resulted from this setting.

4) Maximum boost pressure was set to realize the same maximum mass air flow thus compensating different charge air temperatures.

RESULTS

Figure 8 shows the mean effective pressures and the brake specific fuel consumptions for the conventional turbocharged engine and the engine with VATN turbocharger at full load. The engine with VATN turbocharger is superior in regard to mean effective pressure and hence torque and power over the whole speed range. In the lower engine speed range this is a result of the fully closed vanes and the small loss of the ball bearing system leading to most effective usage of the exhaust gas energy.

In the upper speed range of the engine the vanes can be turned into the optimum position to reach at a positive pressure differential across the engine for maximum engine performance.

Study of the specific fuel consumption at very low engine speeds shows a marked disadvantage of the engine fitted with the VATN turbocharger. This seems to be a result of the fully closed vanes leading to excessive back pressure and bad scavenging. In the upper speed range the benefits of the VATN technology in regard to fuel efficiency are clearly demonstrated.

In Figure 9 the full load boost pressures and back pressures of both engine turbocharger systems are plotted. It shows that the VATN turbocharged diesel engine has a high back pressure because of the fully closed vanes at very low engine speeds, yet the fully closed position is necessary to generate propulsion energy to drive the compressor. At higher engine speeds the pressure differential across the engine is positive, because there is a surplus of energy in the exhaust gas making possible an optimum position of the vanes.

Figure 10 shows that there are advantages for the VATN turbocharger engine system at part load too. There is an improvement in fuel efficiency at 2000 rpm and 3000 rpm over the mean pressure range from 2 bar to the respective maximum. These graphs are proof that the use of the VATN technology in combination with the control unit described earlier is leading towards a better scavenging and hence better fuel efficiency.

Figure 11 shows the fuel consumption map of the 5 cylinder Diesel engine with VATN turbocharger. It can be noticed that there is just a small region of relatively bad fuel efficiency close to the full load operation at 1800 engine rpm. The remaining part of the map is showing good figures for the fuel consumption.

These good results measured on the engine dynamometer under stationary conditions were to be supplemented by driving tests. Because of the better torque characteristic and the lower fuel consumption significantly better driveability and fuel economy of the test car of the 3000 lbs inertia weight class fitted with the VATN turbocharger engine system were expected.

For the evaluation of the driveability a typical acceleration process of day to day urban driving was selected. As the criterion for the fuel economy the 1/3 rule for the consumption values in the ECE testing procedure and the consumption figures at constant driving speeds of 90 km/h and 120 km/h was taken.

For in-vehicle testing of driveability and fuel economy both engine turbocharger systems were installed into the identical car. For example the time for a typical acceleration from 28 km/h to 48 km/h in third gear was measured. The values are given in Figure 12. The car fitted with the VATN turbocharger system reaches the final speed of 48 km/h 0.9 sec. earlier than with the standard system. Related to the distance this means that the car with the VATN turbocharger system needs roughly 8.6 m less to reach the final speed of 48 km/h. Such a performance reserve can be of big importance in urban traffic, for instance during a passing manoeuvre or at cross roads.

Better transient response can just be optained using the completely different charging systems pressure wave charger or positive displacement chargers driven by the crankshaft of the engine.

The fuel economy tests showed an improvement of 11 % using the VATN turbocharger technology instead of the standard equipment.

SUMMARY AND CONCLUSIONS

In this report the influence of different charging systems on power and fuel efficiency of passenger car Diesel engines is shown and discussed. One of the results is that the turbocharged Diesel engine is still especially interesting because of its good fuel efficiency. Therefore, a Diesel engine fitted with a variable area turbine nozzles turbocharger incorporating ball bearings instead of friction bearings was tested against the same engine fitted with the standard turbocharger.

Test results showed that the Diesel engine with the unconventional turbocharger was superior in both stationary and dynamic performance and led to an improvement in fuel economy under given testing procedures. For mass production some problems regarding durability, packaging, and manufacturing costs need to be solved.

REFERENCES

(1) P. Walzer, K.-D. Emmenthal,
    P. Rottenkolber:
    Aufladesysteme für Pkw-Antriebe,
    Automobil-Industrie, 4/1982

(2) M. Rautenberg, A. Mobarak,
    M. Malobabic
    Influence of Heat Transfer Between Turbine and Compressor on the Performance of Small Turbocharger
    International Gas Turbine Congress, Tokyo, 1983

(3) N. Gašparović
    Gasturbinen, Probleme und Anwendungen
    VDI-Verlag, 1967

Linking of Charger and Diesel Engine

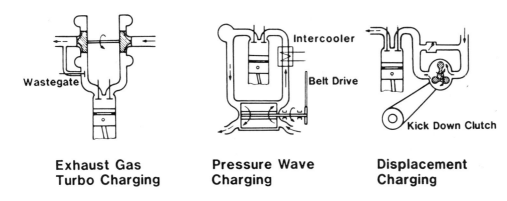

Exhaust Gas Turbo Charging

Pressure Wave Charging

Displacement Charging

Fig 1　Charged engine systems

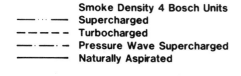

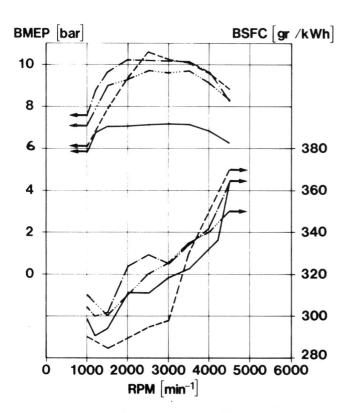

Fig 2　1.6ℓ diesel engine charged by different systems

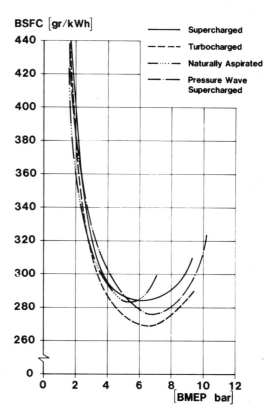

Fig 3　Load range fuel consumption curves at 2000 r/min of the 1.6ℓ diesel engine

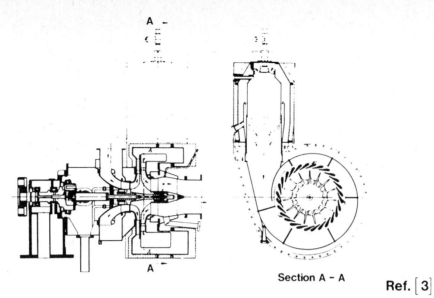

Fig 4 VATN turbomachinery

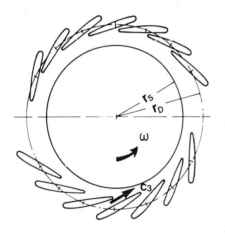

Vanes in Closed Position

Vanes in Fully Open Position

Fig 5 Turbine guide vanes in two different positions

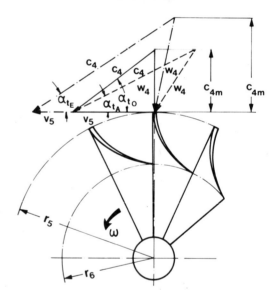

Velocity Triangles

t_O = Time Before Acceleration
t_A = Time Starting Acceleration
t_E = Acceleration Finished

Fig 6 Turbine wheel during an acceleration process

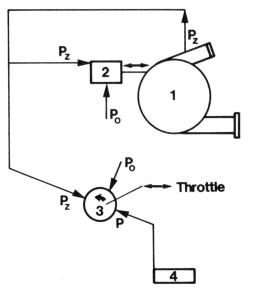

1. VATN – TC
2. Actuator
3. Rotary – Valve
4. Vacuum – Pump

Fig 7   TC control system

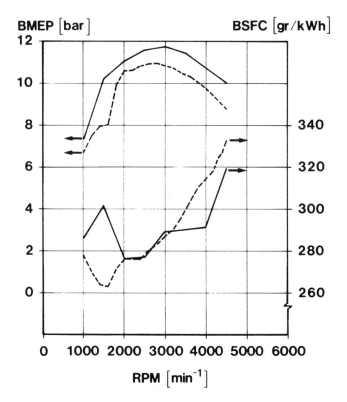

Fig 8   Full load lines of the turbocharged five-cylinder diesel engine

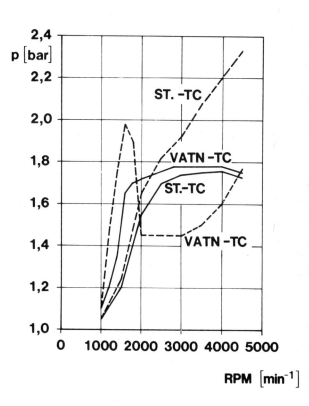

Fig 9   Pressure curves of the turbocharged five-cylinder diesel engine

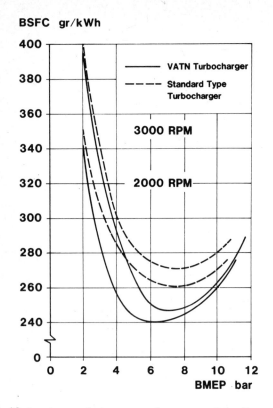

Fig 10 Load range fuel consumption curves of the five-cylinder diesel engine

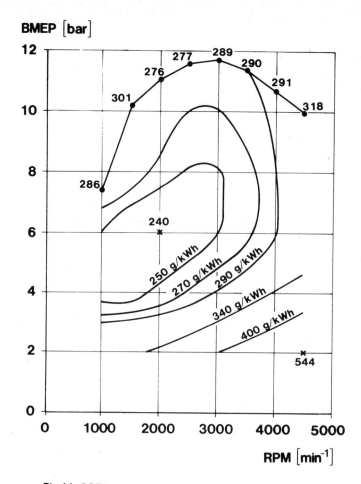

Fig 11 BSFC map of the five-cylinder diesel engine with VATN-TC

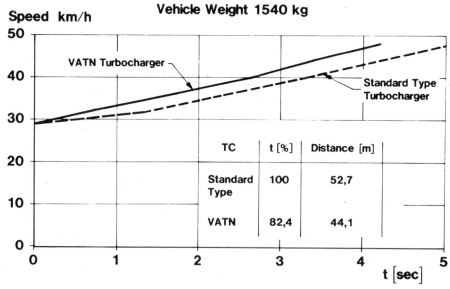

Fig 12 Acceleration process from 28–48 km/h in the third gear

C447/84

Control strategies for a chain drive CVT

U EGGERT, Dipl-Ing and H D SCHNEIDER
Ford Werke AG, Cologne, West Germany

SYNOPSIS

Control strategies for a continuously variable transaxle (CTX) project, which have been conceived and developed for Ford Fiesta/Escort vehicle application, are described.

Various strategies for start-up clutch engagement and ratio change are shown.

Actual test results of vehicle performance and fuel economy are given to verify the predicted benefits of these control strategies.

I INTRODUCTION

CVT's, with their wide ratio spread and inherent capability of continuously varying the transmission ratio, provide the opportunity of operating the engine in the most efficient region and thus improving fuel economy.

Ford are jointly developing, with Van Doorne Transmissie, a CVT in front wheel drive (FWD) transaxle configuration for small car usage. The so called CTX 811 design (see Fig. 1) comprises wet clutches for forward/reverse start-up, a planetary gear set for reverse, the Van Doorne's compression belt drive with primary and secondary pulley, an oil pump, a reduction drive and the final drive with differential unit.

Hydraulics are employed for start-up clutch control, ratio control and belt clamping load generation.

Electronic logic on top of the base hydraulic controls is also under investigation. This allows the application of CVT control strategies specially conceived and developed to improve performance, fuel economy and driveability of the vehicle.

II CLUTCH CONTROL STRATEGIES

The hydraulic control system provides an automatic clutch engagement similar to a centrifugally controlled clutch. The clutch engagement speed varies on the throttle angle/torque demand between 1200 and 2200 RPM. Whenever the driver takes his foot from the accelerator during part throttle start, the clutch will be fully closed (lock-up).

With electronic controls the clutch engagement can be controlled depending on transmission oil temperature for a shockless cold start and to provide sufficient clutch pressure for fast speed changing.
In this way the clutch engagement characteristic can be altered from a "soft clutch" with a relatively long sliptime to a "short clutch" with minimum sliptime for fast acceleration and better fuel economy.

Other features which are provided by electronic clutch controls are as follows:

- different clutch engagement on forward and reverse start-up

- increasing engagement speeds with cold engine to provide adequate fast cold idle

- clutch sliptime limitation and immediate clutch release to dampen jerks on sudden load changes and to avoid any damage to the belt drive system

- clutch disengagement in coast for free wheel effect

- controlled slip on coast.

The electronic logic can also provide a shockless clutch engagement when the clutch has to be closed again for acceleration after freewheeling.

III RATIO CONTROL STRATEGIES

The CTX powertrain has to be matched to different engine characteristics with a CTX calibration for the best compromise between fuel economy and driveability. With the hydraulically controlled CTX, this calibration is fixed for one particular engine application and must be defined as follows.

Fig. 2 shows a typical engine map with its fuel island distribution and the control strategy lines for a hydraulically controlled CTX 811 (HRC) and two different electronic ratio control lines (ERC Economy- and Sport-Mode).

The CTX-811 with its wide ratio coverage of 5.5 or 5.85 is able to get close to an optimum fuel economy line which connects the areas of best specific fuel consumption. To achieve optimum fuel economy it is therefore required to set the minimum operating speed as low as possible and to establish a throttle angle to engine speed relationship into the ratio control cam so that the fuel islands are met.

Unfortunately an ideal control strategy in terms of minimum fuel consumption does not always meet the line of a good driveability. Fuel islands located close to the maximum torque line require a typical "low speed/high torque" operation, which may result in noise and vibration concerns. Subjectively there is at first no response when then accelerator pedal is depressed, as only the throttle angle is changed without change of the engine speed. Thus a single hydraulic control strategy would always represent a compromise between fuel economy and driveability.

The use of electronic controls provides the opportunity to take more advantage of the continuously variable drive potential by using selective control strategies which are tailored to their specific purpose and engaged by a switch on the gear lever.

As shown in Fig. 2 there is a "Sport" electronic ratio control strategy for best possible performance at the expense of fuel economy and an "Economy" ERC strategy to provide optimum fuel consumption with possible driveability deficiencies.

Fig. 3 shows the CTX control strategies "Sport" and "Economy" as a function of engine speed and throttle angle.
Driving the "Sport" Mode (S-Mode) the engine will reach its top speed of 5200 RPM at wide open throttle (WOT) compared to the "Economy" Mode (E-Mode) with 4000 RPM at WOT due to a kickdown limit.
Engine response especially at part throttle is considerably different between the two Modes "E" and "S".

At a slight throttle opening of 40 degree (see Fig. 3) the engine will run at 3300 RPM in S-Mode providing a quick engine response, whereas the engine speed will be lowered to 2000 RPM in E-Mode to achieve better fuel economy and to reduce the interior and exterior noise level at the expense of agility during a vehicle acceleration.

The ratio control diagram of the CTX 811 is represented in Fig. 4 as a function of vehicle speed vs. engine speed between "low" and "high" ratio. The ratio lines of a five speed manual transmission are included for comparison.

At road load start the clutch will be closed at 1200 RPM and the engine speed rises at "low" ratio up to the minimal ratio control speed. Then the CTX ratio will be changed from "low" until "high" is reached.

From this operating point the CTX ratio is in "overdrive" along the road load line until this line is left to provide more power to reach top speed.

Top speed is lower in E-Mode than in S-Mode due to the lower maximum engine speed at WOT.

For a full throttle start the clutch is closed at 2200 RPM and the "low" ratio will be left at 3000 RPM in E-Mode, but at 4800 RPM in S-Mode which will give the driver a much improved feeling of performance.

Apart from the clutch control and ratio control strategies as described above some more features can be realised with an electronic transmission control, see Fig. 5:

- the kickdown limit in E-Mode can be switched off automatically for emergency acceleration
- E or P-Mode is chosen either by a switch or by the rate of change of the accelerator angle (adaptive mode)
- a third mode between E and P-mode can be added for part-throttle
- for an improved engine response, the accelerator can be electrically connected to the throttle replacing the cable ("drive-by-wire")
- the ratio change speed can be varied during transient conditions between acceleration, coast, braking and idling to achieve better driveability
- electronic line pressure control provides more flexibility to match the required clamping pressure to the individual driving conditions which may result in better transmission efficiency
- emergency driving possibilities ("limp home") can be incorporated.

To include and activate all these advanced features the micro processor of the electronic control system will need input data corresponding to Fig. 5 as follows:

- Engine Speed
  derived from the ignition coil

- Primary Pulley Speed
  picked up by an inductive speed sensor on the primary pulley

- Vehicle Speed
  picked up by an inductive speed sensor on the final drive

- Throttle Angle
  provided by a throttle angle potentiometer

- Mode Signal
  supplied from the mode switch on top of the Selector Lever

- Lever Position
  operated by the driver.

All these alternatives and features need further investigation to identify their individual benefits which have to be balanced against their disadvantages, e.g. greater complexity, higher cost and possibly reduced reliability due to higher number of parts.

IV RESULTS OF COMPUTER SIMULATION AND VEHICLE TESTS

A simulation of the vehicle and powertrain dynamics was carried out using a computer program specially developed for a CTX transmission controller. Vehicle/test bed measurements were carried out to investigate engine and transmission efficiencies and the transient behaviour of the transmission. The results were used as input to the mathematical model of the powertrain operation.

Fig. 6 shows the "Simulated Customer Fuel Economy" cycle (vehicle speed vs. time) simulated by the computer program for a vehicle with CTX 811 and hydraulic/electronic ratio control. This so called "Attendorn" cycle was developed by Ford as a "Real World" cycle and is a combination of "City" (Fig. 6a), "Country" (Fig. 6b), and "Highway" (Fig. 6c) cycles.

The ratio control diagrams (Fig. 7) show the computer simulation results for the initial part of the "City" cycle up to 120 seconds. The ratio change characteristic line is shown for a vehicle with electronic ratio control in S- and E-Mode.

Throughout the drive cycle the transmission ratio changes corresponding to changing drive cycle conditions (see arrows).
In S-Mode the engine speed increases up to 2800 RPM during the acceleration from 0 to 80 km/h vehicle speed.
In E-Mode however, the engine speed reaches only 2200 RPM under the same driving condition.

Fig. 8 shows the operating points of the engine during the complete Attendorn "City" cycle in a simplified engine map (see also Fig. 2) for ERC Sport- and Economy-Mode.

The area of the lowest specific fuel consumption is represented by three islands. Also included are lines of constant throttle angle and the required road load torque at the wheels for "high" ratio.

The different operating points for S- and E-Mode are clearly recognizable.

Evaluating these diagrams it can be seen that the operating points of the ERC variant with E-Mode are concentrated in the area of the lowest fuel consumption thus achieving best fuel economy.

This was verified by road test results which were performed with a Ford Fiesta.

Fig. 9 shows test results of the hydraulic and electronic ratio control variants for the complete Attendorn cycle comprising City, Country and Highway parts and the combined figure.
All tests were performed with the same vehicle, engine and transmission. Only the valve body control system was changed.

Based on the fuel consumption of the hydraulically controlled version (100 %), the ERC version in E-Mode indicated a 12 % fuel economy improvement during the Attendorn City cycle. Even the combined fuel consumption figure shows a 4 % advantage for the ERC version in E-Mode, whereas the ERC version in S-Mode could barely improve compared to the fuel consumption results of the hydraulic system version.

Generally the fuel consumption could only be marginally improved. The more dynamic elements a drivecycle represents, the more benefit can be obtained by an ERC compared to a hydraulic control system.
But, for these early tests with a phase I development stage car, it was more important to show how an improvement in economy and driveability compared to the hydraulically controlled version could be achieved.
Some further introduced features did not prove to be efficient, others were promising so that further investigations will have to be carried out.

V OUTLOOK

By carefully investigating ERC it will be possible to enhance the full potential of a CVT to improve economy and driveability as compared with the present control system.

It is understandable that not all the features proposed in this paper can be installed in a normal production car due to cost restrictions in particular for this size of car.

But there will be quite a lot of useful and promising alternatives which should be investigated and developed for application in any future advanced CVT design.

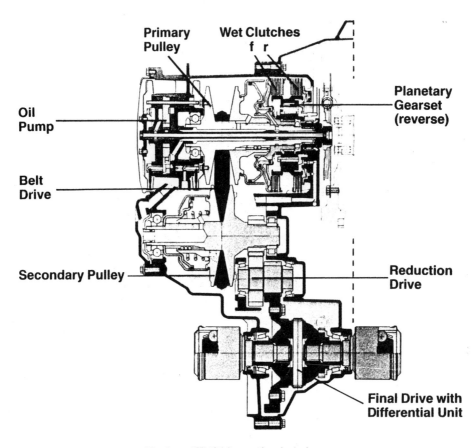

Fig 1 CTX811 wet clutch design

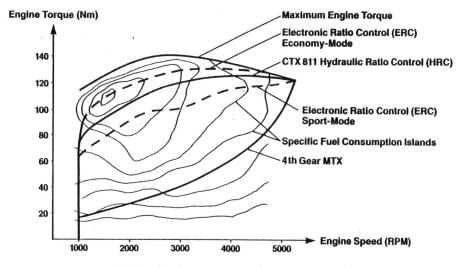

Fig 2 Engine map – ratio control strategy lines

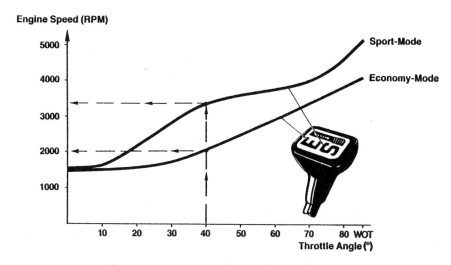

Fig 3 Drive modes Economy-Sport

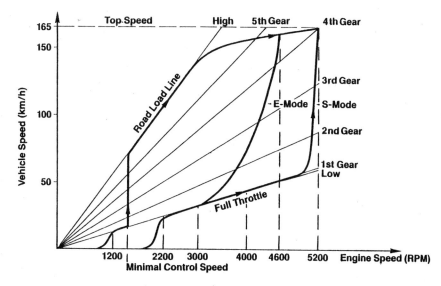

Fig 4 Ratio diagram

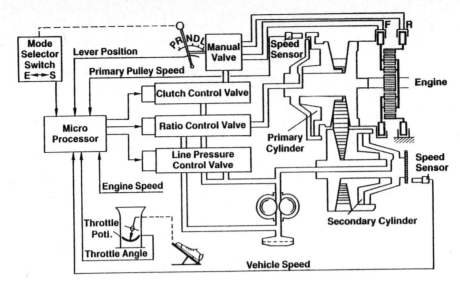

Fig 5   Full electronic control schematic

City-Cycle

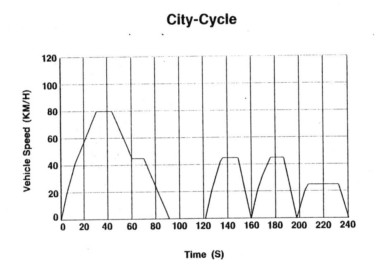

Fig 6a   Ford Attendorn cycle

Country-Cycle

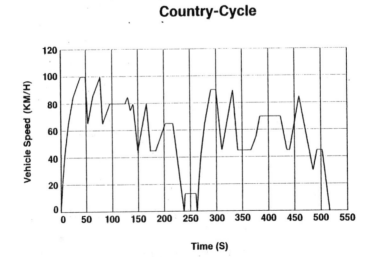

Fig 6b   Ford Attendorn cycle

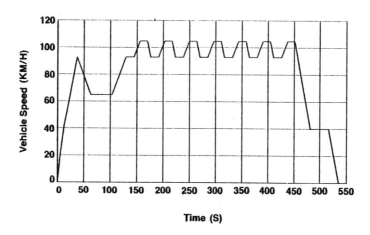

Fig 6c   Ford Attendorn cycle

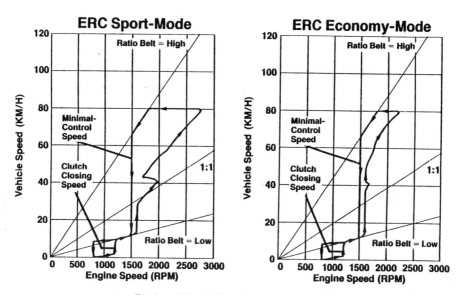

Fig 7   Simulation Attendorn city-cycle

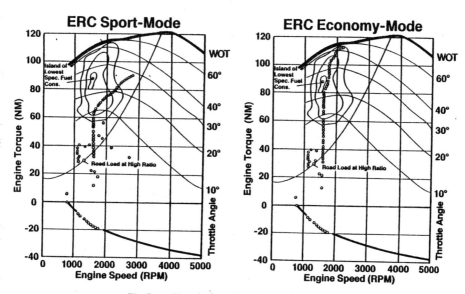

Fig 8   Simulation Attendorn city-cycle

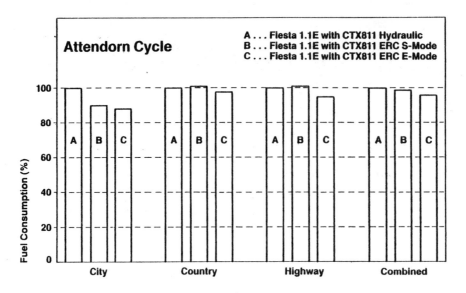

Fig 9 Test results — Attendorn cycle

C431/84

SAE 841303

Fuel economy with small automatic transmissions

R P JARVIS, PhD, CEng, MIMechE
Automotive Products plc, Leamington Spa, Warwickshire

SYNOPSIS Economical performance depends on matching the transmission to the engine and vehicle. The best type is chosen from considerations of overdrive, ratio step and vehicle usage. A particular sort of synchromesh transmission is described and details are given of the use of computer prediction and rig testing to improve the fuel economy.

1 INTRODUCTION

This paper is concerned with an automatic transmission for small and medium sized vehicles and deals particularly with the basis for the choice of design and the detail matching of the transmission to the engine and vehicle. The considerations involved would be of general application but one particular design of transmission is described as a basis for the exercise.

Whatever aspect of vehicle design one considers, fuel economy is a prime requirement. Modern engines are capable of giving low specific fuel consumption figures provided you operate them at a speed and throttle setting not too far away from their preferred working point. It is the job of the transmission to modify the drive ratio in order to achieve this.

Traditionally, the top gear and final drive ratios for a car were chosen to give the ultimate maximum speed, with the result that the engine spent a large proportion of its time at high speed and low torque under ordinary cruising conditions. Significant improvement in economy is achieved by providing overdrive gear ratios. This may be illustrated using a typical fuel consumption mapping for a small car engine (1,3 litre) as shown in Fig 1. A particular operating condition was chosen - in this case 90 km/h cruising - and this implies a certain power requirement from the engine. According to driving ratio, the engine could operate anywhere along the locus shown. The scale of ratios is based arbitrarily on 1.0:1, being the ratio for maximum speed. By reading off and interpolating the specific consumption values, it can be calculated that 17 per cent less fuel is used at 0.6 overdrive ratio than at 1.0:1.

Merely to provide an overdrive ratio is not sufficient, it is also necessary to ensure that the transmission is in the optimum ratio at all times. To obtain full advantage the transmission should change ratio for any significant change in torque demand and this would be too onerous for a driver using a manual box, so the transmission should be automatic. The choice of type will depend on the spread of ratios and the number of steps.

2 GEAR RANGE AND RATIO STEP

Unfortunately, the bigger the range of ratios one tries to provide, the greater becomes the complexity of the mechanism to achieve it. If you restrict the investigation to the study of Fig 1, it appears that a very 'tall' overdrive may be needed but, if it is used too infrequently, it would not be worthwhile. Information is required on actual vehicle usage in order to quantify the potential savings.

An automatic gearbox car was provisioned with instrumentation and a tape recorder to determine the duty cycle when driven over various routes. For the present purpose, vehicle speed and drive shaft torque are the important parameters.

By computer processing, the time spent at various increments of torque and speed was determined for each type of test route. This data was then combined for a typical mix of routes comprising 12.5 per cent urban, 38.5 per cent cross-country, 35 per cent motorway and 14 per cent alpine driving. The result was presented as a joint probability plot of speed and torque as shown in Table 1. These results are a measure of how the driver wishes to drive down the road and are not dependent on the type of gearbox used to do it.

For each box of the table the minimum fuel was determined, assuming the gear ratio for best economy. The range of ratios was considered to be divided into increments and then, by taking the aggregate fuel used for each ratio increment, the final result achieved was a histogram of fuel used against optimum ratio (Fig 2). Ignoring the small percentage when the torque converter was stalled, it can be seen that the gear range could be spread from 3.6 to 0.3.

This wide range and high overdrive is a typical result for this sort of computer exercise, but how real is this requirement?

The fuel consumption variation is a fairly shallow minimum. Fig 3 can be imagined as being a cross-section through the specific consumption contours taken along the constant power locus shown in Fig 1. If the overdrive ratio were to be limited, the increase in fuel consumption would be quite small until the state point got a considerable distance from the minimum. Allied with the fact that the excessive overdrives are only demanded occasionally, one finds that limitation of the ratio to a practical value causes only a slight increase in consumption. The calculations on which Fig 2 was based were repeated for a number of cases in which the available overdrive was successively restricted. The increase in total fuel used was evaluated and plotted as Fig 4. Limiting the overdrive to 0.7 only increased the fuel used by 2 per cent.

Similar considerations apply to the ratio steps. If you have finite steps instead of a continuously variable ratio, any departures from the ideal being equally distributed, the fuel consumption increases very little until you get to quite substantial steps such as 1.4:1, as may be seen in Fig 5.

3 GEARBOX TYPE

At present, passenger cars use gearboxes of two types - manual gearboxes, involving layshafts and synchromesh engagement of gears are the cheapest. Automatic gearboxes generally have a torque converter and a smaller number of ratios derived from epicyclic gearsets using friction bands or clutches. These are more expensive.

Stepless transmissions are the subject of much development work but the market penetration is, as yet, small. Continuously variable transmissions of the Perbury drive or belt drive types are reasonable efficient in the transmission element itself but usually require high clamp loads to transmit the torque and big displacements to change ratio, which leads to the need for a high pressure, large capacity pump for the control circuit. The power losses in such a pump can exceed the 2 per cent savings available in going from a stepped transmission to a continuously variable one.

It is, therefore, proposed that fuel economy requirements can be well satisfied by a gearbox with a number of fixed ratios and reasonable ratio steps. A six-speed box with a ratio step of 1.4 can cover the range of 4:1 to 0.7:1 very adequately.

In choosing the type of mechanical gearbox to use, economy in manufacture must be considered as well as the fuel economy in use. Torque converters make for pleasant driving but they degrade the fuel economy. Lock-up clutches add to the complication and you cannot have the advantages of economy and shift quality simultaneously. Epicyclic gearboxes tend to become increasingly complex as the number of ratios is increased, so a simpler alternative was sought. This leaves us with a synchromesh transmission.

Today's novelty is micro-processor control, and this makes it possible to take up from rest on a clutch and to control the gear-change sequence automatically. By modifying the box to use two clutches, power-on gear-changes are made possible.

The design of a layshaft box makes it relatively easy to add extra ratios. This concept appears best to satisfy all the requirements:
large ratio spread, small ratio step, fully automatic, low parasitic losses and low cost.

4 MECHANICAL DESIGN

This paper is mainly concerned with the basis for a transmission design, and details of the type of transmission I have just mentioned were well described by Mr. Webster in his paper to the S.A.E. Congress in 1981 (1). Nevertheless, it may be helpful to give a brief description here, as a basis for the discussion which follows.

Fig 6 shows the basic design of a four-speed box for a transverse engined vehicle. It can be seen that the gears have been divided into two groups and re-arranged in position, and an extra clutch provided to drive the second group. To this is added a hydraulic system, pump and actuators, to select the gears and operate the clutches, under computer control.

When drive is selected, the clutch is disengaged and first gear engaged. As the engine speed rises in response to an opening of the throttle, the dry clutch progressively engages. When the vehicle speed for changing into second gear is reached, second gear is also engaged. Then the wet clutch engages and the dry clutch disengages simultaneously in a controlled manner to give smooth transition of the drive to second gear. First is then disengaged. The drive alternates between the clutches as it changes into third and top gear.

Fig 7 shows a possible layout for a six-speed gearbox for a front-engined, rear-wheel drive vehicle.

5 COMPUTER PREDICTION

When the general design of the gearbox has been accepted, parameters such as actual ratios and gear-change speeds have to be optimised and, for this, a computer simulation of the vehicle is very helpful. The data required include the details of the vehicle, details of the engine performance in terms of part-throttle torque curves and specific fuel consumption mapping, gearbox details and, finally, driver demand, which will probably be standard cycles such as EPA or ECE15.

The program goes through the cycle second by second. At each interval it determines the vehicle speed and torque necessary to overcome drag and provide acceleration. Assuming the previous gear, throttle opening is determined and then the shift pattern is checked to see whether this will induce an upshift or downshift. Having determined that the correct conditions apply, the specific and actual fuel consumption is calculated for that second. Due allowance is made for inefficiencies and losses and, when the cumulative fuel used is calculated, it is also subdivided into that due to each cause, e.g. acceleration, drag, pump loss, etc.

The fuel used can also be considered as subdivided in another way. Remembering Fig 1 with its locus of constant power, any given driver demand could be met by a particular minimum fuel consumption if an ideal ratio had been chosen. This ideal may be modified by other factors, such as a minimum engine speed. By comparing the actual fuel used with this modified ideal it is possible to compute the amount of fuel wasted due to imperfect choice of ratio. This is a very useful measure when, for instance, comparing an automatic gearbox with a manual, and shows how much of the available gain has been achieved.

Table 2 reproduces actual computer print-out for two cases. This is not the second by second analysis of the cycle one is sometimes given but a summary of the final result. It is designed to be intelligible and to require no further processing before an assessment can be made.

To those not familiar with such an analysis, the results will not be surprising except, perhaps, for the amount of fuel used when the throttle is closed. This is designated 'tick-over and over-run' and amounts to 28 per cent in this particular cycle, which simulates urban driving.

The overall consumption figure of 8.62 litres/100km for the automatic transmission compares favourably with 9.19 litres/100km for the manual transmission. This reflects the advantage of the automatic control where the part-throttle gear-change points can be selected for best economy. The effect is particularly demonstrated by the significant reduction in the fuel wasted term from 14.6 per cent for the manual to 7.6 per cent for the automatic.

Practical confirmation of these results was obtained on a different vehicle, for which in urban driving a six-speed automatic returned 8.7 litres/100km compared with 10.3 litres/100km for a four-speed manual gearbox.

6 MEASUREMENT OF LOSSES

The reduction of gearbox losses is always an important factor in improving fuel economy. It is not enough to have chosen the best concept in terms of minimum hydraulic control power and avoiding hydro-dynamic components with slip losses. It is still necessary to reduce residual losses due to other causes such as oil churning or seal friction. Thus, the purpose of measuring the losses is two-fold:-

(a) To identify what aspects of the design can be improved to reduce the loss and quantify improvements made during development.

(b) To provide the necessary data for the computer prediction already discussed.

The actual measurement of efficiency is always difficult. If it is done directly by measuring speed and torque at the input and output, the losses are going to be calculated as a small difference of large numbers and the accuracy is degraded. A method has been proposed of measuring the losses only as heat rejected by using a water jacket (2). This appears capable of high accuracy but there are long time constants associated with the thermal mass and it may prove tedious in a development context.

A large part of the development objective can be met by motoring the gearbox without output load and measuring the drag reaction. This permits the determination of all the speed dependent losses and probably includes all those where improvements may be made most easily.

An illustration of these is shown in Fig 8 for cold drag loss due to churning and seal friction. From these results it is possible to decide on a filling level that does not cause unacceptable churning whilst maintaining adequate lubrication. It points the way to further improvement, possibly by using spray lubrication and a lower level.

7 CONCLUSION

This paper has shown what a wide spread of information and subsequent analysis is called for in arriving at the specification of a transmission and its controller that will be well matched to the vehicle. Performance and economy compete when deciding on ratios and gear change strategy, and a computer simulation is a convenient way to establish the best compromise.

The information needed includes design data of the vehicle and detailed performance and fuel consumption data for the engine. Recorded data of actual vehicle use is also required. The test cycle chosen for the simulation will depend on whether the transmission is intended for the American or European market.

The work involved in theoretical analysis is considerable, but less than the practical testing of many variants, and provides a sound basis for the transmission specification. Experimental work is, however, vital to refine the mechanical details and improve all aspects of performance, with efficiency particularly important to overall economy.

Bringing together the best choice of concept, the careful matching to the vehicle and detailed mechanical refinement, a fully automatic transmission with a better fuel economy than a manual transmission has been achieved.

ACKNOWLEDGEMENTS

The author thanks the Directors of Automotive Products plc for permission to publish this paper and acknowledges the inspiration derived from all colleagues who have participated in the work on the gearboxes described.

REFERENCES

(1) WEBSTER, H.G. A Fully Automatic Vehicle Transmission using a Layshaft Type Gearbox. SAE Paper No 810104, SAE International Congress and Exposition, Cobo Hall, Detroit, Michigan, February 23rd - 27th, 1981.

(2) KATOH, H. et al. Exact Measurement of Power Loss in Automotive Transmission and Axle for Fuel Economy. SAE Paper No 830573, SAE International Congress and Exposition, Detroit, Michigan, 28th February - 4th March, 1983.

Table 1   Joint Probability of Speed and Torque as Percentages of Total Time

Gearbox Output Torque (Nm)	20	40	60	80	100	120	140	160
400–500	0.2	0.2	0.1	0				
300–400	0.3	0.5	0.9	0.8	0.1			
200–300	0.5	0.8	1.2	2.0	2.3	1.3	0.1	
100–200	0.9	2.0	3.0	4.0	3.5	5.2	2.8	2.4
0–100	3.6	3.6	6.2	5.4	7.9	35.9	1.2	1.3

Road Speed (km/h)

Table 2   Fuel Consumption Program Print-out, ECE 15 Duty Cycle

<u>Transfer Gear Type Automatic - Six-speed - Dry Clutch</u>

Overall consumption: 8.62432 litres/100 km

Calculated Fuel Requirement: .0857016 litres

which may be apportioned as follows:-

Fuel needed	Percentage	Litres
Overcome combined drag	31.8076	.0272597
Accelerate mass	23.6814	.0202953
Accel. trans. inertias	.676549	.000579814
Net accel. of engine	3.02287	.00259065
Drive gearbox oil pump	3.39508	.00290964
Allow for gear losses	6.51658	.00558482
Tick-over and over-run	28.004	.0239998
Take-up and slip	1.68559	.00144458
Gear-shift energy losses	1.21086	.00103773
	100.001	.085702

Fuel wasted: .00652111 litres, i.e. 7.60908%

<u>Four-speed Manual   -   Standard Vehicle</u>

Overall Consumption: 9.19545 litres/100 km

Calculated Fuel Requirement: .091377 litres

which may be apportioned as follows:-

Fuel needed	Percentage	Litres
Overcome combined drag	40.6904	.0371816
Accelerate mass	22.4313	.0204971
Accel. trans. inertias	.647158	.000591354
Net accel. of engine	2.84658	.00260112
Drive gearbox oil pump	0	0
Allow for gear losses	5.42476	.00495699
Tick-over and over-run	26.2646	.0239998
Take-up and slip	1.6956	.00154939
Gear-shift energy losses	0	0
	100	.0913774

Fuel wasted: .0133422 litres, i.e. 14.6012%

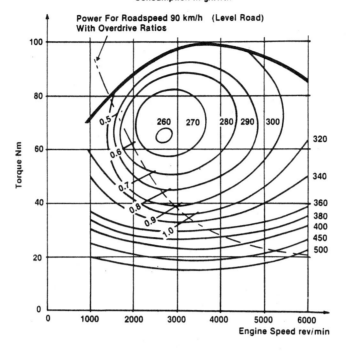

Fig 1   Specific fuel consumption for a typical 1.3 litre engine

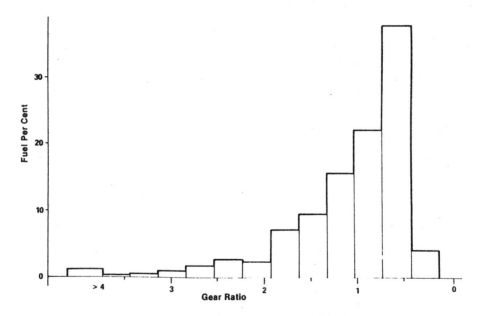

Fig 2   Fuel used versus gear ratio

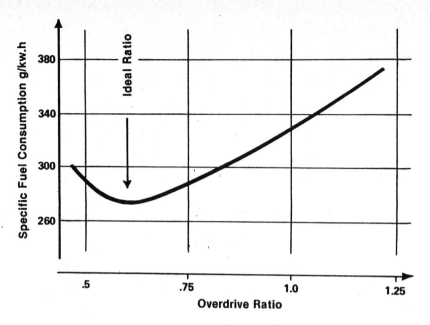

Fig 3 Specific fuel consumption versus overdrive ratio at a fixed power

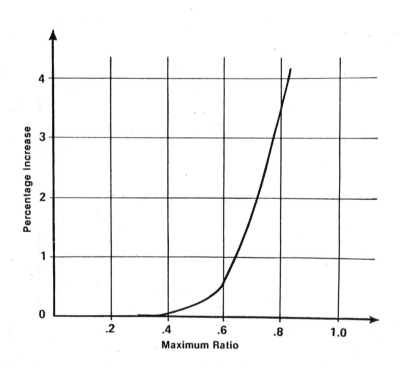

Fig 4 Increase in consumption due to limitation of overdrive ratio

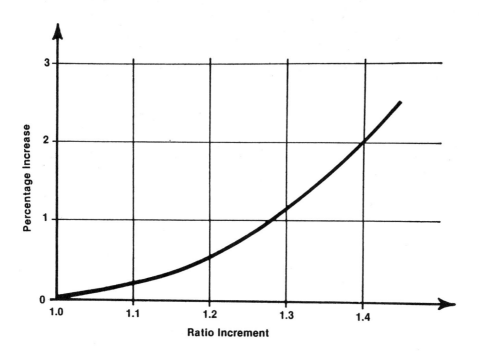

Fig 5   Increase in consumption due to ratio increment

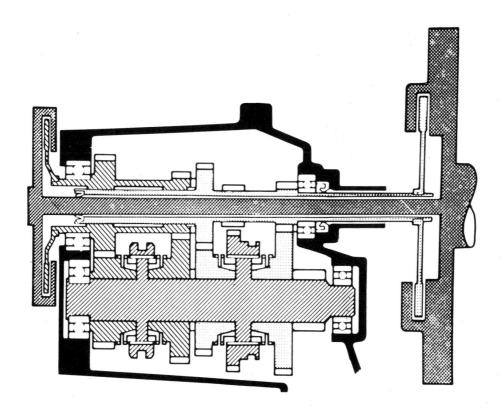

4 2 3 R 1

Fig 6   Typical layout for a four-speed gearbox used with a transverse engine

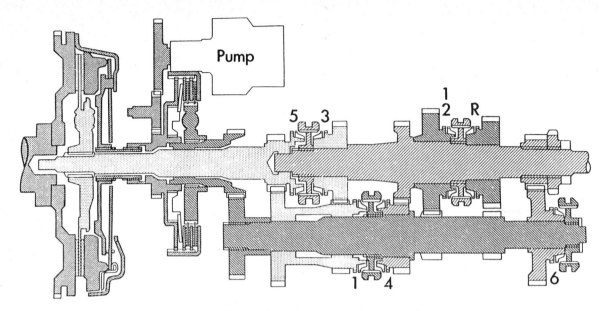

Fig 7 Typical layout for a six-speed gearbox used with an in-line engine

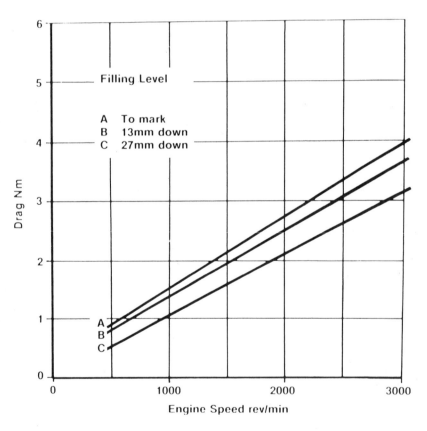

Fig 8 Gearbox drag versus engine speed for different filling levels at 27°C with a gear ratio 1.87:1

C420/84

Mercedes-Benz automatic transmissions for passenger cars

H J FÖRSTER, Dr-Ing, FSAE
Formerly with Daimler Benz AG, Stuttgart, West Germany

SYNOPSIS Transmissions have a great influence upon the operating mode of the engine and by this also upon car performance and fuel consumption. The determining factors are analysed and reassessed and recommendations for the transmission design are given.

INTRODUCTION

The continuously increasing demand for automatic transmissions and the necessity to provide adequate production facilities gave us in the early 70's the chance to develop a complete new range of automatic transmissions for our passenger cars.

Because such an opportunity does not arise too often, we started with a new evaluation of the transmission's influence on power, performance and economy, the results of which, together with the description of the new transmissions MB-W4A 040 and MB-W4A 020, are presented.

ANALYSIS

The new transmissions had to cover the full model range with the data of table 1:

- Power range from 50 kW to 200 kW
- Engine torque range from 120 Nm to 450 Nm
- Engine speed range up to 7000 RPM
- Vehicle mass fully loaded from 1600 kg to 2200 kg, trailer 1200 kg to 1500 kg
- Standard arrangement of the power train: engine in front, driven rear axle
- Lower weight, smaller dimension, lower cost

Table 1, layout data for automatic transmission

Since we built the first automatic transmission in the 60's, we hold in our company a basic philosophy regarding transmissions which, once again, we had to adhere to:

- The engine power has to be made available to the greatest possible extent as propulsion power for the vehicle.
- Special operating conditions, even if rarely encountered, have to be taken into consideration.
- Good performance and fuel economy have to be combined in such a way that the result will be accepted by the driver and thus be effective in real traffic.

- Foolproof automatic control with the possibility to be overriden by the driver.

Because the new compact Mercedes, model 190, was already under development, high emphasis was put on low weight and small cross-section dimensions to narrow the distance between the two front seats. The necessity for 2 sizes of transmission has been acknowledged at an early stage of the investigation. The 2 versions, however, should be interrelated in production as far as possible.

Of course we again began with a study about the ability of CVTs to meet the various demands. After a thorough detailed investigation of CVTs available, their possibilities and their restrictions, we decided, however, finally again in favour of the 'classical' version, that is converter plus multistep gearbox. Besides the highly valued absence of gearshifting in the CVT, no other advantages over the stepped gearbox were found, but many drawbacks regarding efficiency, adaptability to a large power range, weight, bulk volume and cost became evident.

In my opinion, neither car function nor fuel saving demand a CVT.

There may exist design concepts like the front drive with transversally mounted engine and transmission where a large distance between two shafts has to be bridged, where special CVTs offer a favourable design.

With the standard arrangement of engine, transmission in the front and the rear axle driven, that is to say our concept, the greatest objection against CVTs, however, is the negative answer to the question: Is it reasonable to take a transmission type which has to work hard all the time when there do exist other transmission types, which do not have to operate at all for more than 90 % of the time? (The CVT we favoured in this study was of the chaindrive type because of its rather good efficiency).

The basic decision in favour of the converter-multiratio-transmission taken, we concentrated the further analysis upon its potentiality of further improvement.

In transmissions with torque converter the output torque is the product of the converter turbine torque multiplied by the ratios of the gearbox. Therefore not only the amount of the overall torque ratio but also its allotment to converter and gearbox have to be selected.

The hydraulic torque converter is of course a CVT by itself. However, besides its unsurpassed qualities as regards smoothness, reliability and autocontrol of its torque ratio, its efficiency is not good enough to be accepted for the whole working range. Therefore, the well-known 2-phase converter is preferred for use with automatic transmissions for automobiles. Most of the torque multiplication is then done in the mechanical gearbox. The converter substitutes normally only one speed, seldom more.

We finally decided to dispense altogether with the contribution of the torque converter to torque multiplication during operation and retained it only because of its qualities as starting help and vibration damper. Whereas such a decision will enlarge the necessary range of the gearbox, with the possible consequence of more speeds, it gives the freedom of selecting the torque converter according to these starting conditions alone.

The significance of this is shown in figures 1, 2 and 3.

Fig. 1 gives the characteristics of two different torque converters.

At converter A emphasis has been put upon its converter phase which shows a higher torque ratio μ and higher efficiency η over most of the speed ratio ν of turbine to pump. Converter B on the other hand has a much higher power factor λ, especially in its coupling phase.

These two converters will be matched with the engine, fig. 2. Because this has to be done with the power available at the converter pump, the power demand of all the auxiliary drives like air conditioning compressor and oilpumps for power steering, hydropneumatic suspension and of the automatic transmission as well has to be subtracted. Therefore, the power and the efficiencies indicated in fig. 2 are lower than the genuine engine values. Fig. 2 is based upon the 5 litre SI engine with fuel injection which develops a maximum power of 170 kW at an engine speed of 5000 RPM. After the deduction of 15 kW for all the auxiliary drives, the available power of 155 kW and the speed of 5000 RPM is set as unity in the following graphs.

Consequently, following the basic ideas, converter B has a larger diameter than converter A (290 mm to 268 mm). The stallspeed of converter B is as low as 1900 RPM compared with 2300 RPM of converter A. The main difference, however, is the very small converter range of converter B compared with converter A, with the result that converter B has already reached the efficiency of 0.9 when converter A starts working. When converter A (coupling phase) has reached 90 %, converter B works already at 97 %.

Fig. 3 shows the consequences for output power-efficiency and power losses measured at the turbine output. The higher output of converter A in the converter range has to be paid for by higher losses and lower efficiency. The power losses at full load are the product of the higher input power of converter A and its lower efficiency at a given speed, shown in the lower part of fig. 3. The power losses at road load are proportional to (1 - η), because the same speed demands the same output power. The differences between the two layouts are particularly great in the coupling phase, this range working more than 90 % of the time. I see a further advantage of a layout with regards to converter B in the fact that the decision to introduce a lockup clutch, which eliminates the smoothness of the hydraulic drive, can be postponed. It is my personal opinion that the lockup clutch should be kept and stay in reserve as long as possible as the 'last battalion' for the fuel-saving battle. We therefore decided on a layout shown with converter B, without lockup.

After relieving the torque converter from the task of multiplying torque for the driving conditions, the gearbox alone has to fulfil all the demands.

There exists a very simple equation to determine the necessary overall ratio I of the mechanical gearbox. With α for the grade angle, f as coefficient of the rolling resistance, v_o the maximum possible speed, $(P/G)_{eff}$ the specific power available at the wheels and φ the speed factor, the overall ratio of the gearbox I is given by

$$I = \frac{(f + \sin\alpha) \cdot v_o}{(P/G)_{eff} \cdot \varphi} \approx \frac{\tan\alpha \cdot v_o}{(P/G)_{eff} \cdot \varphi}$$

I is proportional to the climbing ability and the maximum road speed and inverse proportional to the effective specific power and the speed factor φ. The speed factor $\varphi = n/n_o$ relates the speed of the transmission input in top gear, at which the drag curve of the vehicle (transferred to this point) equals the maximum engine power, to the engine speed at maximum power.

$\varphi = 1$ means that the maximum speed can be driven.

When the vehicle climbing ability is determined by the working point of the maximum engine power, the additional torque produced by the engine elasticity and/or by the torque converter remains as a torque reserve for start-off-acceleration.

In the layout of a planned vehicle maximum power P_o, the weight G and the maximum speed v_o are given. Then only the 2 factors: the climbing ability $\tan\alpha$ and the speed factor φ remain to influence the overall ratio I of the transmission.

THE FACTOR TAN α

When planning for normal operation at sea level alone, engineers are easily prepared to sacrifice a little of high climbing ability. Particularly in high-powered vehicles the climbing ability seems to be so abundant, that high acceleration has a higher priority. The differences between different cars are great. There exists a much better criterion to determine the overall transmission ratio I, based upon critical driving circumstances. We found that the crucial working condition is crossing the Alps, when overloaded cars are towing their caravans over passes (Großglockner road e.g.). Therefore, we selected the total torque ratio I to meet these conditions. This has the further advantage that the conditions are about the same for all vehicles.

For this purpose the specific power $(P/G)_{eff}$ has to be carefully calculated, considering all the power-reducing factors like accessory drives, losses in the power train and the drop of engine power due to decreasing air density with increasing altitude. Few engineers realize the importance of these effects. Average air density decreases by about 10 % per 1000 m (valid only up to 3000 m); however, it is the indicated engine power which suffers by this. All power demands for engine friction, auxiliary and accessory drives remain about the same.

In our case we have an indicated power of about 196 kW. The indicated power at the altitude of 2000 m is then

$$Pi,_{2000} = 0.8 \cdot Pi_o = 157 \text{ kW.}$$

The effective power at this height then amounts to

$$Pe_{2000} = 116 \text{ kW} = 68 \text{ \% of } Pe_o$$

With the same efficiency of 90 % between the engine and wheels from the 170 kW of the engine at sea level only 104.4 kW remain to drive the vehicle in 2000 m altitude. As regards the total weight which has to be moved, it is composed of the fully laden vehicle (if not overloaded) plus the weight of the caravan.

In our case the vehicle mass fully loaded and the caravan mass amount to

```
   2140 kg = 20 993 N
+  1500 kg = 14 715 N
   3640 kg = 35 708 N
```

Taking into consideration these extreme conditions, the climbing ability can be reduced to the values existing in reality, e.g. 12 % $\cong \sin \alpha = 0.12$. The coefficient for rolling resistance is 0.015 and for the first draft φ is set to unity. The maximum speed at sea level is $v_o = 62.2$ m/s $\cong$ 224 km/h.

Therefore the effective specific power at 2000 m is only

$$(P/G)_{eff} = \frac{104\ 400}{35\ 708} = 2.92 \text{ W/N}$$

The overall ratio I then is

$$I = \frac{0.135 \cdot 62.2}{2.92} = 2.88$$

Similar calculations have to be done for all the car models covered by the two transmissions.

THE SIGNIFICANCE OF THE SPEED FACTOR φ

When the overall ratio of the gearbox is calculated with a speed factor $\varphi = 1$, the whole operating range from the steepest grade chosen, up to the maximum possible speed v_o is covered. However, as has been well-known from the very beginning of automobile technology, in that case the engine will not work at its best efficiency. This can be seen in fig. 4, which is the diagram of a modern SI engine, standardized to the point of maximum power (index 0), in which curves of constant efficiency are drawn. Efficiency and specific fuel consumption are related

$$b_e = \frac{84}{\eta} \text{ g/kWh.}$$

The operating line $\varphi = 1$ is far away from the best efficiencies indicated as η_{opt}. Fig. 4 represents the working conditions with converter locked.

When the converter is always working, this diagram has to be modified by the converter diagram, which is done in fig. 5. This map represents turbine power versus turbine speed and the efficiency curves include the losses in the torque converter. A comparison of figs. 4 and 5 shows that the deterioration of the engine efficiency by the converter remains small over a large part of the whole diagram, the shape of the η_{opt} line has changed within the converter range.

Now the speed factor φ has been varied from unity in both directions and the corresponding operating lines, which are approximately cubic parabolas, have been plotted. They represent the drag curve of the vehicle on flat terrain at sea level transformed to the transmission input = turbine output, values of φ smaller than unity approach the operational line to the optimum efficiency, values greater than unity remove it. (The driveline losses are added to the drag power of the vehicle.)

The results are to be seen in fig. 6, where for the selected values of $\varphi = 0.7; 0.85; 1.0; 1.2$ the efficiencies are plotted versus speed, again standardized with maximum speed. To demonstrate also the effect of the working converter, the efficiencies for driving at road load are given for the locked up condition (fig. 4) as well. The efficiency is increasing with decreasing values of φ. $\varphi = 0.7$ is close to the optimum. The negative effect of the working converter is small but increases with smaller values of φ. At $\varphi = 0.85$ 1 percentage point, at $\varphi = 0.7$ 2 points are lost. The absolute maximum possible (in theory) is the η_{opt} line without any losses. With $\varphi = 0.7$ the maximum speed is reduced by about 8 % the operation limited in the range of the low speeds to $v/v_o = 0.3$. The curves indicate the fuelsaving reserve which could be mobilized by a lockup-clutch or smaller values of φ. The higher losses, also of the best CVTs known, unfortunately always overcompensate for these theoretical gains. If $\varphi = 1$ is taken as reference line, about 10 to 15 % fuel can be saved by using the 0.7 line, or 7 to 10 % will be lost by selecting $\varphi = 1.2$.

There remains the question why so many studies promised high fuel savings by operating the engine in the right way, and praised the high fuel-saving potentialities of CVTs. When built and actually used, the fuel-saving effect was mostly much smaller than anticipated and hoped for. There exist 3 reasons for such a discrepancy:

- The SI engines, on which the studies were based, often showed a dramatic deterioration in efficiency when operating outside the optimum, which is no longer the case with modern SI engines (but still bad enough).

- The losses of CVTs with a wide range have very often been underrated.

- However, there is a second question to be answered:
  If the fuel-saving effect of layouts with $\varphi = 0.7$ is well known since many decades, why has it not been applied and is apparently rediscovered in our days? The answer lies in the driver's acceptance of such a device. The significance of the last point can be understood by fig. 7 which shows the reserve power available without any ratio changing just by opening the throttle (shown only with converter locked). It can easily be seen that this power reserve for instant acceleration and climbing drops drastically for only a small saving on fuel. Regardless of the maximum power of the engine, the performance of the car becomes poor and gives the feeling of a lazy car. Because frequent gearshifting by hand is inconvenient and tiresome, drivers use the overdrive ratios less often and the fuel consumption is often greater than before. That means for defining the factor φ and with it the operation line:

> Approach the optimum efficiency as far as possible while maintaining an acceptable power reserve for acceleration and climbing. This decision has to be made anew for every car-engine-transmission combination.

In the example shown in figs. 4 to 7 the operation line $\varphi = 0.85$ represents such a compromise for high-power cars. Fuel consumption is close to the optimum and there remains an acceleration reserve of about 1 m/s².

After we had calculated the overall ratio I and chosen an appropriate speed factor φ for each model, we determined the power range and the ratio of the two transmissions to

W4A 020   50 ÷ 100 kW   I = 4.3
W4A 040   100 ÷ 200 kW   I = 3.6

In doing so, we were fully aware that such a decision made a 4-speed transmission compulsory also for the high-power vehicles, but it was still possible to cover the enlarged range of the smaller version with 4 speeds, too.

TOP GEAR DIRECT OR OVERDRIVE?

After the overall range has been fixed and also the number of gears, necessary to cover the range, has been chosen, the ratios of the steps must be selected. In doing this, it is of great importance to decide whether the top gear should be direct or have an overdrive ratio. Being fully aware that all (or most) of our competitors favour the overdrive version, we nevertheless did decide on a 4-speed gearbox with direct top gear. There are 3 reasons for this decision:

- All calculations, and all tests in the field as well, show that passenger cars drive 90 to 95 % of their time in top gear. The actual percentage depends on the specific power of the car, the topography of the terrain it is operated in, and the share of stop-and-go. If the drivetrain arrangement makes it possible it seems only reasonable to us to have no gear working and running in the transmission for more than 90 % of the time in order to avoid losses, noise and wear.

- Reduction of power losses, which arise from 3 sources when using overdrive:
  The almost permanent running of the planetary gearset forming the overdrive ratio, together with the losses of the disengaged direct clutch and the freewheeling unit.

- The higher speed of the propellershaft (ventilation, joints, bearings) and the greater ratio in the final drive of the rear axle.

According to our calculation and measurements these additional losses may easily reduce the efficiency in top gear about 2 to 3 %. The direct top gear in the transmission demands a rather direct ratio of the final drive, which on the other hand gives the designer the chance to reduce the size of the rear axle housing.

Of course, if the transmission engineer is compelled to use the already existing 3-speed automatic and the old rear axle drive too, he has no other choice than to add an overdrive before or behind the old unit. But if there is a possibility, which we had, to begin with the centerline, the optimum design for standard drives is the transmission with direct top gear and ratio 1. With this the approximate gear ratios which should be obtained by the transmissions could finally be set.

The performance results are shown only for two extreme models, the smallest one, the 190 D with 53 kW and the largest one, the 500 SE with 170 kW.

Fig. 8 gives the acceleration (climbing) ability versus carspeed at sea level, the cars having their normal testing weight (2 passengers). It can be seen that the 500 can reach the same top speed in 3. and in top gear as well. Top gear is the economical gear with lower acceleration, 3. gear the power gear for performance. Shifting is done automatically up to about 130 km/h, then by kick down. With the gear ratios chosen and the characteristic of the engine a very good approximation to full power, hyperbola is reached.

The rather low top speed of the 190 D demands a speed factor $\varphi = 1$ to reach 158 km/h. The sacrifice in fuel saving can be accepted because the Diesel engine has a higher part load efficiency and less improvement can be gained by values of the speed factor smaller than unity. Also the small acceleration reserve in top gear recommends such a layout. The superiority of the 500 in speed, acceleration and climbing is evident.

Fig. 9 then compares the working condition at an altitude of 2000 m, both cars fully loaded and towing their caravan. On a 12 % grade starting and driving is possible for both vehicles. Because the gear ratio has been calculated with the point of maximum power, the converter helps to bring the 'tractor-trailer-train' into motion.

LAYOUT OF THE GEARBOX

There exist many arrangements of planetary gear sets to get 4 forward and one reverse speed, top gear direct, and many of them we analysed. In the end we found that the principle of our previous 4-speed automatics - in production well over 10 years - has many advantages regarding efficiency, building gear ratios and simplicity. Fig. 10 shows this principle with the symbols of Wolf where each circle represents a planetary gear set, the 3 lines are the 3 shafts which are connected to each other as shown.

- 2 groups of planetary gear sets, which form first and second speed by multiplication of the two single ratios.
- In third speed the first group is blocked and only one planetary gear set is working.
- Both groups are blocked in top gear.

The principle of the ratio-forming arrangement, however, was the only thing left from the old design, because better devices to fulfil the different demands were developed. Fig. 11 shows schematically the final solution.

The 2 groups, the first a Ravigneaux set and the second a simple planetary gear set are side by side with very short connections. The turbine wheel drives the rear sun gear of the first group. In first gear the ring gear of the first group and the sun gear of the second are held stationary by the same brake B2. Clutch K2 and/or the freewheel unit are engaged. In second, clutch K2 is disengaged. With brake B1 and brake B2 applied, the front sun gear of the Ravigneaux set and the sun gear of the rear set are grounded. In third gear brake B1 is opened, clutch K1 engaged. The first group is blocked and the ratio is formed only by the rear set. In top gear in addition to clutch K1 also clutch K2 is engaged, ratio 1. In reverse, brake B3 grounds the carrier of the Ravigneaux set and at the same time the ring gear of the rear set which are interconnected. Clutch K2 is engaged which makes the sun gear the input into the rear set.

MB AUTOMATICS W4A 020 AND W4A 040 AS DESIGN

In fig. 12 both transmissions are shown in their final realization. It can be seen that both sizes are identical in design, identical in all length dimensions so that they can be machined on the same tooling facilities, but that the diameters are different. As a result the smaller version weighs about 10 kg less. Of great importance is the reduction of the housing diameter by 18 mm because it allows a smaller transmission tunnel so that the front seats could be arranged more closely together, which was of importance for the seating comfort in the compact Mercedes.

Because of the wide power range demanded for the 2 transmissions, 3 different sizes of torque converters with diameters of 0.245; 0.270; 0.290 m with very similar characteristics have been developed, the middle one of which is used with both transmissions. Both versions have a rear pump which is disengaged during normal driving.

The servo units are 2 multidisc clutches, 1 multidisc brake (reverse) around the Ravigneaux set and 2 brake bands arranged around the clutches.

After parallel development of an electronic control we finally decided again on a genuine hydraulic version. The reason is very simple. Whereas there is no doubt that the electronic solution permits more factors to influence the shift program and the power shifts, the remaining hydraulic circuit does not get much smaller. The electronic control, therefore, is an addition with solenoid-valves, outside wiring, boxes as well as connections and it increases the costs substantially. Therefore as long as a genuine hydraulic control can fulfil the demands it seems reasonable to keep it.

To cut production cost, the number of machined parts is reduced to the absolute minimum. Only the gear wheels and some shafts have survived. Great care was taken to avoid hollow shafts, parts which are so familiar from transmissions composed of various planetary gear sets. The only hollow shaft, the reactor hub, is an extruded part.

Many parts are made of diecast aluminium, fig. 13, like the main transmission housing, which forms one piece with the converter housing, the front cover with the piston of brake B3, insets and pistons of all clutches and brakes, the valve plates including all the valves (also in aluminium) and the end cover at the rear.

Butmost interesting are parts stamped from sheet-metal. The design not only incorporates many of this type but also some really sophisticated ones, fig. 14.

There is of course the torque converter where only the reactor wheel is of cast aluminium. Other stamped sheet-metal parts are: the carriers for the planetary sets, the splined hubs for the inner disks of the clutches K1, K2 and brake B3, the centrifugal parts of the governor and the oil pan. The most sophisticated parts are the housings for clutch K1 and K2, fig. 15, which at the same time form the drums for the corresponding brakes B1 and B2. Each of these is produced in 28 steps from the sheet, slots and splined included, fig. 16.

By this production technology many parts, though different in size, require only different dies for casting or stamping.

MINIMIZING PARASITIC LOSSES

This is the everlasting and never-ending battle for every transmission engineer. For further progress also minute steps have to be considered. The positive results concerning the losses in the converter have already been mentioned. With the speed factor of 0.85 the converter in its coupling phase works at road load mostly better than the 98 % efficiency line (fig. 5). Because the ventilation losses of the converter cannot be neglected at high speeds, and because they also increase with the 5th power of the diameter, it is important to adapt this diameter to different engine sizes.

The greatest 'contribution' to the mechanical losses of transmissions comes from the oil pumps. The secondary pump is, therefore, mechanically disengaged when not in use (most of the time) and the primary pump is further improved by reducing dimensions and by a better profile of the teeth. As already mentioned, in third gear (less than 6 % of all operating time) only 1 planetary set is working, in top gear (more than 90 %) everything is running blocked and the friction losses of the disengaged servo units (2 brake bands and 1 disk brake) are kept extremely low. Even the oil flow for lubrication has been reduced in top gear to reduce its share of the losses. Compared with their predecessors the mechanical losses are less than half the former values, fig. 17. The total transmission losses have to include the hydraulic losses of the converter (if not locked), which are most of the time under 2 % when driving at roadload. Transmission losses expressed in percentage, however, have no real meaning because only some factors are proportional to the input power.

Whoever is working in this field will appreciate the improvement.

Naturally all losses cost fuel. However, automatic transmission can help tremendously to improve fuel economy in actual traffic, because it will shift in the most economic gear ratio (an appropriate shift program assumed) whereas many drivers avoid shifting to overdrive and prefer a driving mode with high performance without shifting. In my opinion automatic transmissions are, therefore, a prerequisite to maintain a low fuel consumption in actual traffic (perhaps not during tests).

SUMMARY

A philosophy for the 'correct' matching of car, engine and transmission which also considers the feeling and the behaviour of the driver is demonstrated by the example of the Mercedes-Benz automatic transmissions. The ratio of the 2 transmissions, which have to cover the wide power range from 50 to 200 kW, is adapted to the different power/weight ratios of the extreme combinations. Both transmissions are of identical design but different in their radial dimensions. They have 4 speeds, with direct topgear and their design represents the most advanced production methods.

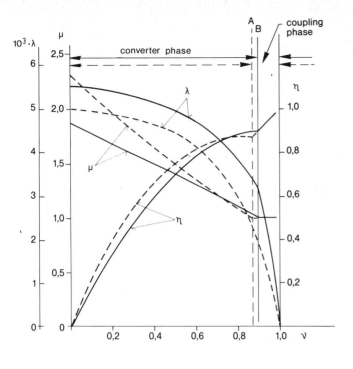

- - - converter A, emphasis upon converter phase
——— converter B, emphasis upon coupling phase

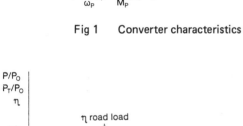

Fig 1   Converter characteristics

Fig 2   Engine map with the converter A and B from Fig 1; ratio of diameter $D_B / D_A = 1.08$

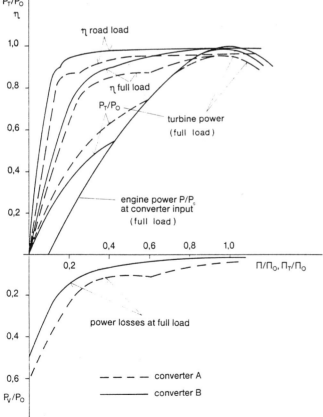

Fig 3   Power, efficiency, power losses at turbine output

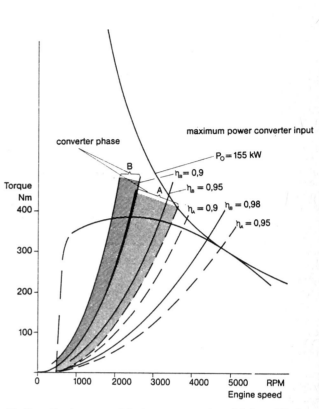

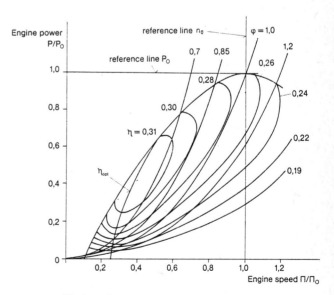

Fig 4   Engine power at converter input

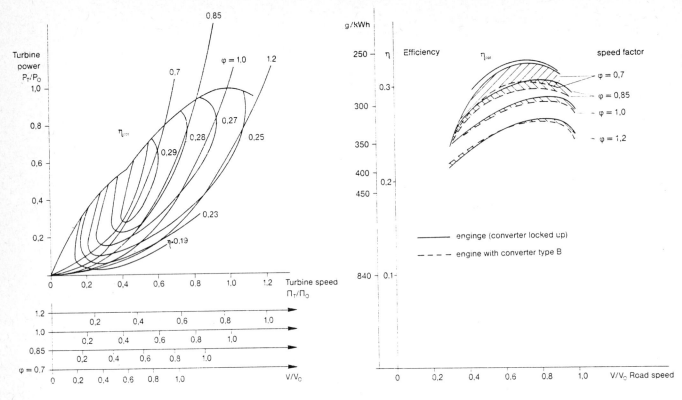

Fig 5 Engine + converter, turbine output converter B. Road load for various speed factors φ

Fig 6 Efficiency at road load; various speed factors φ

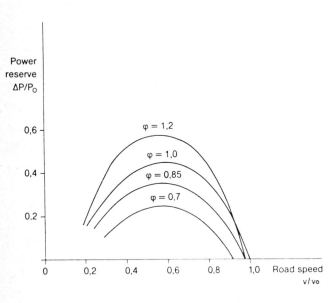

Fig 7 Power reserve in top gear for accelerating and climbing (without converter)

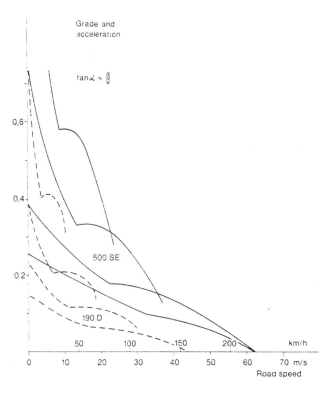

Fig 8 Driving conditions at sea level; vehicles with two passengers

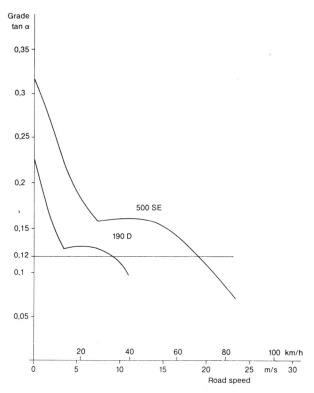

Fig 9   Driving conditions at an altitude of 2000 m vehicles fully loaded, towing a caravan, 1. gear

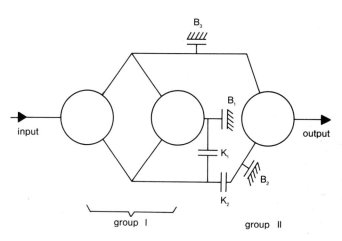

Fig 10   MB-Automatic, principle of planetary gear set arrangement shown in symbols according to Wolf

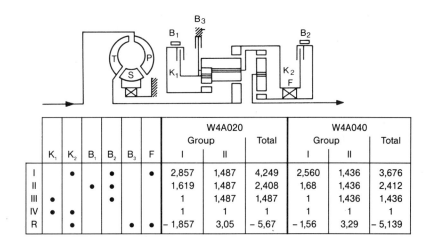

	K_1	K_2	B_1	B_2	B_3	F	W4A020 Group I	W4A020 Group II	W4A020 Total	W4A040 Group I	W4A040 Group II	W4A040 Total
I		•		•		•	2,857	1,487	4,249	2,560	1,436	3,676
II			•	•			1,619	1,487	2,408	1,68	1,436	2,412
III	•			•			1	1,487	1,487	1	1,436	1,436
IV	•	•					1	1	1	1	1	1
R		•			•	•	−1,857	3,05	−5,67	−1,56	3,29	−5,139

Fig 11   MB-Automatic W4A 020 and W4A 040, principle arrangement and ratios

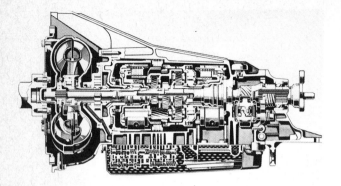

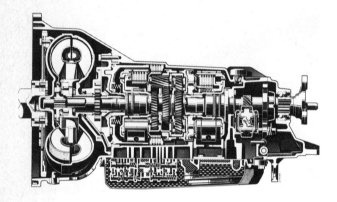

Fig 12   MB-Automatic transmissions W4A 020 and W4A 040

Fig 13   MB-Automatic, essential diecast parts
    1 Main case
    2 Front cover
    3 Reactor
    4 Thrust washer (converter)
    5 Clutch and brake pistons
    6 Spring retainers
    7 Connection cylinder
    8 Valve body
    9 Servo covers
  10 Servo pistons
  11 Rear pump housing
  12 Extension housing

Fig 14   Sheetmetal stampings MB-Automatic W4A 040
    1 Clutch housing K1
    2 Splined hub for the inner disks brake B3
    3 Splined hub for the inner disks clutch K1
    4 Governor
    5 Planetary carrier, Ravigneaux set
    6 Oil pan
    7 Splined hub for the inner disks clutch K2
    8 Rear planetary carrier welded to output shaft
    9 Clutch housing K2

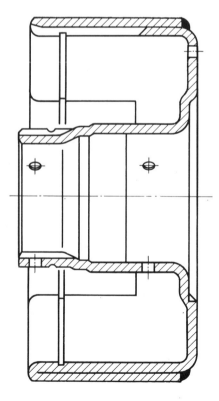

Fig 15   Clutch housing K2

Fig 16 Essential stamping steps in the production process of the clutch housing K2

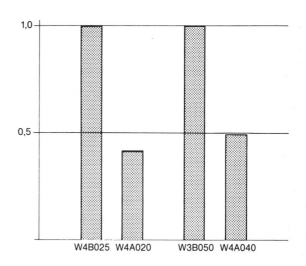

Fig 17 Reduction of mechanical losses of the new transmissions. The value of the predecessors taken as unity

C445/84

SAE 841305

The Ford research dual mode continuously variable transmission

T R STOCKTON, MSE
Ford Motor Company, Dearborn, Michigan, USA

ABSTRACT

The familiar continuously variable steel belt transmissions being introduced for automotive use are "straight" CVT's, where total speed ratio span is limited to capability of the variable speed belt and pulley system. Meeting start-up performance and gradeability requirements penalizes the final drive or axle ratio resulting in the transmission reaching the maximum overdrive ratio prematurely.

The Ford Research front wheel drive Dual Mode CVT employs a torque converter for start-up in addition to a full range of CVT operation as a means of improving both smoothness and fuel economy.

The torque converter and a separate chain drive power path from the engine to the variable speed unit are arranged in parallel. This combination provides an approximate 30% effective increase in ratio coverage to 7:1. Vehicle fuel economy is thus improved an estimated 5-9% during highway driving because the final drive ratio can be reduced 30%. Vehicle start-up through the torque converter is inherently smooth.

A 4 x 4 system is also derived from the base FWD, having a bevel gear power take-off near the vehicle center.

INTRODUCTION

This paper describes the application of a torque converter to an automotive CVT as a means of extending overall ratio range for improving fuel economy, while also providing excellent start-up feel. The limitations of speed ratio span of candidate variable speed belt and pulley systems without some scheme of ratio augmentation causes an area of vehicle operation where the transmission ratio remains in the max. overdrive position and consequently the engine operates at speeds above the optimum for fuel economy.

Transmission design and control strategies, with vehicle test results of the Dual Mode CVT concept have contributed to the understanding of trade-offs between fuel economy improvement and design factors.

SYSTEM CONCEPT AND KEY FEATURES

The Dual Mode CVT employs a torque converter for start-up in addition to a full range of CVT operation as a means of maximizing both smoothness and fuel economy. This is a two-axis system, shown schematically on Fig. 1. The torque converter and variable speed drive pulley are located on the engine centerline with the driven pulley, the planetary forward-reverse reduction gearing and differential on the drive wheel axis.

The design cross section (Fig. 2) is sized for 136 N-M torque capacity. The prototype powertrain assembly shown on Fig. 3 has been installed in a 1.6L Ford Escort for test and evaluation. Specifications for the dual mode CVT are listed below.

Ratios (N_1/N_0)

Chain	2.241
Belt	
Max. Underdrive	2.42
Max. Overdrive	.41
Range	5.90
Final Drive	
Forward	3.871
Reverse	3.452

N/V In Escort
(Engine Rev./Min./Vehicle Speed Km/Hr)

Converter Mode	80.03
Belt Mode	
Max Underdrive	86.44
Max Overdrive	14.76

Torque Capacity (Engine)   136 N.M.

Dimensions

Pulley Radii	33.27-80.52 mm
Center Distance	226.82 mm

Torque Converter
Stall Ratio   2.1
Stall K $\dfrac{\text{Engine Rev/Min.}}{\sqrt{\text{Engine Torque N.M.}}}$ 202

The steel "V" belt is an experimental development provided by the Morse Chain Division of Borg Warner Corp.

The torque converter chain drive power path is arranged in parallel with the variable speed unit. This combination provides an approximate 30% effective increase in ratio coverage to 7:1 overall for the transmission. The final drive ratio can be accordingly reduced 30% without compromising performance and gradeability.

Vehicle start-up through the torque converter is inherently smooth. Near the converter coupling point a power transfer clutch is engaged during acceleration, transferring power flow from the converter to the variable speed belt drive. The chain transfer drive ratio is set approximately equal to the pulley underdrive ratio so that at heavy-to-full throttle the transfer clutch is approximately synchronous with the input side of the variable ratio pulley drive, which is in its maximum underdrive ratio position. At light-to-medium throttle, the reduced converter speed ratio results in a small engine speed drop after the shift.

As vehicle speed increases above the power transfer point, the pulley system strokes from maximum underdrive towards maximum overdrive position. An overrunning clutch unloads the converter in this mode. It also permits a conventional forced downshift and a rapid change in ratio for passing maneuvers. Smoothness of the upshifts and downshifts is assured by the overrunning clutch on the chain transfer drive sprocket.

Hydraulic pressure to activate the three clutches, to provide the pulley clamping force and to charge the converter and lubrication system is provided by a variable displacement vane type oil pump, which is engine driven through a quill shaft.

CONTROL STRATEGY

The primary objective of the control system is to force the engine to operate at the most fuel efficient combination of RPM and torque for any power level. Figure 4 is a map of the 1.6L engine for the vehicle, showing a family of constant RPM lines plotted aginst B.S.F.C. and horsepower. The lowest B.S.F.C. point for each of these fish hook curves is identified by the corresponding throttle opening. Thus for any power demand level the control must seek the optimum RPM. The vehicle speed can vary at constant engine RPM to accommodate hills or accelerations. At closed throttle the engine speed drops to the max. overdrive value, and continues to drop with vehicle speed to idle RPM. The transmission then downshifts into converter mode.

Figure 5 shows the relationship between the 1.6L engine and vehicle speeds in an Escort for various throttle positions. N/V lines for the Ford ATX high gear and equivalent straight CVT in maximum overdrive are included for reference. Assuming a minimum engine speed of 1200 RPM, it can be seen that at road load condition, the dual mode CVT upshifts to maximum overdrive at 80 km/h. As vehicle speed increases the CVT gradually downshifts to meet road load power requirements. Note that the straight CVT reaches limiting overdrive ratio positon at a much lower vehicle speed.

During a typical start-up acceleration at 20° throttle, engine speed will rise to 2100 RPM and fall to 1700 RPM at the power transfer point. It will then hold constant RPM and power to 103 km/h. At larger throttle angles the engine speed change at the power transfer point diminishes toward zero, consistent with increasing converter speed ratio. At full throttle, engine speed in the CVT mode holds constant to maximum vehicle speed. At this point the variable pulley system has shifted from maximum underdrive to approximately 1:1 ratio.

It can be seen that the CVT is under full variable control at all vehicle speeds but that ratio range beyond that available would be of little value for this particular engine and vehicle combination.

In Fig. 6 the desired engine RPM vs. throttle opening is conveniently approximated by a straight line. Referring back to Fig. 4 the engine RPM is not overly critical for best economy and precise control of ratio in the belt mode is not necessary. This also simplifies control of the transfer clutch. As shown on Fig. 7 the clutch apply and release schedules are straight line functions of belt output pulley RPM, or road speed, and throttle angle. Above 70° throttle the forced downshift is made easily available for performance. The corresponding upshift line must be located slightly higher to maintain hysteresis between shifts. A broader hysteresis band exists at lower throttle angles, to minimize upshift-downshift business.

BELT LOADING

Belt clamping load is provided by hydraulic pressure acting upon each of the two moving pulley cones. The required belt loading to insure efficient operation and to prevent slip is directly controlled only on the lower, or output pulley. Clamping on the upper or input pulley is a resultant equilibrium force, which varies as a function of ratio and torque. These characteristics are shown on Fig. 8 for max. input torque, along with operating radii of the pulleys. Note that clamping load of the output pulley is higher than the input pullley in underdrive, but lower in overdrive. Ratio is downshifted by bleeding oil from the input pressure cavity. Conversely for upshifting oil is forced into the cavity until the desired ratio is reached. The oil quantity in the cavity is trapped during constant ratio. System line pressure must be held sufficiently above the ratio control pressure to provide adequate ratio change response.

CONTROL SYSTEM

Overall function of the transmission is electronically controlled. An Ultra PAC 500 computer was used to permit rapid reprogramming for developmental changes. The interface between electronics and the hydraulic control is comprised of four linear actuators with position feed back potentiometers, and four pressure transducers, with magnetic sensors for engine and output pulley speeds.

The hydraulic control schematic is shown on Fig. 9 and the electronic-hydraulic diagram on Fig. 10. System line pressure ranges from 552 kPa to 1930 kPa for the 1.6L engine. The linear actuators consist of a 12 v. DC motor which drives a ball screw to compress a spring to the height required. Spring load on the spool valves establishes pressure on the main regulator, belt load, and transfer clutch. The ratio control actuator serves to control flow in and out of the input pulley cavity, see Fig. 11. This system was selected at the outset because no suitable commercial actuators were available. Although numerous development problems were corrected, this type of linear actuator would not be considered for production because of high cost.

FUEL ECONOMY

At the time of writing this paper fuel economy testing was commencing and no results were available. Computer projectons indicated a 25% Metro and Highway improvement over the base Escort, with ATX automatic transmission. Compared with a straight CVT with no extension of ratio range beyond the variable speed belt and pully system, there was no Metro improvement but up to 9% Highway improvement. Due to uncertainty of the projections the level of improvement is not yet established.

CVT 4x4 DERIVATIVE

Historically, 4x4's have been RWD adaptations, but such systems penalize the complexity of a base FWD.

A modest cost 4x4 drive line is derivable from some transverse FWD powertrains. However, the "U" drive transaxles, such as the dual model CVT are much more favorable for packaging because the rear drive shaft can be conveniently located near the vehicle centerline.

A bevel gear set is integrated with the FWD final drive gearing to provide the longitudinal drive shaft take-off for driving the rear axle.

On the premise that all new FWD powertrain designs should consider 4x4 option, an "either-or" modification to the dual mode CVT was derived (Fig. 12). As a result the base transaxle would serve either 2x4 or 4x4 installations.

In the interest of minimizing parasitic drag losses, spiral bevel gearing operating in ATF is employed. This not only eliminates the gear tooth sliding losses inherent in hypoids, but avoids the need for heavy E.P. lubes.

The above system is potentially a cost/weight effective means of providing a 4x4 option for small transverse FWD powered passenger cars, vans, or pick-up trucks, and is compatible with various rear axle drive schemes.

A vehicle test bed installation is being evaluated.

CONCLUSIONS

1. The torque convertor is an effective means of extending the overall ratio range of a CVT.

2. A potential improvement in EPA Highway fuel economy results from reducing the final drive ratio of a CVT.

3. Extending CVT ratio range beyond the optimum value for a particular vehicle offers no benefit in fuel economy.

4. A modest cost and favorably packageable power take-off to drive the rear axle for a 4x4 option is deriveable from the Dual Mode CVT.

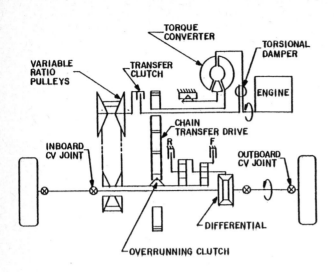

Fig 1 Dual mode CVT schematic

Fig 3 Dual mode CVT prototype with 1.6ℓ engine

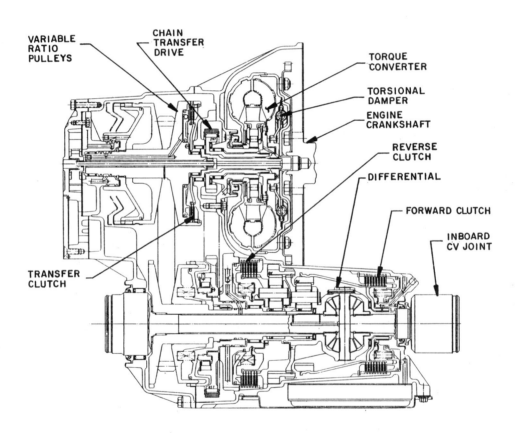

Fig 2 Dual mode CVT

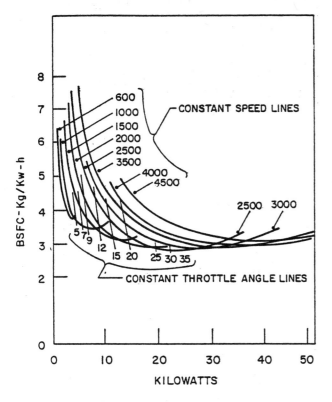

Fig 4 1.6ℓ engine BSFC versus power

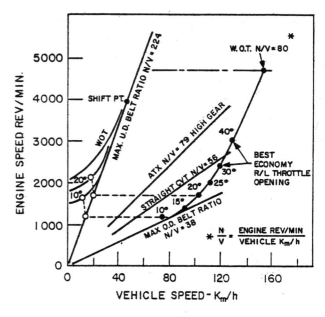

Fig 5 Engine speed versus vehicle speed

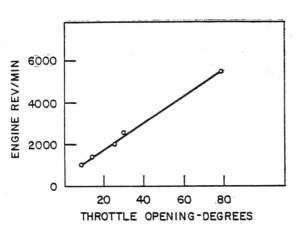

Fig 6 Engine speed versus throttle opening — CVT range

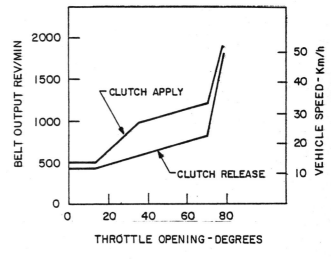

Fig 7 Transfer clutch control schedules

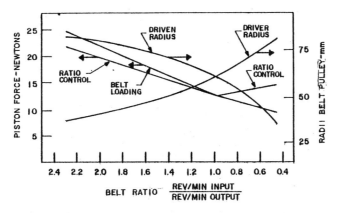

Fig 8 Belt and clamping forces — 136 Nm engine

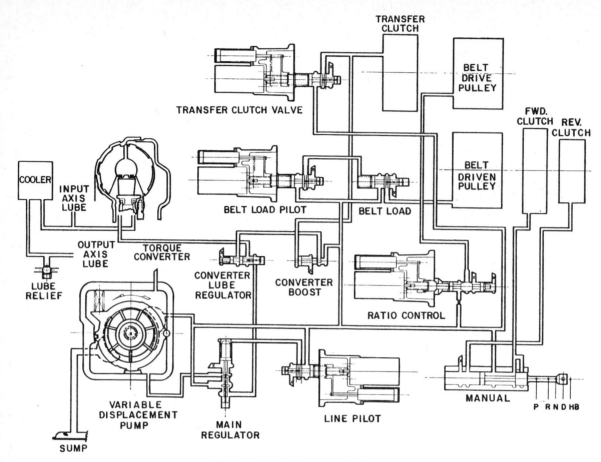

Fig 9 Hydraulic control schematic

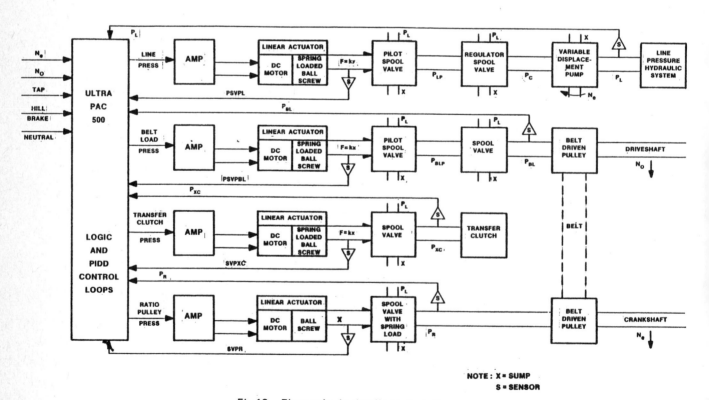

Fig 10 Electronic–hydraulic control diagram

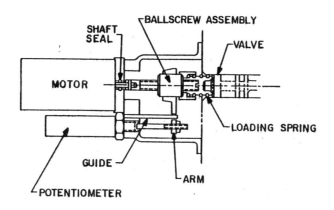

Fig 11 Actuator assembly

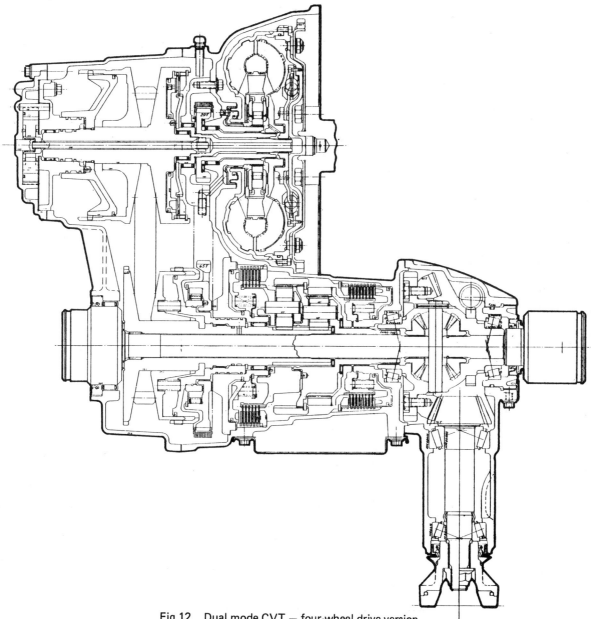

Fig 12 Dual mode CVT — four-wheel drive version

SAE 841306

The design of an engine-flywheel hybrid drive system for a passenger car

N A SCHILKE, MS, ME, A O DeHART, L O HEWKO, MS, EM, C C MATTHEWS, MS, ME,
D J POZNIAK, MS, ME and S M ROHDE, PhD
General Motors Research Laboratories, Warren, Michigan, USA

The quest to improve the fuel economy of General Motors passenger cars has led to the investigation of an engine-flywheel hybrid powertrain at the GM Research Laboratories (GMR). An engine-flywheel system was designed for a compact car and its performance was predicted analytically. The system was estimated to achieve an improvement in EPA Composite fuel economy of 13% over a 1984 production compact car. This margin of improvement was judged insufficient to justify the complex drivetrain, and, therefore, a prototype system was not built. However, the current potential of engine-flywheel hybrids for GM passenger-car applications has been defined, and the technology areas requiring additional research attention have been identified.

1. INTRODUCTION

Interest in using flywheel power for vehicle propulsion goes back many years. Oerlikon flywheel-powered buses were in service in Switzerland as early as 1953 [1]*. But the size, weight, complexity and safety aspects of flywheel power precluded wide-spread use, especially in passenger cars. The re-emergence of the flywheel as a possible vehicular energy source is attributable to technological advances in flywheel systems, continuously variable ratio transmissions (CVT's), and microprocessor-based control systems.

The incentive for considering the addition of a flywheel to a vehicle drivetrain is its ability to store and regenerate energy, thereby providing potential for fuel economy improvement. This is particularly true when driving on schedules with many stops, e.g., urban driving. Braking is an essential vehicle function, but the energy involved in stopping the vehicle is normally wasted. Additionally, the widely varying speed and load demands in urban driving result in fuel economy losses because the engine is often not operating at its best efficiency. As pressure increases for fuel economy improvements, so does the possibility of utilizing more complex vehicle drive systems for more complete use of the available energy.

Numerous government-sponsored studies have addressed the fuel economy potential of flywheel-hybrid vehicles [2-8]. GMR's renewed interest in these vehicles stemmed from analytical studies [9,10] and from in-house research which preceeded those studies. Conclusions from this research indicated that the potential gain in fuel economy for hybrid passenger cars could be as high as 20-30% [9]. They also indicated that the fuel economy of the hybrid powertrains is very sensitive to system losses and to the control strategy used for energy management.

All of the above experiences with flywheel-based hybrids suggested the need to establish the viability of practical hybrids in general, and the advisability of their use in passenger cars in particular. These considerations have led to the research summarized in this report.

2. DISCUSSION

The objective of the current project was to design an engine-flywheel-hybrid drive system incorporating the most advanced practical technology and to integrate that system into a compact car. The more global objective was to define the engine-flywheel-hybrid technology as related to passenger cars and as applicable to future General Motors products.

The design targets for the project are shown in Table I. The vehicle selected was a 1982 front-wheel-drive compact car, in the 3 000 pound EPA inertia weight class. It was projected that the installation of the hybrid drive system would add one EPA inertia weight class (125 pounds) to the mass of the vehicle. Not only would additional mass, beyond 125 pounds, have an adverse fuel economy impact, it would also require re-examination, and possible redesign, of the chassis. These considerations led to the 3 125 pound inertia-weight-class target.

The 1985 federal emissions standards were adopted as the program target. Vehicle performance targets were selected to match a production compact car in all aspects, i.e. wide-open-throttle acceleration in mid-speed range and high-speed range, part-throttle acceleration, and general performance "feel". The top speed was selected as 135 km/h (85 mph).

Sustained gradeability may be a problem with hybrid vehicles, because reduced-power engines are typically employed to maximize fuel economy. The engine power (size) reduction is made possible by using the flywheel to reduce peak engine-power demands for acceler-

---

* Numbers in brackets designate References listed at end of report.

ation. For this project, it was decided to use a reasonable, but not overly restrictive, gradeability target. A minimum speed of 88 km/h (55 mph) on a 6% grade was selected--because over 95% of the grades on U.S. highways are 6% or less.

Table I Hybrid Vehicle Design Targets

VEHICLE: 3 125 pound Inertia Weight Class

EMISSIONS: Meet '85 U.S. Federal Standards
HC 0.41 grams/mile
CO 3.4   grams/mile
NO_x 1.0   grams/mile

PERFORMANCE: Match Production Compact Car

TOP SPEED: 85 mph

SUSTAINED GRADEABILITY: minimum of 55 mph on 6% grade

2.1 Concept Definition

Conceptually, the fuel economy gains for a hybrid system can be achieved by:

1. Utilizing a smaller than conventional engine and operating it more efficiently, i.e. closer to wide open throttle: The flywheel can be utilized to reduce the peak power demands of the engine and/or the CVT can be utilized to operate the engine more efficiently than with a conventional step-type transmission.

2. Turning the engine off during idle and unpowered decelerations: The presence of a second energy source (the flywheel), which can provide power for the accessories, makes this feasible.

3. Utilizing regenerative braking to charge the flywheel, thereby regaining the vehicle kinetic energy that would otherwise be dissipated as heat during braking.

In order to choose the best configuration, alternate systems were conceived, configured kinematically, designed to a component-specification state, modeled and analyzed [10]. This process involved the generation of detailed component models and the integration of these models to simulate specific driveline configurations. GMR CVT and hydraulic pump performance test data, as well as test data obtained from GM divisions, were used in the simulation studies. A key element in defining a propulsion system was the close interaction between the mechanical design and simulation activities. Through these interactions, accurate models of specific hardware designs were developed. Likewise, hardware constraints were incorporated into the analyses. The results of simulation studies were then used to identify potential problem areas, sources of energy loss, and desirable hardware modifications and configuration changes. Thus, the hybrid drive system evolved in a synergistic manner.

The initial design analysis effort centered on a "two-mode," tri-axial hybrid system. This system offers considerable flexibility in operating characteristics and control schemes, but it is also very complex. Subsequently, a simpler, so-called "one-mode" hybrid evolved and the concept definition became a comparison and choice between the two-mode and one-mode drivetrain configurations.

2.1.a <u>Two-mode hybrid analysis</u>

The two-mode hybrid configuration, designed in several possible versions, features a relatively high-energy (250 W-h), high-speed (22 300 r/min) flywheel system capable of providing flywheel-assisted vehicle acceleration up to 100 km/h. To obtain flywheel assist over such a large speed range requires a wide CVT ratio range. The required ratio range is obtained by traversing the CVT ratio range twice, i.e. two hybrid modes are used. The kinematics for the two-mode arrangement are shown in Figure 1. The drivetrain is shifted kinematically--what was the CVT output becomes the input, thereby enabling the CVT ratio range to be traversed a second time. In addition to the two hybrid modes, there is a Low Mode where the flywheel is declutched and the engine provides power to the vehicle through the CVT, and a Flywheel-Charge Mode to facilitate engine charging of the flywheel.

The flywheel is charged utilizing vehicle kinetic energy (i.e., during regenerative braking) or the engine, or both. In addition, the flywheel speed is required to be in a specified band as a function of vehicle speed such that the system maintains nominally constant kinetic energy. The engine is run intermittently to charge the flywheel--even at constant vehicle speed. Upon completion of flywheel charging, the engine is declutched from the transmission and stopped. The flywheel is then used to power the vehicle. During those periods in which the engine is off, accessories are powered by the flywheel.

Although conceptually simple, the hardware required to implement the two-mode propulsion system turned out to be complex. For example, ten clutches and six chains had to be included in the two-mode transmission design. All of these elements were incorporated into a simulation of the two-mode system. As expected, the losses associated with these elements adversely impacted fuel economy.

The simulation programs developed for the two-mode hybrid were used in a multitude of parametric and optimization studies. Specifically, the effects of gear ratios, CVT ratio range, engine size, flywheel speed range, and energy management strategy were studied. These studies gave valuable insight into the tradeoffs involved in the design of a hybrid vehicle. For example, minimum engine size in a two-mode hybrid is determined by gradeability requirements. The minimum engine size obtained in this manner is smaller than that needed to meet acceleration performance requirements. Utilizing a smaller engine, however, necessitates a higher flywheel minimum speed constraint as a function of vehicle speed. The consequent flywheel losses compromise the fuel economy of the hybrid. Hence, a 1.8 L engine was chosen for the two-mode hybrid. With this engine choice, the predicted hybrid fuel economy was 33 mpg on the EPA Urban Schedule, 34 mpg on the Highway Schedule, and 33 mpg Composite.

2.1.b One-mode hybrid analysis

As the simulation and design activities proceeded, it became increasingly evident that the two-mode hybrid was mechanically very complex and had high parasitic losses. These losses were primarily the result of open clutch and chain losses, and flywheel rotating losses. For example, about 40% of the energy available for regeneration at the wheel-road interface would be dissipated as flywheel losses. In addition, the benefit of charging a flywheel using the engine and thus "passing" energy through the transmission twice became questionable. Finally, results obtained for a CVT-equipped vehicle underscored the benefits associated with the use of a CVT, i.e., the ability to operate the engine efficiently. Hence, a simple hybrid design was sought which would not have high parasitic losses, but would retain the desirable features of an engine-flywheel hybrid: the ability to run the engine at (or near) minimum brake specific fuel consumption (BSFC), the opportunity to turn the engine off during unpowered deceleration and idle periods, and the recovery of energy which would normally be dissipated in braking. The one-mode hybrid concept emerged as the most favorable choice since it incorporates all of these features.

Figure 2 shows a schematic of the one-mode hybrid. This hybrid system is essentially a CVT with the addition of a flywheel. During normal cruising, this driveline functions as one with a CVT, i.e., the engine can be operated near, but not necessarily at, minimum BSFC. Upon braking, vehicle kinetic energy is used to charge the flywheel and/or power the accessories, but in this version there is never any engine charging of the flywheel. The charged flywheel is used subsequently to power the accessories and to accelerate the vehicle. After the flywheel energy is (essentially) depleted, the engine is restarted and the vehicle reverts to the engine-CVT mode. The flywheel system was designed for a maximum operating speed of 12 000 r/min for an energy level of 60 W-h.

2.1.c Comparison of the one- and two-mode hybrids

As indicated in the previous section, the one-mode hybrid is mechanically simpler than the two-mode design and has lower parasitic losses associated with the transmission/flywheel systems. The one-mode design, however, does not allow the engine to be operated at minimum BSFC at all times. Hence, the average engine efficiency for the one-mode design is about 5 percentage points less than for the two-mode design. These factors result in the almost comparable fuel economies of the two designs:

	Fuel Economy (mpg)		
	Urban	Highway	Composite
One-Mode Hybrid:	34	34	34
Two-Mode Hybrid:	33	34	33

A sensitivity study was conducted to examine the relative merits of the one- and two-mode designs. Figure 3 shows the results of that study. The two-mode transmission has more clutches and chains than the one-mode, and it is considerably more sensitive to losses from those elements. Likewise, the two-mode design has a moderately high-speed flywheel and is extremely sensitive to flywheel losses. These results favor the one-mode hybrid design, and the choice was made to proceed with the one-mode configuration for the passenger car drive system. In summary, as compared to the two-mode design, the one-mode design was: predicted to provide comparable fuel economy, mechanically less complex, smaller, lighter, less costly, easier to package, functionally simpler, used a lower-speed flywheel system, and appeared to require the shortest development time.

2.2 Drive System Design

Once the one-mode hybrid had been selected, considerable work was done to select components and configure the drivetrain to satisfy vehicle packaging constraints. Every effort was expended to use available components, and to ensure that new components were practical and producible.

The design chosen for final development is shown in the schematic diagram of Figure 4. The design is based on a variable-sheave, metal-belt CVT driving the front wheels of the vehicle through an engine clutch (which is necessary for starting, because the CVT does not go through zero ratio), and suitable reduction gearing, including forward and reverse gears. Added to this engine-CVT portion of the drivetrain is a fixed ratio chain and suitable gears and clutches to connect the flywheel and accessory drive portions of the system. This hybrid arrangement is of the parallel type in which the engine and flywheel are coupled to the vehicle in parallel via the CVT. In addition, the accessories can also be driven from the CVT. There is a two-speed drive arrangement for the accessories which reduces the accessory power requirements by slowing the accessories at higher vehicle or flywheel speeds. With this drivetrain, the accessories can be powered by the flywheel, by the engine or, on deceleration, by the vehicle. This enables the engine to be turned off whenever it is not required to power the vehicle and when there is available vehicle or flywheel power for the accessories.

It was an engineering challenge to design the hybrid system so that it could be packaged in the engine compartment of an existing compact car. This objective was met with no structural modifications to the vehicle chassis. The placement of the system components is shown in the schematic side view of the engine compartment in Figure 5. The axes for the engine, flywheel, axle and accessories are indicated in the figure.

A hydraulic system was designed to fulfil all of the operational requirements of the hybrid transmission system. The main goal was to design a microcomputer-controlled, electro-hydraulic transmission control system utilizing advanced hydraulic technology which would satisfy the unique hydraulic requirements of a hybrid transmission, minimize hydraulic losses, and ensure a reliable, fail-safe system. The final system is innovative, yet it is within current automotive design practices and manufacturing capabilities. The hydraulic system receives electrical command signals from the central electronic control system. Nine pulse-width-modulated (PWM) solenoids provide the electronic-hydraulic interface and respond to these signals by creating appropriate hydraulic pressure control signals. These pressure signals, in turn, actuate hydraulic valves, which operate the mechanical driveline components such as slipping clutches, on-off clutches, the lubrication and cooling systems, and the CVT ratio control system. Working pressures are provided by a pressure regulator module which also modulates the line pressure according to the level of transmitted torque. Several advanced hydraulic concepts were utilized in the design of the electro-hydraulic control system, some for the first time in automotive transmission practice. Examples of these are: an electro-hydraulic interface using PWM solenoids, a two-speed pump drive, and a dual-pump-system pressure supply using a fixed and a variable-displacement pump in tandem. To increase control accuracy and further reduce system losses, the function valves were operated by pilot pressure using a three-level pressure source, cooling oil was provided to clutches only when they were slipping and, where possible, valve leakages were directed toward lower pressure sinks.

2.2.a Driveline analysis

Engineering calculations were performed to define component size, durability and strength. Detailed analyses of the dynamics of the complete driveline system were also included. Specifically, analyses included gears (planetary, helical, spiral bevel), fixed-ratio chains, clutches (on-off type, over-running, slipping), hydraulics (including valve design, orifices, actuators, and pumps), shafts, splines and various other subsystems. Where possible, accepted GM and SAE automotive design practices were used and existing computer programs, available at GMR, Hydra-matic Division and other divisions, were employed.

An analytical program was developed to address the dynamic stability of this multi-mass system. This program can be used to analyze up to 50 individual masses, connected with springs, dash pots, variable-ratio drives or slipping devices, to calculate the transfer function and provide dynamic analysis of the system, or any portion thereof, including dynamic stresses.

2.3 CVT Development

A major effort was directed toward analysis and development of the metal-belt CVT element that is the heart of the hybrid transmission. There are a number of operational requirements imposed on the CVT in this design, which are unique to hybrid operation and which required special treatment. These unique requirements are: power reversal (e.g., regeneration) during operation, faster-than-conventional rates of ratio change, and complex functional requirements (e.g., an asynchronous shift). All of these are interactive, which makes the simultaneous attainment of acceptable operating levels difficult, particularly because the highest possible operating efficiency must also be achieved.

Analyses showed that the important considerations for the application of the variable-diameter-sheave, metal-belt drive, for either hybrid or conventional passenger cars, are the magnitudes of the axial forces required for both steady-state and transient operation, and the belt efficiency. Both factors have a major influence on the fuel economy achievable with such a drive. For steady-state operation, the axial sheave forces and belt efficiency are functionally dependent on the magnitude of the traction coefficient, i.e., the ratio of the transmitted torque divided by the product of the belt pitch radius and the applied axial force. Analytical models and experimental data have shown that the maximum belt efficiency occurs when the axial sheave force is at or near the minimum axial force required to prevent slip. This minimum axial force is determined by the maximum available traction coefficient. Alternatively, the traction coefficient determines the output torque capability for a given applied axial force.

Experimental data have confirmed the dependence of belt performance on traction coefficient. Because the measurements include bearing and windage losses not accounted for in the belt analytical model, a correlation of the data was obtained and an empirical model was used in the modeling and simulation of vehicle performance. For the hybrid CVT, the sheave axial forces are developed hydraulically. Using the metal-belt analytical model, predictions were made of the hydraulic pressure required in each sheave for belt ratio maintenance, and experimental data confirmed the predicted pressures. Not only did these measurements confirm the validity of the sheave-to-sheave force relationships, but they also suggested a ratio maintenance strategy which would maximize belt performance and life, while minimizing the losses within the hydraulic system.

The ratio-control strategy, summarized in Figure 6, is to operate the belt at a traction coefficient approximately 20% lower than the maximum traction coefficient at slip. This constant-traction-ratio strategy is equivalent to operating at an axial force which is a fixed percent greater than the minimum permissible axial force. Because the belt efficiency variation with traction coefficient is relatively flat and increases in the region of incipient belt slip, the belt losses would be minimized. Further, the hydraulic system requirements would be minimized by reduction of the hydraulic pressure to the lowest possible value, but still providing a margin of safety for torque or pressure fluctuations.

For the hybrid design, the required rates of ratio change are expected to be twice as large as required in a conventional CVT-equipped passenger car. Further, regeneration, during which vehicle kinetic energy is transferred to the flywheel, represented an additional operational requirement for the transmission. During the transition period from propulsive to regenerative power flow, a sudden change in the magnitudes of the sheave axial forces is required to avoid belt slip and maintain high belt efficiency. The required rates-of-change and regeneration required a detailed study of CVT ratio dynamics.

A belt test fixture, shown in Figure 7, was used for dynamic testing. The fixture was designed for use with the largest capacity metal belt available, and has a variable center distance for evaluation of the various ratio ranges using belts of different length. The sheave piston area ratio can be varied over a large range for evaluation of the effect of area and area ratio on rate of ratio change. The fixture uses two flywheels--one represents the vehicle inertia and the second the hybrid-vehicle flywheel. The two flywheels are connected to the belt test fixture via in-line torque transducers. This same facility can be used to test slipping clutches, which are also required in the hybrid vehicle driveline. The double-flywheel belt test fixture was used in the development of ratio maintenance devices, strategies and control interfaces.

Other development efforts were directed at reducing the energy consumption of the CVT hydraulic system, and included improvement of the variable-displacement pump, and the development of a dual-pump supply system. The dual-pump system consists of a small fixed-displacement, high-pressure pump, used to meet the normal high-pressure needs, with assistance, when necessary, from a variable-displacement pump which normally only supplies the low pressure needs. The dual-pump system was motivated by the higher pressure requirements of the hybrid hydraulic system relative to conventional transmissions. A research facility was developed for evaluation of both the conventional pumps and the novel dual-pump system proposed for the hybrid drivetrain, and theoretical analyses of the pump were used to give the development guidance.

To evaluate the dual-pump concept, fixed-displacement, high-pressure pumps, sized to meet normal high pressure needs, were tested and a model was incorporated into the simulation software. Simulation results predict a fuel economy advantage with the dual-pump system at higher line pressures. In addition, a dynamic model of the dual-pump system helped to define proper system design values, and to identify instabilities, thereby defining an appropriate operating philosophy.

2.4 Engine Development

The objectives of the engine development activity were to: select the best engine in terms of fuel economy characteristics, subject to the constraints of sufficient power, acceptable exhaust emissions and acceptable packaging; develop operating philosophies and control algorithms; and devise and develop a fuel shut-off/restart system. Manpower limitations and time constraints restricted the engine selection process, but an international survey was conducted before an available GM engine was selected.

The selection process was complicated by the fact that some of the engine selection criteria were dependent on the overall system configuration and control strategy, which, of course, were not defined at the start of the project. In the two-mode system, sustained gradeability is the major engine-selection criterion, because the flywheel is sized to satisfy peak power requirements. The one-mode system, however, has more severe engine operating requirements. Consequently, when attention focused on a one-mode design, some redirection and intensification of the engine development effort was required. Only the one-mode, engine-selection process will be discussed. It should be kept in mind that a different hybrid configuration could lead to different engine selection criteria. The criteria used in the engine selection process are delineated in Table II, along with the factors that were examined in the evaluations relative to each criterion. Also, availability was a major consideration because of project time constraints. From a technical viewpoint, power rating was the initial screening parameter.

Table II

SELECTION CRITERIA	SELECTION FACTORS
Adequate Power	Power, Torque
Minimum Size, Weight	Length, Head Material
Readily Available	Supplier Location
Adaptable to TBI/EFI*	Induction Configuration
Good BSFC	Plug Location, Chamber Geometry
Low Noise	Perceived Level
Low Emissions	Certification Data, Crevice Volume
Good Durability	Documented Problems
Suited for Hybrid Use	Intake/Exhaust Location, Driveability, Cooling, etc.

*TBI ≡ throttle body injection, EFI ≡ electronic fuel injection

In any hybrid drive system with an energy storage device, the engine, or primary power producer, must be sized assuming no availability of stored energy. By determining the engine-power requirement in this manner, one accounts for situations wherein charging (and use) of energy storage is either not possible or not advisable, for example, during cold starts, during sustained steady speed operation, or during sustained operation up a grade. For this system, gradeability and wide-open-throttle (WOT) acceleration were examined using simulation programs, and the results were used to determine the peak engine power requirement. As indicated in Figure 8, with the design targets, a minimum power of 55 kW was required. Note that acceleration performance dictated peak engine power for the one-mode hybrid system. Allowing for an engine power excess of 20% fixed the engine peak power range at 55 kW to 65 kW.

Over 100 engines in the appropriate power range were evaluated using the other criteria listed in Table II. The GM 1.8 L OHC Family II engine was selected because it satisfied all of the selection criteria and was immediately available.

As indicated in Table I, the design targets for emission control were the 1985 federal standards. The emission control approach selected was to use closed-loop, stoichiometric air-fuel ratio control, moderate EGR, and an oxidizing-reducing catalyst. It was anticipated that the intermittent engine operation, associated with the engine-flywheel hybrid vehicle application, would affect exhaust system temperatures, thereby influencing catalytic converter performance. Vehicle studies on the Federal Test Procedure with intermittent engine operation, simulating that expected in the hybrid system, showed that it was possible for the catalyst temperature to drop below the minimum reaction temperature for a contemporary, aged catalyst. Analytical modeling studies, using the approach described in [11], showed that the catalyst temperature could be held above the minimum reaction temperature if the distance between the exhaust manifold and the catalytic converter was about one meter or less.

Another concern, relative to emission control and fuel economy of the hybrid system, was the development of a fuel system for engine shut-off and restart. Experimental vehicle studies indicated that a standard throttle-body-injection (TBI) system would not provide adequate control for a hybrid application, since there was inadequate engine response restart. Also, it was desired to have a more positive fuel control, to prevent fuel storage in the intake manifold and the resultant penalties in fuel economy and hydrocarbon emissions. The fuel supply system selected used electronic port fuel injection (PFI), with the electronic controls to implement the system developed as part of the hybrid.

Engine testing was done for the hybrid project to obtain BSFC data for the engine operating conditions expected to be encountered, i.e. high load at both low and high engine speeds. The initial engine tests were conducted with TBI because PFI hardware and controls were not available at the start of the program. These data were used in the hybrid simulation program.

2.5 Flywheel System Development

The design of the flywheel system was also complicated by the fact that two distinctly different drive systems, i.e, the one-mode and two-mode, were considered in the project. During the initial phase of the project, when attention was focused on the two-mode hybrid, a high-speed (23 000 r/min), moderate-capacity (250 W-h) flywheel system was completely defined and partially designed. As the program evolved, the flywheel system finally selected and designed was of relatively low-speed (12 000 r/min) and low capacity (65 W-h). Most of the flywheel systems considered, and the final system selected, were based on the unique GMR Cymbal Flywheel concept.

The Cymbal flywheel consists of a stack of thin discs, each formed in the shape of a cymbal. The advantage of this is illustrated schematically in Figure 9 where tangential stress distributions are compared for a flat disc and a cymbal disc. A flat disc has a maximum stress, at its inner radius, which is about twice as large as for a cymbal disc. The reason for the reduced peak stress in the cymbal is that it tends to flatten out with increasing speed, thereby generating a compressive stress to counter-act the increasing tangential stress. The ability of the Cymbal concept to operate at lower stress levels enabled the use of common steel, thereby precluding the necessity to use more exotic materials of unproven practicality in flywheel applications.

The issues addressed in the flywheel design were: attainment of the required energy capacity, minimization of mass and volume, selection of flywheel orientation, definition of vacuum level to minimize windage loss, selection of a seal or seals, selection of bearings, and definition of the containment and housing structure. Overall considerations in the design process were a desire to have a low-hazard and low-cost flywheel system.

The gear ratio for the flywheel drive must be kept low enough to be acceptable in terms of current design practices, including considerations such as noise, but yet high enough to ensure maximum energy recovery. Also, to obtain the required maximum energy capacity, there is a trade-off between flywheel speed and flywheel mass. Perhaps it is instructive to point out that the best flywheel system is one with minimum mass operating at extremely high speed in a system having no-cost vacuum, with no-loss bearings and no-loss seals. Unfortunately, no-cost and no-loss components are not available, and all losses increase dramatically with increased operating speed.

A 12 000 r/min maximum flywheel speed was selected as a reasonable compromise of all factors. This included consideration of the 5.6:1 CVT ratio and a 6:1 flywheel gear ratio. The flywheel inertia needed for 65 W-h energy capacity at 12 000 min is 0.3 kg-m^2. This corresponds to a rotating mass of 22 kg obtained with (18) 1 mm thick cymbals of 384 mm diameter.

The choice of flywheel orientation in the vehicle must consider possible gyroscopic effects on vehicle handling [12]. The gyroscopic phenomena were studied analytically, and it was predicted that the 0.3 kg-m^2 flywheel, which had an angular momentum of 381 N-m-s at 12 000 r/min, would have minimal effect on vehicle handling in any flywheel orientation. A flywheel with higher angular momentum might affect vehicle dynamics, particularly if mounted with its axis in the horizontal, transverse orientation. In this drive system, the horizontal, transverse flywheel orientation was favored because it eliminated the need for a right-angle drive. Since this was the worst flywheel orientation from a handling viewpoint, even though the analysis had predicted no problem, experimental tests were also conducted. A test vehicle was equipped with an electrically-driven flywheel system, as shown in Figure 10. The test flywheel was mounted as closely as possible to the planned mounting location in the hybrid, and was designed to match the angular momentum of the hybrid flywheel system. The vehicle tests, which were done at the General Motors Proving Grounds, also showed that the flywheel gyroscopic moments had a negligible effect on vehicle dynamics. With these results, the hybrid flywheel-axis orientation was fixed as horizontal, transverse.

With the basic flywheel system defined, the detailed design of the various flywheel sub-systems and components was addressed. Following is a brief overview of each of the sub-system development activities.

2.5.a Cymbal discs

Spin tests conducted on a series of cymbal-shaft-retaining ring assemblies indicated that clamping rings are required for stability of the disc-shaft assembly, and that a steel sleeve on the hub would be required. Detailed stress calculations for cymbals of different angles, and with various cymbal-to-shaft interference fits, indicated that the cymbal-to-shaft interference fit could be as low as 0.18 mm to ease flywheel assembly. The prediction was verified in testing, as long as a clamping ring was employed in mounting the cymbal discs to the shaft. The minimized-interference-fit of the GMR Cymbal concept turned out to be the over-riding advantage of the Cymbal Flywheel in the low-speed hybrid application. The stress levels in the cymbals are very low for this flywheel assembly so that the 12 000 r/min flywheel design has a factor of safety over six.

Our studies showed that inexpensive sheet steel, coupled with the novel GMR Cymbal design, provided the most cost-effective approach for automotive flywheel applications. Most of the flywheels tested were fabricated of SAE 4130 sheet steel, and some development work succeeded with the lower-cost HSLA sheet steel. Laminate thicknesses of 0.6, 0.9 and 1.0 mm were successfully fabricated into cymbal constructions.

Of interest during the design of the flywheel assembly were the frequencies and modes of vibration of the cymbal discs. Vibrational spectrum analysis was used to determine the mode shapes and frequencies of vibration of the system. The first bending mode of the discs occurred at 177 Hz, the second diametral mode at 194 Hz, and the third mode at 235 Hz. These frequencies occur near the operating speed of the flywheel assembly, but they are not considered to be a problem because of the friction between the cymbals.

2.5.b Bearings

A bearing-loss rig, having an in-line torque meter, was built to measure the losses in a number of candidate bearings. Ball bearings, needle bearings and floating-ring bearings were tested under a variety of load and lubrication conditions. Needle bearings and ball bearings exhibited comparable losses which were about 25% less than for floating-ring bearings. This result led to the final design of a ball and needle bearing in combination as shown in Figure 11--the ball bearing provides needed thrust-load capability, while the needle bearing provides the needed radial stiffness for the system and has minimal space requirement.

2.5.c Seals

Initial seal tests were based on a flywheel speed of 26 000 r/min. Seals considered for the high-speed flywheel operating in a 100 kPa vacuum were: (1) a viscoseal; (2) a GMR spiral-groove face seal; (3) a spiral-groove seal; and (4) a ferromagnetic seal. Test results were obtained for all of the high-speed seals. When the flywheel design speed was reduced to 12 000 r/min, a standard face seal became feasible. The losses for this seal (oil-lubricated on the interface outside diameter), were 30 W at 6 000 r/min and 80 W at 12 000 r/min. Leakdown rate for this seal was 0.027 kPa/h--the vacuum loss to be replenished by a vacuum pump.

2.5.d Housing

The flywheel housing must provide adequate support for the containment ring, resist piercing in the event of a flywheel burst, resist the large forces due to the system vacuum, have minimum windage loss, fit within the available space, and have minimum mass. A two-part housing design was developed with an aluminum main housing having a sheet metal ribbed cover. It was found that a conical steel cover, shown in Figure 12, that is 2.25 mm thick with radial stiffener ribs would support the force of over 8 kN due to the system vacuum. The axial and radial clearance between the housing and flywheel rotor were selected to minimize windage loss at maximum speed. Analysis of losses showed that the side clearance should be about 3.5 mm while the radial clearance should be set at 7 mm.

2.5.e Vacuum system

Flywheel housing pressure is important, since operating the flywheel at design speed and at atmospheric pressure consumes about 3 kW in windage loss. Consequently, an inexpensive commercial Gerotor pump was developed to provide the necessary reduced housing pressure. This pump required 30 W of drive power. The power consumption can be reduced further by cycling the vacuum pump on and off as required to maintain the desired vacuum level. With a broken-in system, vacuum levels of 3.32 kPa were achieved. The calculated windage loss at this pressure is about 90 W.

2.5.f Containment

One of the advantages of the GMR Cymbal flywheel design is that, in the event of a burst, the individual fragments deform readily upon impacting the containment. Consequently, much of the energy is absorbed in bending the flywheel fragments rather than in damaging the containment. In addition, the total energy to be absorbed with the Cymbal flywheel design--assuming the flywheel fractures into thirds--is about 5% of a solid flywheel. A series of candidate materials and constructions was tested, and a four-layer wrap of 1 mm thick SAE 1090 steel was selected for the containment. The direction of wrap was selected to wind onto the rotor in the event of rotor contact. Flywheel burst tests at 32 000 r/min, where a single disc was purposely failed, showed little damage of the wrapped containment ring.

2.5.g Flywheel assembly tests

While each of the subsystems was analyzed and tested separately, the synergistic effects could not be evaluated until a complete assembly was tested. The operating friction was measured to be about 300 W without break-in, or fine-tune adjustment of the various subsystems. The predicted minimum friction, based upon broken-in components and optimum adjustment, was 245 W, with distribution as follows:

Bearings	75 W
Face Seal (1)	80 W
Windage	90 W
Total	245 W

Some additional power consumption should be alloted for the vacuum pump, but in the preferred cyclic-operation mode, the vacuum pump power would be minimal (30 W).

2.6 Control System Development

The control system of an engine-flywheel hybrid vehicle is critical in determining the operating characteristics and efficiency of the vehicle. This control system must coordinate the flywheel, engine, transmission, and brake system operations and failure-mode procedures. The goal was to develop an electronic system which would control the propulsive and braking energy of a flywheel hybrid vehicle to satisfy performance and driveability requirements, maximize fuel economy, recover braking energy, maintain transparency of system operation to the driver, and ensure the attainment of applicable emissions constraints.

The approach was to define the controller functions and appropriate controller hardware, and to develop mathematical models of the systems being controlled, as well as to identify and address critical control issues. This information was used to develop the appropriate control algorithms, software, and hardware.

2.6.a Hierarchical Developmental Control System

The control system development began in parallel with the mechanical design activity, before a specific mechanical configuration had been selected, and while the energy management strategy was an open issue. Hence, control system flexibility was a necessity from both the hardware and software standpoints. A user-friendly interface was desirable so that algorithm development could proceed rapidly in a high level language.

A schematic of the hierarchical developmental control system is shown in Figure 13. This system consists of a coordinator, or host controller, and separate engine and transmission controllers. A major advantage of the hierarchical structure is that it facilitates subsystem-controller and mechanical-hardware development, e.g., the coordinator/engine controller and the coordinator/transmission controller can be developed and tested independently.

Figure 14 shows the Intel 8086-based, 16-bit microprocessor for the coordinator which was chosen primarily on speed and software considerations. Phoenix Digital-GM function control modules (FCM) were selected for the subsystem controllers.

2.6.b Subsystem controllers

A shared-memory interface was developed which allows interconnection of up to four FCM controllers to a multibus host machine. Realtime software in the satellite machine handles the specifics of the particular controlled object (engine, transmission, brakes, etc.), with overall coordination done at the host level.

To allow development and calibration of separate subsystems without the host controller, a separate operating and software system was developed. This allows the operator to communicate with the portion of memory normally accessed by the host, and to monitor operation of the inter-connected host-satellite system.

Special purpose I/O hardware and software are required to interconnect the subsystem controllers to either engine or transmission. This was completed for the engine satellite controller in three versions: TBI, grouped PFI and sequential PFI. All provide control of idle air, EGR, spark timing, and engine start-stop.

2.6.c Hydraulic system modeling

A complex hydraulic system forms the interface between the hybrid transmission and the electronic control system. To determine the feasibility of the proposed hydraulic system design, e.g., its stability and response characteristics, and to evolve realistic transmission control algorithms, critical hydraulic subsystems were modeled. In addition to their use in this program, these models are also applicable to other automotive systems and have demonstrated the utility of applying analytical methods to transmission hydraulic-system design.

The dynamics of the dual-pump hydraulic supply system are dominated by a variable-displacement vane pump. A dynamic model for the variable-displacement pump was developed, considering its detailed geometry, line pressure, "hydraulic lock" and leakage. The model explains some of the pump's characteristics which have been observed experimentally but not previously understood. Results from the pump model agree favorably with steady-state experimental data.

The hydraulic system design incorporates a low-volume, fixed-displacement pump; a large-volume, variable-displacement pump; and a valve body which couples the fixed- and variable-pump flows according to pressure-control commands and load requirements. The valve body also regulates the output of the variable-displacement pump. Since the characteristics of the dual-pump hydraulic supply system may determine the behavior of the entire hydraulic system, it became essential to: (1) derive a dynamic model for the system, (2) identify critical areas and design parameters, and (3) study the dynamics of the proposed design.

A dynamic model for the dual pump/valve body system was developed to enable the computation of valve motions, and the pressures and flows in the hydraulic lines. Effects such as fluid compressibility, flow forces (steady-state and transient) on individual valves, internal-volume variations (in particular, inside the variable-displacement pump), and leakages past valve lands were included in the model. Computer simulations using this model predicted that the proposed design was feasible and had a high sensitivity to valve damping. Tradeoffs between leakage (i.e., efficiency) and response time were determined. Moreover, use of the model led directly to a number of proposed design changes to ensure stable, responsive operation. These changes were incorporated into the final design of the hydraulic system.

2.6.d Simulator Develoment

A simulator was used to develop control algorithms and study the response of the propulsion system to external inputs. The simulator, which operates on the GMR engineering computer system, emulates the engine, transmission and vehicle. To accomplish this, a lumped-parameter dynamic model of the one-mode hybrid was developed. This model consists of a 13-degree-of-freedom driveline model and a dynamic engine model based on manifold gas dynamics.

A Lagrangian formulation was used to derive the equations for the driveline model. All hydraulic clutches in the transmission, including the reverse clutch and roller clutches, are included. Clutch lock-up is treated by eliminating the corresponding degree-of-freedom using matrix condensation techniques. The final system consists of 22 state variables which include 13 speeds, seven shaft deflections, CVT ratio and engine-manifold pressure. Test data for the engine were used in the engine simulation. Production service brake characteristics were used with a front brake shut-off switch for flywheel regeneration.

2.7 Regenerative Braking System Development

Regenerative braking in a front-wheel-drive vehicle should be done using only the two front wheels to maximize energy recovery. To determine the braking deceleration rates that would be acceptable and to assess flywheel gyroscopic effects, analytical vehicle handling studies were performed and subsequently validated by vehicle experiments. The studies concerned braking-in-a-turn under controlled conditions of deceleration, turn radius and vehicle velocity. Measurements included the degree of deviation in vehicle motion from the motion intended by the driver. Both the analytical and experimental parametric study, with varying deceleration rates, road conditions, braking options,and flywheel "on" or "off" conditions, were done by the Vehicle Handling Group at the GM Proving Ground. Both studies showed that braking via the two front wheels only on a dry or wet road was not a problem at deceleration rates up to 0.3g, which is adequate for urban driving.

Thus, a flywheel can be incorporated into a passenger vehicle, and regenerative braking via the two front wheels can be done without adverse safety ramifications due to vehicle handling problems. The proposed regenerative braking system design, shown schematically in Figure 15, is still subject to final development and in-vehicle evaluation. However, the braking system design does appear viable and was configured with the following design features and objectives:

- maximize regenerative braking to maximize energy recovery,
- utilize regenerative braking on the front wheels only,
- retain service brake capacity, so that full brake capacity is available even when regenerative braking is unavailable,
- do not sacrifice brake-system-failure safety,
- facilitate automatic, and unobtrusive, transitions between braking modes, and
- implement automatic operation of either the regenerative or service brake system with a single brake control.

2.8 Overall System Operation

The general operation of the hybrid drive system is described below using information from the simulation studies and considering operation on the EPA Urban Driving Schedule. The vehicle speed-time trace for Cycle 1 of the 18-cycle EPA Urban Driving Schedule is shown in Figure 16, along with the tractive power required to drive that cycle. A comparison of the vehicle speed and tractive power traces shows the correspondence of positive power to increasing vehicle speed, and negative power to decreasing vehicle speed. Negative tractive power represents the energy available for regeneration to charge the flywheel.

Cycle 1 features an engine start and short warm-up period, an initial acceleration, some intermittent moderate accelerations and decelerations, and a final deceleration to rest. The characteristics of the system operation during these various operating modes are delineated in Figures 17 and 18. Figure 17 shows the engine-speed and CVT-ratio variation with time. Figure 18 shows the flywheel speed and power traces. The operation can be described by considering the following operating modes:

2.8.a Cold start

Upon starting from a cold condition, the engine is operated as in current production vehicles, i.e. no abnormal loading during warm-up. Power for the accessories is provided by the engine. The flywheel is not charged during the cold start, but the vacuum pump required to evacuate the flywheel housing is operated, using power from the accessories drive shaft.

2.8.b Intermittent acceleration (without flywheel power)

The initial acceleration of the vehicle is accomplished using engine power only. This acceleration starts 20 seconds after engine start-up and continues for about 10 seconds. The variation of engine speed and CVT ratio during the acceleration are shown in Figure 17. As indicated, the engine speed increases and the CVT ratio moves away from maximum underdrive as the vehicle is accelerated.

2.8.c Initial deceleration

At about 37 seconds into Cycle 1, the vehicle experiences its initial deceleration and, as noted on the tractive power plot of Figure 16, power becomes available to charge the flywheel for the first time. The corresponding flywheel speed and instantaneous power into the flywheel are shown in Figure 18.

2.8.d Intermittent acceleration and deceleration

During the remaining portion of Cycle 1, up to the final deceleration, the vehicle experiences relatively modest speed changes. All tractive power comes from the engine.

2.8.e Deceleration to rest

The final deceleration for Cycle 1, which begins at about 115 seconds, constitutes an opportunity to regenerate a substantial amount of vehicle kinetic energy and, as indicated in Figure 18, the flywheel is charged to a speed of about 7 000 r/min. Part of the energy recovered is used to power the accessories, providing the opportunity to turn the engine off and save fuel. Note that the engine stops at the 115 second point (Figure 17), 10 seconds before completion of the deceleration.

The ability to use the vehicle to power accessories during a deceleration is a major advantage of the hybrid drive system, since vehicle kinetic energy is the cheapest energy available. At the end of Cycle 1, after the vehicle has come to rest, the accessories are powered by the flywheel and the engine remains off. The accessory load for the EPA Driving Schedule consists only of the alternator and the various pumps (water, oil, fuel, hydraulic and power steering), some of which do not operate when the engine is off. However, the accessory load is sufficient to deplete the flywheel energy by 7 W-h between the end of Cycle 1 and the start of Cycle 2. The depletion of flywheel energy to provide accessory power is shown in Figure 19 for each of the idle periods on the Urban Schedule. Two points are noteworthy with regard to Figure 19. First, the flywheel is never charged to its maximum (nominal) energy on the Urban Schedule. Recall that the energy capacity of the flywheel is 65 W-h at 12 000 r/min, so the flywheel only gets to about one half of that level during the Urban schedule. The second point is that with greater accessory loads e.g., in an idle with the air conditioning on, considerably more flywheel energy would be extracted.

The traces of vehicle speed and tractive power vs time for Cycle 2 are shown in Figure 20. The engine-speed and CVT-ratio traces are shown in Figure 21, and the flywheel speed and power traces in Figure 22.

2.8.f Acceleration with flywheel power

The acceleration at the start of Cycle 2 constitutes another important operating mode for the hybrid drive system. As indicated in Figure 22, prior to the start of Cycle 2, the flywheel is providing about 1 kW to power accessories. Then, at about the 162 second point, the vehicle acceleration is started solely with flywheel power. Because the flywheel only had 13 W-h of energy at the acceleration start, it quickly uses up (in about 5 seconds) its vehicle-accelerating capability and the engine must take over. The vehicle speed at this point is 8 km/h. In order to facilitate a match between the vehicle (output shaft) speed and the engine speed, the CVT must go through an asynchronous shift. Because the vehicle speed at the shift point is so low in Cycle 2, the shift is not severe, as shown in Figure 21 where the CVT ratio changes from 0.7 to 0.6. After the engine is brought into operation to provide tractive power, the flywheel speed decays due to parasitic losses. This can be seen in Figure 22 where the flywheel power and speed are shown to continue to decrease following the 162 second point. Similar flywheel power and speed depletion are indicated after an intermediate charging of the flywheel between 180-190 seconds.

2.8.g Fuel economy

Figure 23 shows the breakdown of tractive energy the hybrid system over the complete Urban Schedule. The flywheel contributes only 8% of the total tractive energy. However, as indicated previously, the hybrid drive system enables the use of flywheel and vehicle energy to drive the accessories, and the importance of this is shown in Figure 24. Forty-one percent of the power for accessories is provided by the flywheel and vehicle, thereby substantially reducing the load on the engine.

The efficiency of the energy recovery process during the Urban Schedule is illustrated in Table III which shows that 69% of the vehicle kinetic energy, that would otherwise be lost, is recovered and used to charge the flywheel and power the accessories. Table IV shows that, of the energy stored in the flywheel, 50% is utilized beneficially to power the accessories and to provide tractive energy for vehicle acceleration.

With the energy recovery/utilization characteristics delineated in Tables III and IV, the predicted fuel economy of the hybrid-powered compact car (3 125 pound EPA Inertia Weight) is indicated below in comparison with fuel economies measured for typical 1984 production compact cars (3 000 pound EPA Inertia Weight).

	Fuel Economy (mpg)		
	Urban	Highway	Composite
Hybrid:	34	34	34
1984 Production Car:	25	39	30
% Change for Hybrid:	+36%	-13%	+13%

As indicated, the hybrid vehicle is predicted to have a substantial (36%) fuel economy benefit on the Urban Schedule. On the Highway Schedule, the hybrid vehicle fuel economy is predicted to be 13% below the 1984 production compact car due to the losses associated with the complex hybrid powertrain. Overall, a 13% improvement in EPA Composite fuel economy is predicted for the hybrid vehicle.

Table III Flywheel Charge-Mode Energy Distribution on EPA Urban Schedules
(% of vehicle kinetic energy available for regeneration)

Flywheel	60	69%
Accessories	9	
Charging clutch	9	12%
Service brakes	3	
System losses	19	
Total	100%	

Table IV Flywheel Discharge-Mode Energy Distribution on EPA Urban Schedule
(% of flywheel energy available for use)

Accessories	12	50%
Tractive	38	
Flywheel losses	12	
Discharge clutch	12	
System losses	21	
Flywheel residual	5	
Total	100%	

2.8.h Acceleration performance

The system operation on the Urban Schedule should be fairly typical of city driving, but in actual driving it would not be uncommon to surpass the maximum Urban Schedule acceleration and deceleration rates of 0.15g. For a 0.2 g deceleration from 100 km/h, the flywheel should reach a maximum speed of 11 500 r/min.

The predicted acceleration performance for the hybrid vehicle is indicated in Figure 25 relative to calculations for a 1982 production car. The hybrid predictions are shown for accelerations with a fully charged flywheel and without any flywheel power, since both conditions could be encountered in actual operation of such a system. There is a difference between the two hybrid accelerations under the two flywheel conditions, and it is felt that the difference would be perceptible to the driver, and perhaps bothersome, particularly at the initial acceleration point.

With a fully charged flywheel, the hybrid vehicle accelerates faster than the production car. This can be seen better in Figure 26 which shows the difference in distance traveled by the production car and the hybrid vehicle with a fully charged flywheel. The hybrid vehicle, with a charged flywheel, has an ever increasing lead on the production car. Without flywheel assist, the hybrid vehicle lags behind the production car initially and then, because of the benefit of the CVT, catches and passes it.

3. SUMMARY

A complete engine-flywheel hybrid drive system was designed for a passenger car. This included: the design and specification of all transmission and driveline components; the design, construction and testing of a flywheel system based on the GMR Cymbal-Flywheel principle; tailoring of a 1.8 L OHC Family II GM engine with sequential port fuel injection for hybrid-type operation, including on-off fuel control strategy and hardware; specification of the hardware for a drive-by-wire development control system; conceptualization of a regenerative braking system; and resolution of vehicle handling characteristics. The hybrid system is a one-mode type, wherein the flywheel is charged only via regenerative braking, i.e., there is no engine charging, and the drivetrain is not shifted kinematically, as is the case for systems with larger flywheels. The performance predicted for the hybrid system, with or without a charged flywheel, is comparable to a production car. The fuel economy predicted for the hybrid system was compared to a 1984 production compact car. The hybrid was predicted to achieve a significant gain in urban fuel economy, but could not match the projected highway economy for the production car. Overall, the hybrid was predicted to achieve a 13% improvement in EPA Composite fuel economy.

The predicted fuel economy improvement for the hybrid system did not justify construction of a prototype for experimental validation. However, the definition of the state of flywheel hybrid technology for passenger car applications brought this project to a successful conclusion by delineating those areas requiring additional research.

4. ACKNOWLEDGMENT

The authors thank all of the personnel from GMR and Hydra-matic Division who contributed to this project--it was a true team effort. Among the many GM managers who were supportive of this effort, Mr. John D. Caplan and Dr. William G. Agnew deserve special recognition.

5. REFERENCES

(1) "The Oerlikon Electrogyro - Its Development and Application for Omnibus Service," Automobile Engineer, December 1955.

(2) R. R. Gilbert, et al., "Flywheel Drive Systems Study," Lockheed Missiles and Space Co., Report LMSC-D246393 to EPA, July 1972.

(3) "Feasibility Analysis of the Transmission for a Flywheel/Heat Engine Hybrid Propulsion System," Mechanical Technology Inc., November 1971.

(4) J. Helling, et al., "Hybrid Drive with Flywheel Component for Economic and Dynamic Operation," Institute fur Kraftfahrwesen Technische Hochschule, Aachen, Proc. 3rd International Electric Vehicle Symposium, February 1974.

(5) A. A. Frank, et al., "The Fuel Efficiency Potential of a Flywheel Hybrid Vehicle for Urban Driving," Proc. 11th Intersociety Energy Conversion Engineering Conference, September 1976, pp. 17-24.

(6) G. Larson and H. Zuckerberg, "Hybrid Heat Engine Propulsion for Urban Buses," Proc. 4th International Symposium on Automotive Propulsion Systems, April 1977.

(7) E. Behrin, et al., "Energy Storage Systems for Automobile Propulsion: 1978 Study," prepared for U.S. Department of Energy by Lawrence Livermore Laboratory, December 15, 1978.

(8) F. Surber, et al., "Hybrid Vehicle Potential Assessment," Vol. I-X, prepared for U. S. Department of Energy by Jet Propulsion Laboratory, September 30, 1979.

(9) S. M. Rohde and N. A. Schilke, "Fuel Economy Potential of Heat Engine/Flywheel Hybrid Vehicles," SAE Publication P-91, 1981.

(10) S. M. Rohde, T. R. Weber, R. J. Coleman and L. O. Hewko, "Computer Simulation of Vehicular Propulsion Systems," Report No. GMR-4463, General Motors Research Laboratories, September 2, 1983 (ASME Computers In Mechanical Engineering, March 1984).

(11) G. F. Robertson, "A Study of Thermal Energy Conservation in Exhaust Pipes," SAE Paper 790307, February 1979.

(12) A. T. McDonald, "Simplified Gyrodynamics of Road Vehicles with High-Energy Flywheels," Proceedings of the 1980 Flywheel Technology Symposium, Scottsdale, Arizona, October 1980.

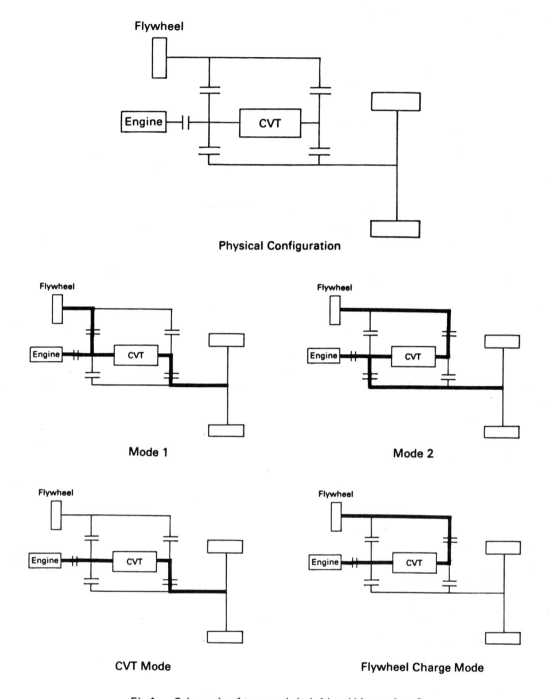

Fig 1 Schematic of two-mode hybrid and kinematics of driving modes

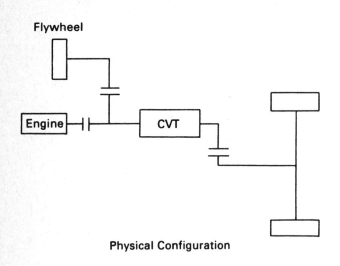

Physical Configuration

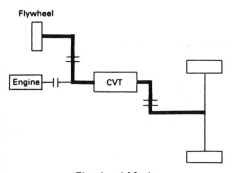

Flywheel Mode

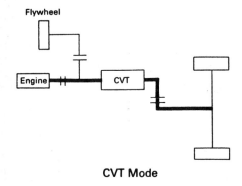

CVT Mode

Fig 2　Schematic of one-mode hybrid and kinematics of driving modes

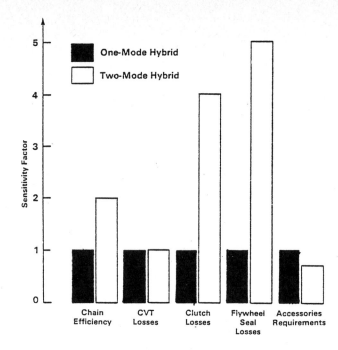

Fig 3　Relative sensitivity of hybrid fuel economy to system losses

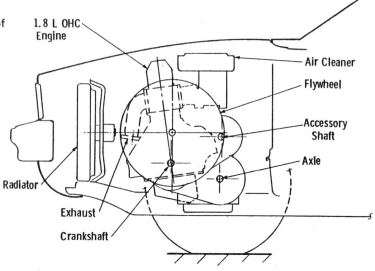

Fig 5　Vehicle installation of hybrid drive (left-side view)

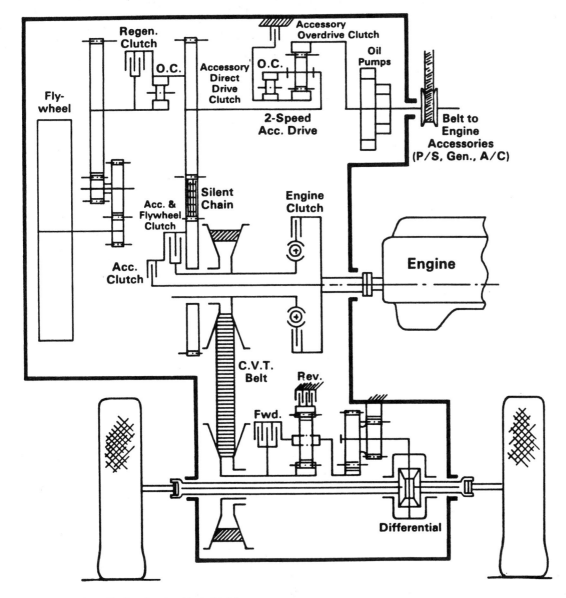

Fig 4 Engine-flywheel hybrid drive system

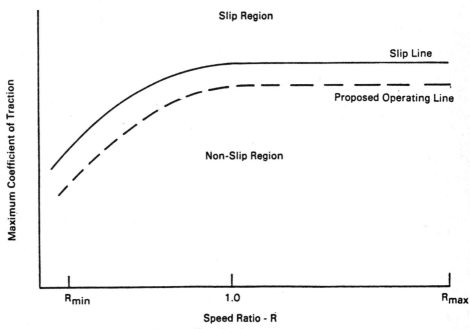

Fig 6 Proposed coefficient of traction strategy for metal-belt CVT

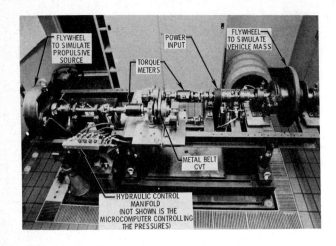

Fig 7 Photograph of belt-test fixture for dynamic testing

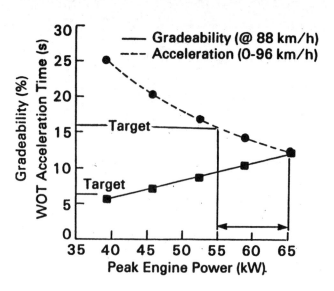

Fig 8 Vehicle performance versus peak engine power

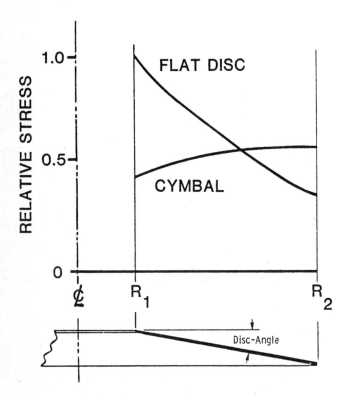

Fig 9 Relative stress for flat and cymbal-disc flywheels operating at the same speed

Fig 10 Engine compartment of gyroscopic/vehicle handling vehicle

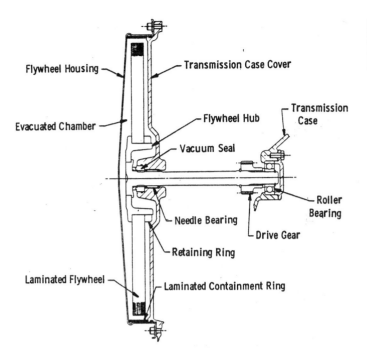

Fig 11 Schematic of flywheel system

Fig 12 Assembled flywheel in housing

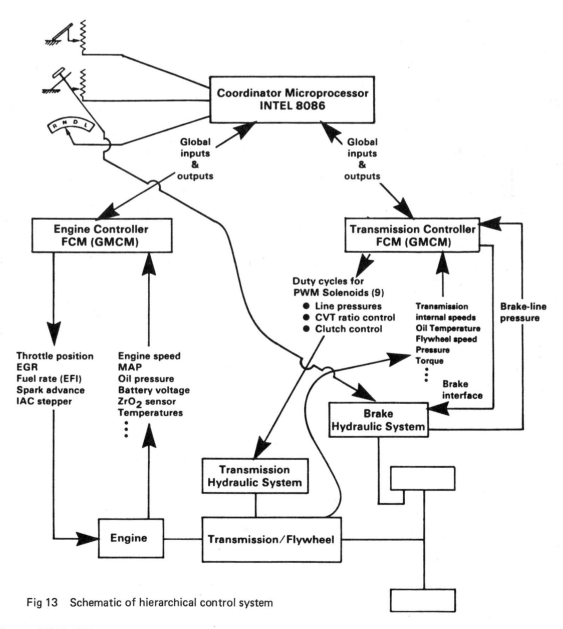

Fig 13 Schematic of hierarchical control system

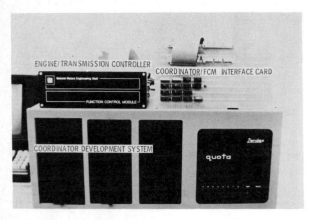

Fig 14 The electronic control development system

Fig 15 Schematic of proposed brake system

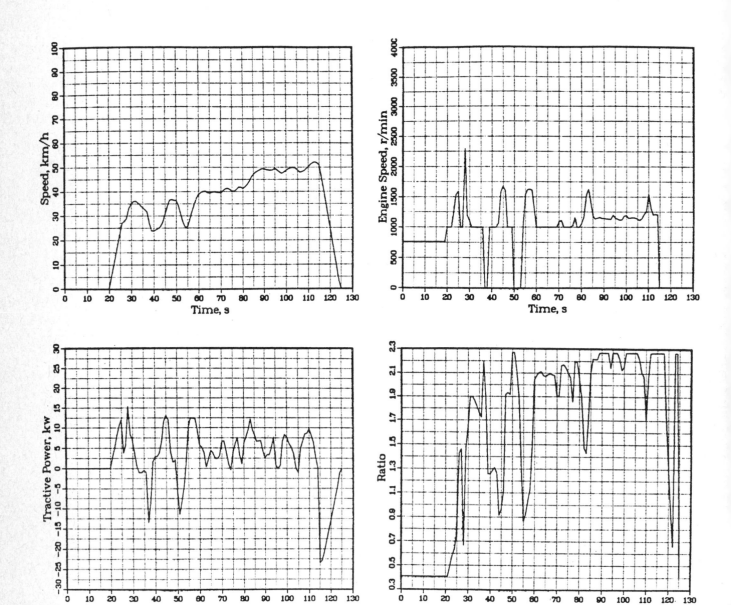

Fig 16 Vehicle speed and tractive power versus time plots for the first cycle of the EPA urban driving schedule

Fig 17 Engine speed and CVT ratio versus time plots for the first cycle of the EPA urban driving schedule

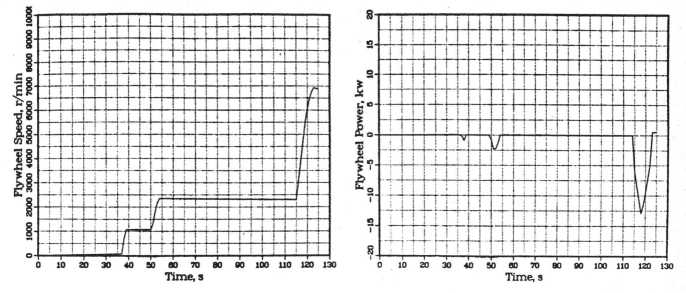

Fig 18 Flywheel speed and flywheel power versus time plots for the first cycle of the EPA urban driving schedule

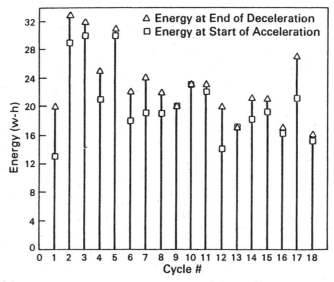

Fig 19 Energy at end of decelerations and beginning of accelerations for the EPA urban driving schedule

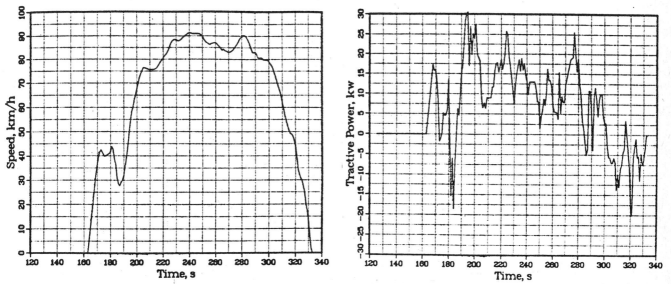

Fig 20 Vehicle speed and tractive power versus time for the second cycle on the EPA urban driving schedule

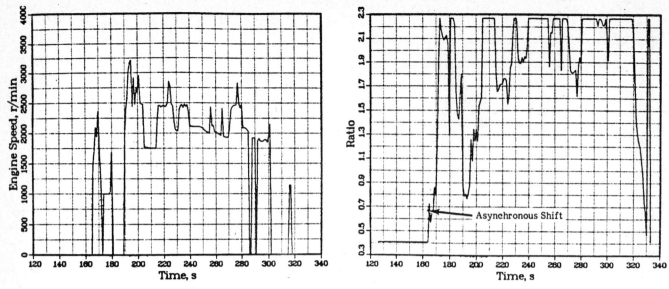

Fig 21 Engine speed and CVT ratio versus time for the second cycle of the EPA urban driving schedule

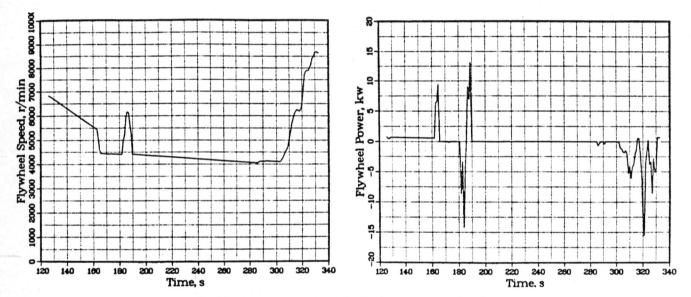

Fig 22 Flywheel speed and flywheel power versus time for the second cycle of the EPA urban driving schedule

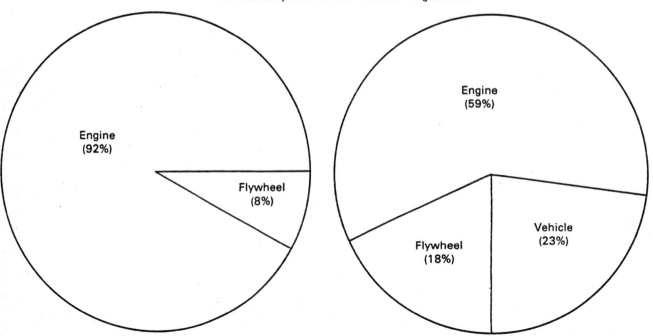

Fig 23 Distribution of motive energy; EPA urban driving schedule

Fig 24 Distribution of energy used to power the accessories: EPA urban driving schedule

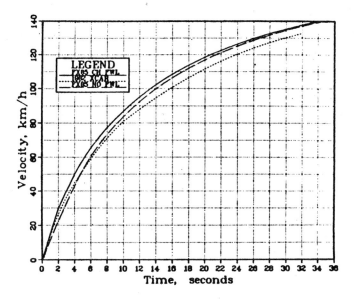

Fig 25 Wide-open throttle performance curves for the hybrid with and without flywheel charge and for a 1982 production car

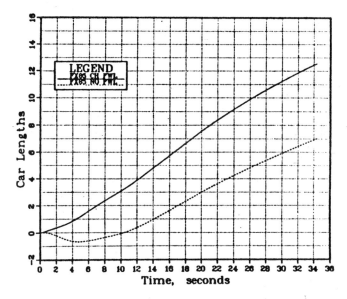

Fig 26 Relative distance travelled versus time for hybrid and for production car

C423/84

SAE 841307

The 'Unomatic' transmission

G L FALZONI
Fiat Auto SpA, Turin, Italy

ABSTRACT Work carried out by FIAT on metal belt C.V.T.s since 1974 is illustrated briefly and the electronic control transmission developed for a 2 liter displacement engine is presented.
Finally, the results are provided of fuel economy tests carried out on car with 5 speed transmission and with electronic control C.V.T.

1 INTRODUCTION

So far the main purpose of fitting cars with automatic transmissions has been that of easing driving with respect to cars with manual transmissions; this, above all, where road and traffic characteristics require frequent actuation of both clutch and transmission.
On the other hand, the very good reliability and operating characteristics of modern synchromesh transmission have strengthened the concept that the automatic transmission should be considered a "surplus" for the driver's comfort and a costly optional intended primarily for cars of medium -- high class.
Now, however, because of the problems that the car designer has to tackle as a consequence of the high cost of oil, the automatic transmission is viewed under a completely different light.
Paradoxically, whereas a few years ago penalization of fuel economy through the use of automatic transmissions was accepted and justified, today, the new automatic transmissions are considered as a means to reduce fuel consumption.
If this goal is reached, "new generation" automatic transmissions will be increasingly adopted not only on medium-high class cars but also on medium-low class cars first and on economy class later.

2 A TRANSMISSION TO "MANAGE" THE ENGINE

As known, I.C. engines can provide the same horse power values but with consumption that varies considerably depending on the R.P.M. rates at which the power is delivered. Figure 1, referred to the "UNO" 1,3 litre, 70 HP engine is a clear demonstration of this concept.
It is possible to see, for example, that 20 HP can be produced between 6 and 10 l/hr depending on the engine R.P.M. rate.
It is therefore possible to plot for each engine the curve at which power should be delivered to minimize consumption. The dash line curve in Fig. 1 shows minimum consumption.
To deliver the same power at different engine R.P.M. rates means that this is done with different throttle openings. Figure 2 shows, always for the above engine, the power curves with constant throttle opening.
It is evident that 20 HP can be delivered by the engine with 45 to 100% openings depending on the R.P.M. rate.
When the power output has been set, two parameters must be monitored to minimize consumption.
1) Engine R.P.M. rate and transmission type and mode of use.
2) Throttle opening.
With regard to the transmission type and mode of use it is interesting to observe Fig. 3 which shows the histogram of the ratios to be used in the European urban cycle to minimize consumption.
It is evident that transmissions with high ratio "spread" are required to cover the wide range of ratios.
In the case of the UNO, the spread of the 5-speed transmission is as high as 5.
For many operating conditions it would, however, be necessary to further increase the total ratio.
This cannot be done because when the lower ratio (13.37) is set to ensure the required performance, the higher ratio (2.67) and therefore the operating range shown in figure are also bound.
In theory, such operation range could be extended using a six-or higher-speed transmission. It is, however, easy to realize that gear shifting would be so frequent as to be unacceptable, above all in urban and suburban areas; the driver should, in fact, pay more attention to driving maneuvers and it would therefore be difficult for him to use the transmission with the same care.
It follows that to minimize consumptions it is necessary to automatize use of the transmission thus easing the driver's task.
In our opinion, the most effective way to solve the problem is the use of a continuously variable transmission that, when suitably automated and controlled, provides the best overall result in terms of fuel economy and driver's comfort.

3 "UNOMATIC" BELT-DRIVE CONTINUOUSLY VARIABLE TRANSMISSION

3.1 General

The continuously variable transmission based on a belt fitted between variable-diameter pulleys is well known but its use on cars became a reality only after Van Doorne Transmissie had designed and manufactured a metal belt.

This belt has permitted a reduction of transmission dimensions to values compatible with space available in the engine compartment. It was above all possible, even though with such small units, to transmit through a single belt the high power produced by present engines.

The Van Doorne belt consists essentially of a series of thin trapezoidal metal elements (about 1.5 mm thick) held in place by two special loops. These consist, in turn, of an assembly of ten thin (about 0.15 mm) concentric loops fitted one inside the other.

Thanks to this configuration the loops meet both the flexibility and mechanical strength requirements necessary for operation of the system fig. 4.

By taking this new metal belt as the basis, FIAT, in 1974, started the first exploratory studies to design a transmission for medium-class cars. In consideration of the encouraging results obtained, a cooperation agreement was signed in 1975 between FIAT and Van Doorne to manufacture metal belt - drive continuously variable transmissions.

A project resulted from the studies and tests carried out on some prototypes and in 1979 150 Ritmo 75 cars were equipped with this transmission.

One hundred of these cars were sold to selected customers and introduced on the market in January 1980; the others were subjected to severe test cycles by the Test Departments to assess, above all, quality and reliability.

The results obtained both from cars sold to customers and from those tested by the Test Departements convinced us that the line followed was the right one and that we had available "the transmission".

"The transmission" because this is not, in our opinion, an alternative to or a replacement of the conventional automatic transmissions, but a means that will permit a reduction of fuel consumption, leaving performance unchanged and at the same time improving the driver's comfort.

It is with these convinctions that FIAT started the project to introduce the CVT on the market.

The choice fell on the UNO because this car proved suited to be equipped with a transmission that greatly enhances its performance and that leaves consumption unchanged with respect to the car with 5-speed manual transmission.

3.2 Transmission Layout

The transmission, fig.5, consists of the following:
a) A front unit in which an epicyclic train, driven by hydraulic multi-disc clutches, has the dual purpose of providing pickup and Forward Drive - Neutral and Reverse.
b) A continuously variable unit in which the driving and driven pulleys, connected by the metal belt, are fitted on two parallel shafts.
c) A final reduction unit with integral differential.
d) A hydraulic control unit with gear pump that delivers oil under pressure to feed the pulleys and lubricate and cool all transmission components.

a) Front unit.
This unit consists of a torsional damper fitted between the flywheel and the input shaft to damper, rapid transients, torque irregularities at the transmission inlet.
The solution adopted has the advantage - by exploiting the Forward Drive and Reverse clutches also for pickup - of eliminating a component such as a hydraulic coupling, or converter or automatic clutches of different design, that in the automatic transmissions has been specifically provided for pickup.

b) Continuously variable unit.
Its ratio varies continuously between 1 : 2.47 in Low and 1 : 0.445 in High.

c) Final reduction unit.
This unit consists of two helical gear sets: the first set is placed on the secondary pulley shaft and on the intermediate shaft respectively (gear ratio is 1 : 1.405). The second set is mounted on the intermediate shaft and on the differential respectively (gear ratio is 1 : 3.842). The differential is of the conventional type and is provided with two constant velocity universal joints to which the drive shafts are connected.

d) Hydraulic control unit.
Consists of a gear pump driven directly by the engine to deliver oil to the valve control unit (fig.6). The group consists of 7 valves which receive oil under pressure from the pump and operate through various signals to adjust pressure and delivery of oil to the following actuators and hydraulic circuits:
- Hydraulic clutches for pickup in forward drive or reverse.
- Primary pulley whose movable half pulley effects the change of ratio.
- Secondary pulley whose movable half pulley compresses the belt depending on the input torque to be transmitted.
- Forced lubrication of the various rotating components.
- Oil delivery to the heat exchanger.

Operation of the above actuators and hydraulic circuits and, consequently, of the transmission, is controlled by the valve unit depending on the signal that this unit receives from:
- The selector on the transmission that - connected and actuated by the bowden and by the hand lever - transmits to the "manual valve" the P-R-N-D-L functions selected by the driver.
- The transmission "Kick down" bowden that - connected and actuated by the accelerator control - transmits to the "throttle valve" the position that the throttle has taken after the driver has depressed the pedal.
- The "Engine pitot tube" inside the transmission that - fitted in a centrifugal chamber rotating at the same speed as the engine - transmits to the "clutch valve" and to other valves of the hydraulic

unit a pressure signal proportional to the engine R.P.M.s.
- The "primary pitot tube" inside the transmission that - fitted in a centrifugal chamber rotating at the same speed as the primary pulley - transmits to the "primary valve" and to the "reverse inhibitor valve" of the hydraulic unit a pressure signal proportional to the primary shaft R.P.M.s.
The pressure differential between the two pitot signals provide to the hydraulic group the exact instantaneous work conditions of the clutches (disengaged - about to engage - engaged).
- The "primary pulley sensor" that - through a metal rod between primary pulley and hydraulic unit - transmits to the secondary valve the exact instantaneous position of the pulley and, therefore, the exact transmission ratio.

4 SPECIFICATIONS

- Input torque between 90 and 125 Nm
- Engines from 1100 to 1300 c.c.
- Ratio spread from 5.8 to 5.55
- Weight of transmission complete with joints, without oil: about 54 Kg.
- Oil cooling through oil/air heat exchanger fitted externally to the transmission
- Controls available to the driver:
a) Selector lever with the following positions:
  P = Park
  R = Reverse
  N = Neutral
  D = Drive   - Car always in Forward Drive with automatic operation between Low and High.
  L = Low   - Car always in Forward Drive operating automatically but mainly on low ratios with high engine braking efficiency
b) Accelerator pedal which directly affects transmission operation:
  - Slightly depressed - The transmission is rapidly set on high ratios with minimum fuel consumption
  - Very depressed   - The transmission "holds" the low ratios longer; pickup and acceleration are good
  - Fully depressed (kick down) - The transmission changes ratios only at 5000 R.P.M. with maximum pickup and acceleration.

5 PERFORMANCE AND CONSUMPTION (1300 cc. VERSION)

On the same car have been installed for test a five-speed manual transmission first and the CVT after. The shift-pattern diagram of the two transmissions is shown in fig. 7.
The results obtained were the following:

- Acceleration from rest of the CVT equipped car is, in theory, poorer than acceleration of the car with manual transmission (0 to 100 Km/h in 12,7 secs against 11,5 secs).

In practice, instead, the positions are inverted because the times indicated for the 5-speed transmission can only be obtained by an expert test driver with very fast starts and gear-shifts made in the order of tenth of seconds.
With the CVT the values shown can instead be obtained by everybody because the only maneuver required is to fully depress the accelerator pedal.
Pickup is much better than for the manual transmission because kick-down is obtained with the CVT when the accelerator pedal is fully depressed.
Finally, fuel consumption are practically the same as for the 5 - speed manual transmission.

6 ELECTRONIC CONTROL OF THE TRANSMISSION

The UNOMATIC transmission just described is provided with a fully hydraulic logic control.
A hydraulic control system, however, has some limitations due to the complex valve unit.
In other words, if we want the transmission to operate with different logics depending on the demand of the driver to fully exploit all the possibilities offered by a CVT, it is necessary to switch to an electronic control system.
In order to verify the possible advantages deriving from the use of a control system of this kind, the control system of a UNOMATIC transmission was changed from hydraulic to fully electronic.
The diagram of fig. 8 shows the control components grouped as follows:
a) Sensors
b) Hydraulic solenoid valves
c) Electronic control module

6.1 Sensors

a) Electromagnetic pick-ups to detect the engine and drive/driven pulley R.P.M.
b) Potentiometer on the carburettor to indicate throttle opening
c) Sensor to detect pulley drive oil pressure.
These are the main sensors; the less important ones are shown in fig. 8.

6.2 Hydraulic Solenoid Valves

These valves are of the proportional type and control:
1) The transmission ratio
2) Pulley operation pressure
3) Engagement of Forward and Reverse

6.3 Electronic Control Module

This module consists of an eight bit microprocessor with integrated custom circuit for the interfaces.

7 RESULTS OBTAINED

Exhaustive research was based at first on considerable theoretical preparatory work and subsequently on bench and car testing to select the control logics.
Research was aimed at defining two logics: one to raise performance and one to minimize consumption

(without penalizing driveability).
An outline is given here only of the work done to minimize consumption. To this end a series of comparison tests were carried out using the same car fitted with:
- A standard production manual transmission (ratio spread 5).
- A CVT (ratio spread 5.55) and a hydraulic control logic with low to high up-shift point at 1750 R.P.M.
- A CVT and an electronic control logic.

For all versions provisions were made to obtain more or less equivalent performances.
These versions and constraints permitted an investigation and evaluation to be made of the parameters which affect consumption and driveability either positively or negatively.

Research has evidenced that the more important parameters are essentially two:
a) Pulley clamping pressure as a function of the torque to be transmitted.
b) Low to high up-shift point.

Whereas the former does not affect driveability, the latter is important to this effect and is directly linked to what stated on page 277, (power can be delivered with consumptions which vary considerably as a function of the R.P.M. rate and valve opening at which power is produced).

A long series of studies and testing was therefore conducted to link the engine R.P.M. parameter (by working on the drive ratio) to throttle opening. And this to avoid any detrimental effect on driveability when power is delivered with minimum consumption at any operating point.

The result of research to fit the CVT with electronic control is summarized in Fig. 9. The full line shows the link that must be obtained between throttle opening and engine R.P.M. to ensure driveability. The dash line shows, instead, the link that must be obtained, always between throttle opening and engine R.P.M. to ensure minimum consumption.

It is possible to see that the compromise reached is fully satisfactory because driveability (full line) is achieved by running the engine practically with minimum consumption (dash line).

The tests also indicated that correct management of the pulley clamping pressure parameter was only possible through the electronic control. Moreover, this was the only means whereby the up-shift point could be lowered considerably.

Thanks to the foregoing, equivalent average consumption were obtained from the manual transmission and from the CVT with hydraulic logics.

Consumptions with electronic control transmission are instead illustrated hereafter along with the two parameters which affect them significantly.

a) <u>Pulley Clamping Pressure</u>
By modulating the pressure according to the torque to be transmitted the ECE 15 cycle consumptions are reduced by 5% (Fig. 10).
At steady speed, instead, consumption is not affected significantly (reductions are of the order of 1 to 2%).

b) <u>Up-shift point from low to high ratio</u>
A decrease in engine R.P.M. in low ratio significantly reduces consumption in the ECE 15 cycle. Such reduction is about 12% when the up-shift point is set at 1000 R.P.M. rather than at the 1750 R.P.M. (See Fig.10).
Car driveability is still good provided the engine has been designed to operate with a C.V.T.
Good driveability was, however, obtained by suitably setting throttle opening in relation to engine R.P.M. rate.
At a steady speed of 60 Km/h reduction of the R.P.M. rate gives an 8% economy gain because the control law imposes operation at a lower R.P.M. rate.
Such influence becomes nil at higher speeds because the transmission always handles the powers in the high ratio.

8 CONCLUSION

Use of the electronically controlled belt-drive CVT permits a reduction of average fuel consumption with respect to the same car equipped with 5 - speed transmission. Moreover, performance penalties are minimum.
Since this transmission is automatic and continuous, driving comfort is also greatly improved.
Finally, by suitably programming the transmission electronic control system it is possible to use the car according to different performance, economy and other logics by simply operating, for example, the accelerator pedal and/or a suitable selector lever.

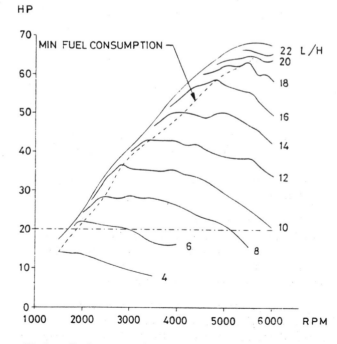

Fig 1 Fuel consumption curves related to engine speed and power

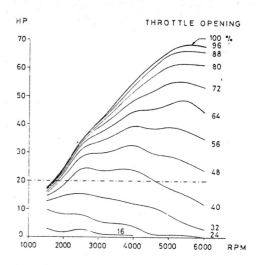

Fig 2   Power curves related to throttle opening

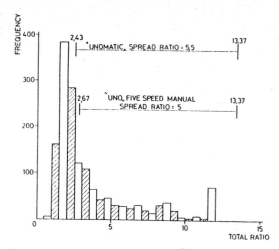

Fig 3   Ratios employed to minimize fuel consumption in ECE 15 cycle

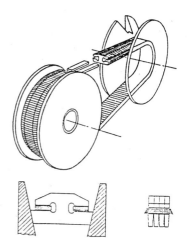

Fig 4   VDT's pulleys and belt system

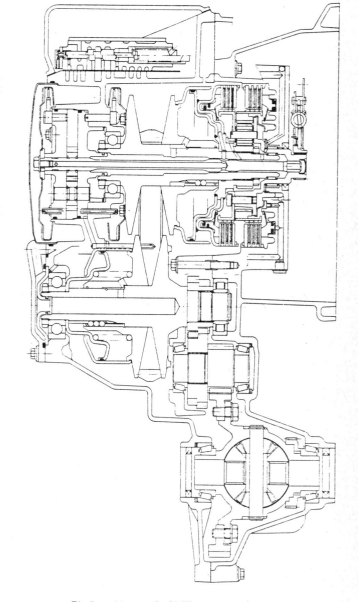

Fig 5   Unomatic CVT's cross section

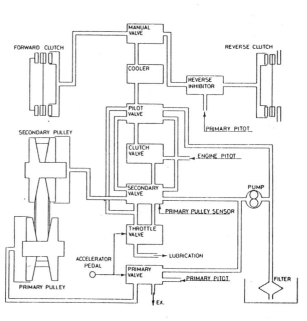

Fig 6   Unomatic CVT's hydraulic schematic

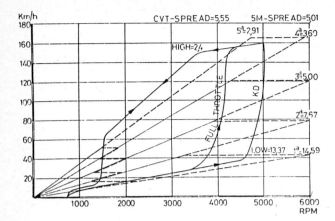

Fig 7 Unomatic CVT's and five speed manual transmission variogram

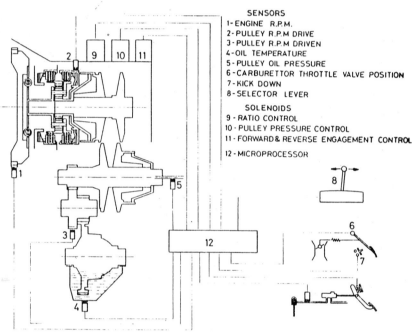

Fig 8 Unomatic's electronic control system

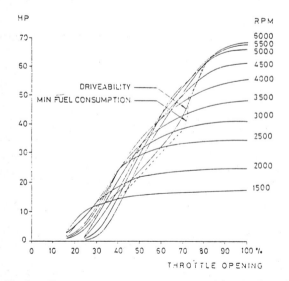

Fig 9 Power curves at constant speed related to throttle opening

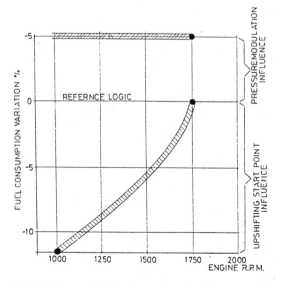

Fig 10 Unomatic's fuel consumption improvements obtainable by means of electronic control

C442/84

SAE 841308

Field experience with a fleet of test cars equipped with Comprex supercharged engines

P REBLING, Dipl-Ing and F JAUSSI, Ing HTL
BBC Brown, Boveri and Company Limited, Baden, Switzerland

The Comprex was developed by BBC to the series production stage in cooperation with various automobile manufacturers.(Ref. 1-6)

In order to subject the consequent design of the pressure wave machine components to realistic operating conditions twenty-seven vehicles of identical type (Fig.1) were put into on-road operation by BBC. The vehicles have been in use since the middle of 1983. All vehicles are powered by a 1.6 ltr. swirl chamber engine which was designed to be turbocharged.

CONTENTS

1. Comprex Installation Design
2. Engine Performance
3. Experience and Data obtained with the Test Cars
4. Conclusions
5. References

1. COMPREX INSTALLATION DESIGN

- The Passat is a very popular car in Europe and in Switzerland in particular. It was at this time the only one in this category having a 1.6 ltr. diesel engine, which could be pressure-charged. The definitive mounting arrangement can be seen in Figs. 2 and 3. Because of the required speed ratio, the alternator shaft was incorporated into the engine/Comprex transmission train; by this arrangement it was possible to use moderate pulley diameters. Since the belt must be insensitive to dirt, slip and the positional errors encountered in installation, a V-belt raw-edged and notched was used. The V-belt is tensioned between the alternator and the Comprex by an additional tensioner pulley (Fig. 4).

- An advantage of this Comprex arrangement is the configuration of the exhaust bend. The relatively large bend radii of the exhaust pipe,in conjunction with its large diameter, helps to keep the internal flow resistances low. By the Comprex itself frequencies up to 500 Hz are reduced to an acceptable level, therefore only high frequencies (500Hz must be damped, the measures are absorption type silencers in the exhaust-outlet system.

- The position of the Comprex air casing permits a short connecting line to the intake air filter housing, which was adapted to the increased air mass flows while retaining a small cross-section.

- Special care was taken not to obstruct accessibility to the parts which require servicing. Because of this arrangement a special exhaust gas manifold had to be designed onto which the Comprex is mounted. The configuration of the exhaust gas receiver was governed by three partly contradictory objectives:

    * Smoothing of the pressure pulsations in the exhaust gas manifold
      → large volume.

    * Small surface to keep the heat loss low
      → small volume.

    * Good accessibility in order that the exhaust gas receiver can be mounted on the cylinder head with stud bolts
      → small volume.

  The configuration, which represents a compromise to account for the aforementioned objectives, is shown in Fig.5. This design also affords easy installation with automatic tools. GGG 40 (DIN) was selected as a casting material. It has proven its stability of shape at high temperatures in many accelerated endurance tests at BBC.

- On the charge air side, depending on application an air-air cooler could be installed between the Comprex and the engine inlet manifold.

- The crankcase vent has to be changed; i.e. blow-by-gas into the intake line is not advisable because the mixture of soot particles, stemming from the internal recirculation (which reduces the NOx), and the oily air would result in inadmissible contamination of the charge air system. It was therefore decided to inject the blow-by gases into the exhaust pipe by means of an ejector. Fig. 6.
  Exhaust emissions measured at full load both with and without blow-by to the exhaust showed no measurable difference.
  To protect the engine against deleterious operating conditions, the supercharging system is continuously monitored. (Fig.7)

- The parts required for the conversion are shown in Fig.8 along with the corresponding parts for the turbocharged version in Fig.9.

The Comprex installation kit without charge air cooling consists of 52 parts and with CAC 55 parts. The corresponding turbocharger installation kit (w/o CAC) consists of 43 parts.

2. ENGINE PERFORMANCE

2.0 TEST BED EVALUATION

The variants mentioned and described below were tested on the same engine on the same test rig; only the adjustments or accessory parts were changed. The objective of these measures was to minimize performance scatter.

All tests were performed on an engine test rig at BBC Brown, Boveri & Co. Ltd. in Baden, Switzerland. The test rig equipment reflects the latest state of the art in the automotive industry.

In the Performance diagrams to follow, charge air pressure and charge air temperature are used as characterising values for the Comprex. Brake mean effective pressure, fuel injection flow rate (BETA), max. cylinder peak pressure, exhaust-gas temperature, smoke number and brake specific fuel consumption in function of engine speed at full load are used as characteristic values of engine.

2.1 ENGINE VARIANTS

The vehicle and the engine are both of production design.

The measurements on the individual engine variants were performed in the following sequence :

2.1.1 TURBOCHARGED ENGINE

The equipment and adjustment, especially that of the fuel injection pump, correspond to the production status 1983 for delivery ex works. The standard engine has no charge air cooling.

2.1.2 COMPREX-SUPERCHARGED ENGINE: OPTIMIZATION WITH STANDARD INJECTION PUMP SETTING

The objectives were not to exceed any of the engine limits, and to demonstrate how by the selection of the supercharger the performance of the engine could be improved.

For simplicity reasons no changes were made to the fuel injection pump from the settings of the turbocharged engine.

The greater injected quantities at particular engine speeds result from the distinctly higher charge air pressure of the Comprex-supercharged engine, de-activating the smoke limiter. The supercharging was limited by the charge air pressure due to the maximum allowable pressure of 130 bars in the swirl chamber.

Excessive loading of the engine was hence avoided in the version without charge air cooling, and thermal loads were even lower in the version with charge air cooling.

The volume of the exhaust gas manifold (between the engine and the Comprex) was optimized. Similar investigations were performed to select the Comprex specification and the speed ratio.

2.1.3 COMPREX-SUPERCHARGED ENGINE: OPTIMIZATION FOR INCREASED OUTPUT

In contrast to the foregoing exercise, here the possibilities of the Comprex are more fully exploited. To achieve this objective, with the versions with and without charge air cooling it was necessary, especially in the lower speed range, to increase the fuel injection flow.

2.2 COMPARISON OF THE INDIVIDUAL VARIANTS

2.2.1 TURBOCHARGED ENGINE COMPARED WITH COMPREX-SUPERCHARGED ENGINE WITHOUT AND WITH CHARGE AIR COOLING

2.2.1.1 BRAKE MEAN EFFECTIVE PRESSURE (Fig.10)

Because of the high charge air pressure the air mass flow to the engine increases and the charge air pressure-dependent fuel supply also delivers a higher fuel flow. For these Comprex-supercharged engines this results in a markedly higher BMEP particularly in the lower speed range.

Above 2400 rpm the Comprex-supercharged engine without charge air cooling with the same pump adjustment has the same values as the turbocharged engines; otherwise, the BMEP of this version is up to 2 bar higher.

2.2.1.2 FUEL INJECTION QUANTITY (MM^3/STROKE) (Fig. 11)

Because of the higher charge air pressure P_2 in the speed range 1000 - 2000 rpm, without an alteration of the fuel injection pump the Comprex version injected a substantially larger quantity/stroke.

In the range of higher engine speeds the quantities became equal again. With the engines with increased output the fuel injection flows clearly increased. This applies especially to the version with charge air cooling.

2.2.1.3 MAXIMUM CYLINDER PEAK PRESSURE (Fig.12)

With none of the five variants is the maximum swirl chamber pressure limit of 130 bar exceeded. However the Comprex-charging gave a higher BMEP than turbocharging, particularly in the lower speed range, resulting in a higher peak cylinder pressure.

2.2.1.4 EXHAUST-GAS TEMPERATURE (DEG.C) (Fig. 13)

The exhaust-gas temperature of all the measurements have only a small deviation. The temperature difference of the measurement : Comprex with charge air-cooler, pump adjustment same as with turbocharger is due to the difference in air-fuel-ratio.

2.2.1.5 SMOKE NUMBER (BOSCH UNITS) (Fig.14)

The shape of all measurements are within the acceptable area.

2.2.1.6 SPECIFIC FUEL CONSUMPTION (gKWh)

FULL LOAD : (Fig. 15)

Over almost the entire speed range the consumptions of the Comprex engines are lower than those of the turbocharged engines. At $n_M=1600$ rpm differences of up to 40 gr/kWh are achieved. Above n_M = 3000 rpm the differences become less. An exception is the Comprex-supercharged engine with unaltered fuel injection pump adjustment and charge air cooling, which also consumes substantially less than the turbocharged engine, on average 10 gr/kWh.

PART LOAD : (Fig. 16)

The fuel consumption map shows the Comprex-supercharged engine to be superior at most part loads. When the BMEP is less than 2 bar the turbocharged engine has slight advantages under steady state conditions.

2.2.1.7 CHARGE AIR PRESSURE (Fig. 17)

Here, the superiority of the Comprex-supercharged engines is most clearly seen. Only above a speed of 3000 rpm do the pressure of the turbocharged engine and the Comprex-supercharged engine approach one another with unaltered injection pump adjustment (for the version w/o CAC).

2.2.1.8 CHARGE AIR TEMPERATURE (Fig. 18)

Due to higher charge air pressure, the charge air temperature of the Comprex versions is also clearly above that of the turbocharged version.

With the Comprex-supercharged engines with charge air cooling the thermal loading of the engine is estimated to be of the same order as that of the turbocharged engine.
In the test rig measurements the cooler efficiency was set at 60%, which is a realistic figure for on-road operation.

3. EXPERIENCE AND DATA OBTAINED WITH THE TEST CARS

3.1 COMPREX PERFORMANCE AND DURABILITY

- From the experience with the test fleet no negative effects of any kind on the drive train could be traced to Comprex-supercharging. This is especially noteworthy for the engine which endured the relatively high pressures and temperatures without damage.

- With one exception no malfunctions of any kind were encountered with the Comprex pressure-wave machine. The only trouble which has occured up to this time is attributable to a foreign body having found ingress into the cell wheel. A deformation of the cell walls on the exhaust gas side was established which affected the intake capacity of the Comprex to the extent that the maximum charge air pressure was reduced by 0.2 bar.

- On the other hand, negative experience was gained with a few of the ancillary parts. This included primarily the tension pulleys whose grease-lubricated-for-life ball bearings had the tendency to seize. By modifying the bearings the problem was solved. In the first vehicles leakage of the wire-reinforced rubber hose connections on the charge air side was experienced. By the use of couplings without wire reinforcement the leakages ceased.

All other components performed the functions assigned to them satisfactorily.

3.2 ROAD TESTING-ACCELERATION, RESPONSIVENESS, CONSUMPTION (Table 1)

The measuring instruments were as are commonly used today in the automotive industry.

The measuring facility was not a test rig, but a specially selected and specially surveyed highway. To keep the measuring errors as low as possible, the instruments were recalibrated for each run.

All test runs were made by the same two persons. The values shown in table 1 are arithmetic mean values from three vehicles per version.

3.2.1 ACCELERATION

The advantages of the Comprex-supercharged engine over the turbocharged version are most clearly demonstrated by the time required for 1000 m : Δt = 2 sec, and in the distance travelled in 4 sec : Δd = 2m.

3.2.2 RESPONSIVENESS

With increasing speed change not only a greater difference between turbocharged and Comprex-supercharged vehicles was established, but also a greater difference between the Comprex-supercharged vehicles with and without charge air cooling. This applies for fourth gear, just as well as for fifth gear.

3.2.3 FUEL CONSUMPTION

3.2.3.1 CONSTANT SPEED (Table 1)

The differences in vehicle performance lie within the instrumentation scatter. Only the consumption of the Comprex-supercharged vehicle with charge air cooling was clearly below that of the other two versions.

3.2.3.2 ROAD-AVERAGE CONSUMPTION OF THE TEST FLEET

The following consumptions were established as an average from all vehicles delivered to users. At the time this paper was written the average distance travelled by each vehicle was 25 000 km (min. : 8000 km. max.: 60 000 km). In our opinion the drivers, as well as the roads on which driving was done, are representative of the national averages. All test cars accumulated a total mileage of 680 000 km.

The average consumption was 6,6 liters/100 km, the lowest consumption being 5,9 liters/100 km, and the highest 7,5 liters/100 km. These differences result from :

Driving style, roads driven, vehicle, vehicle equipment, tires, season etc.

3.3 NOISE MEASUREMENTS (Table 2)

The drive-by noise was measured according to ISO 362 and the local regulations for noise emission of passenger cars.

All values given in table 2 were obtained with three vehicles.

Since the vehicles with five-speed transmissions must be driven into the measuring section in third gear, the corresponding values are valid. The values measured are clearly below the upper limit presently valid for Switzerland 78 dB (A).

4. CONCLUSIONS

Generally speaking, from all the information available the following conclusions can be drawn :

- The selection of the test vehicle PASSAT as a suitable and reliable carrier for the test programm was vindicated.

- The Comprex superchargers have achieved the performance objectives, as well as the expected high operating reliability.

- The vehicle owners are very satisfied with the behaviour and performance of their vehicles.

- By optimizing the adaption of engine and vehicle (transmission, rear axle) to the higher torque of the Comprex-supercharged engine, further improvement could be expected.

5. REFERENCES

1. A. Mayer, G.M. Schruf : Practical experience with the pressure-wave supercharger COMPREX on passenger cars, i. Mech E 1982, C 110/82

2. G. Gyarmathy : How does the COMPREX Pressure-Wave Supercharger work ? SAE 830234

3. A. Mayer, G.M. Schruf : Downsizing and Downspeeding of automotive Diesel Engines SAE 820443

4. Croes N. : Die Wirkungsweise der Taschen des Druckwellenladers COMPREX ; MTZ 40 (1972) Z.

5. G.M. Schruf, T.A. Kollbrunner : Application and Matching of the Comprex Pressure-Wave Supercharger to Automotive Engines SAE 840133

6. G. Zehnder, A. Mayer : COMPREX Pressure-Wave supercharging for automotive diesels - state of the art. SAE 840133

Table 1 Passat comparison: Turbo-Comprex

			Turbo	PWS w.o. CAC	PWS w. CAC
Acceleration	0 - 100 km/h	s	15,7	14,8	14,3
	0 - 400 m	s	19,9	19,2	19,1
	0 - 500 m	s	23,5	22,2	22,2
	0 - 1000 m	s	30,9	32,8	32,9
	Distance during the first 4 sec. measured in METER (m)				
	V_{max} (km/h)		160,1	161,1	162,5
Driveability / Elasticity	4. Gear (i = 0,91)				
	40 - 60 km/h	s	7,1	5,4	5,3
	40 - 80 km/h	s	14,7	11,5	10,6
	40 - 100 km/h	s	21,5	17,2	15,0
	40 - 120 km/h	s	28,9	23,1	22,0
	60 - 120 km/h	s	21,9	19,5	19,0
	5. Gear (i = 0,71)				
	40 - 60 km/h	s	9,8	8,0	7,7
	40 - 80 km/h	s	19,0	15,7	14,7
	40 - 100 km/h	s	28,3	23,9	22,6
	40 - 120 km/h	s	37,9	32,5	30,3
	60 - 120 km/h	s	31,6	28,4	25,5
Fuel consumption	60 km/h	l/100 km	3,5	3,5	3,35
	90 km/h	l/100 km	4,5	4,5	4,35
	120 km/h	l/100 km	6,3	6,3	6,20
	150 km/h	l/100 km	11,1	10,9	9,60
Test-Weight		kg	12 10	12 10	12 15
Tires			185/70 SR	185/70 SR	185/70 SR
Ambient pressure		bar	0,974	0,977	0,976
Temperature		°C	27	20	25
Axle ratio			4,11	4,11	4,11

Table 2: **Drive by noise according to ISO 362**

vehicle with 5-speed gearbox

		Turbo db (A)	PWS w.o. CAC db (A)	PWS w. CAC db (A)
3. gear 50 km/h	left	75	73	73
	right	73	73	73
2. gear 50 km/h	left	81	78,5	78,5
	right	80	77,5	77,5

All values are arithmetic mean values
(3 measurements)

Fig 1   BBC's VW Passat test fleet 'Comprex Diesel'

Fig 2   View of engine compartment

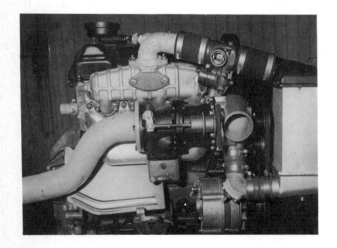

Fig 3   R.h. side view of Comprex installation showing drive via alternator, favourable exhaust layout and air/air intercooler

Fig 4   Front view showing Comprex belt drive

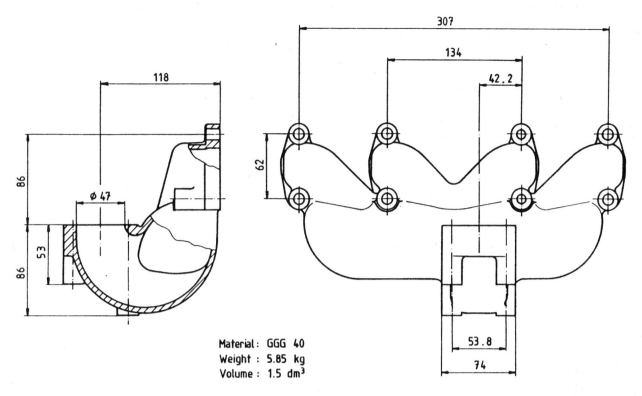

Material: GGG 40
Weight: 5.85 kg
Volume: 1.5 dm³

Fig 5 Exhaust gas manifold — Comprex version

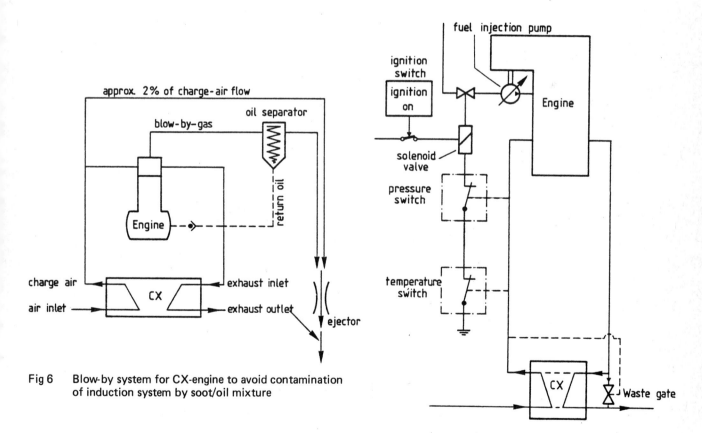

Fig 6 Blow-by system for CX-engine to avoid contamination of induction system by soot/oil mixture

Fig 7 Schematic diagram of engine overload protection system

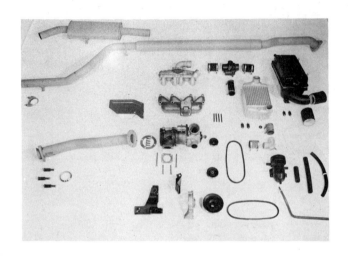

Fig 8   Comprex conversion parts

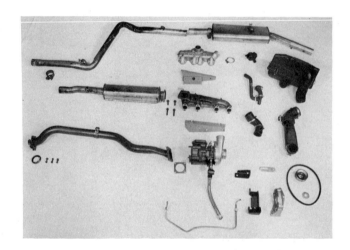

Fig 9   Turbocharger original parts

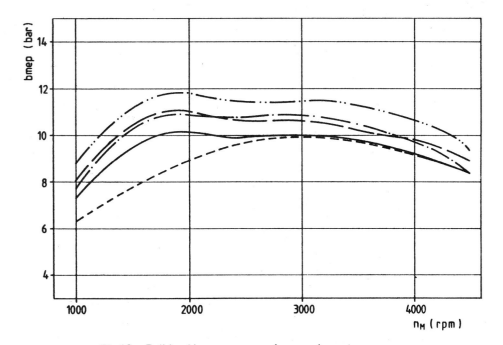

Fig 10 Full load bmep versus engine speed
– – – – – – engine with turbocharger (as-delivered condition)
– – – – engine with Comprex with charge air cooler. Pump adjustment same as with turbocharger
——— engine with Comprex without charge air cooler. Pump adjustment same as with turbocharger
—··—··— engine with Comprex with charge air cooler, optimization for increased output
—·—·— engine with Comprex without charge air cooler, optimization for increased output

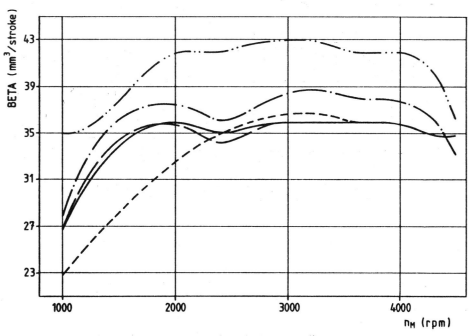

Fig 11 Full load fuel injected quantity versus engine speed

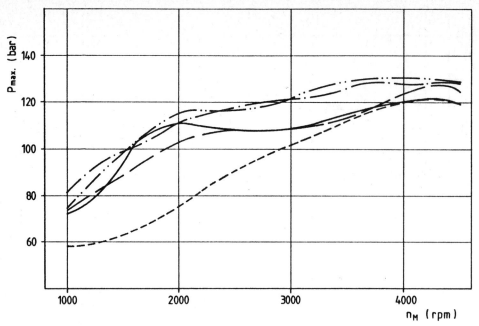

Fig 12 Maximum pressure inside swirl chamber versus engine speed

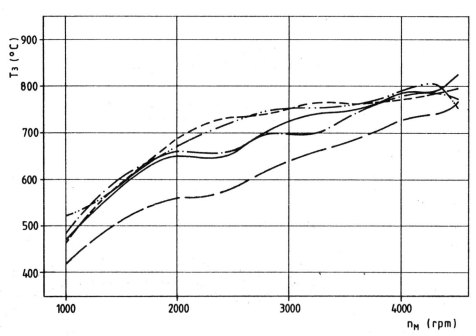

Fig 13 Full load exhaust gas temperature versus engine speed

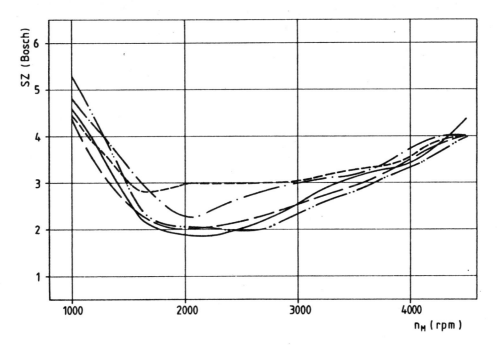

Fig 14 Full load smoke number versus engine speed

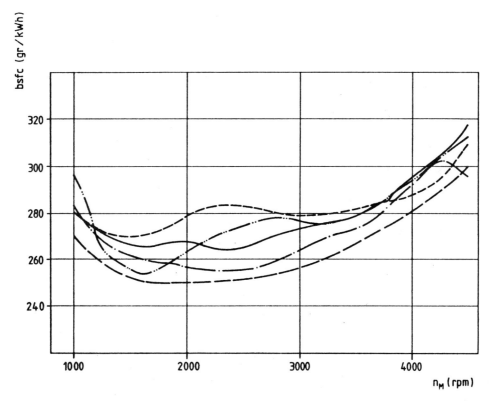

Fig 15 Full load bsfc versus engine speed

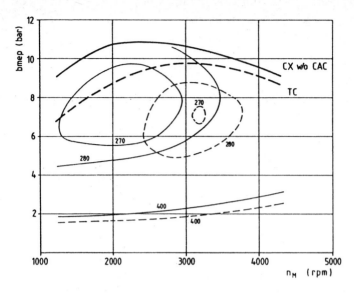

Fig 16 Fuel consumption maps: Comprex and turbocharger

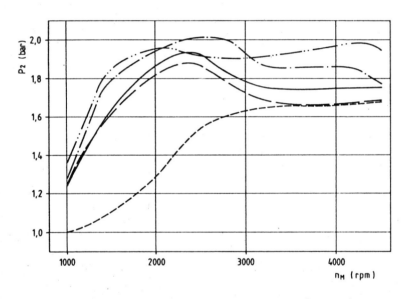

Fig 17 Full load charge air pressure versus engine speed

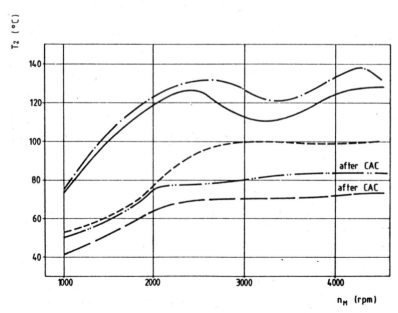

Fig 18 Full load charge air temperature versus engine speed

C437/84

SAE 841309

Development of Mazda fuel-efficient concept car

K YOKOOKU, A NAGAO and H ODA
Toyo Kogyo Company Limited, Hiroshima, Japan

SYNOPSIS There are growing demands for higher-performance yet less-gasoline-consuming cars, thus combining the essential automobile function with fuel conservation in these days of slow economical growth. In view of this background, development was made on a concept car incorporating maximum use of technologies which meet such demands and have high feasibility for production. A 1490cc four-cylinder engine had a number of fuel-saving techniques incorporated, such as those for reducing mechanical resistance of the piston and rings and increasing compression ratio, swirl intensity control system (Dual Induction System), and variable displacement system. Also, in order to bring out a maximum performance of the engine, an electronic system for controlling fuel flow, ignition timing, EGR ratio, and idle speed was adopted. The concept car has been reduced in its inertia weight class from 1000 kg, original production-level, to 875 kg by using light-weight materials. The rolling resistance also has been reduced. Through matching and evaluation including the choice of the gear ratio, the concept car has achieved 30 percent fuel economy improvement over a 1982 model, while keeping a power output of a present production car level. In addition, application of a simulation program has helped identify the contribution of each design factor to the fuel economy improvement.

1 INTRODUCTION

Since the 1973 oil crisis, fuel economy improvement has been a predominant consideration in developing automobiles the world over. In Japan, automakers were forced to trade it off to meet the 1976 exhaust emissions standards but have since been making continued efforts to develop technologies that improve fuel economy while meeting emissions regulations. Regarding the engine, their development efforts have been aimed at improving combustion efficiency and mechanical efficiency and at developing an engine control system to maximize engine performance. As for the vehicle, rolling resistance reduction, aerodynamic drag reduction, and weightsaving have been given major attention. As a result, fuel economy of Japanese cars has been improving year by year, as shown in Figure 1, but it is predicted nonetheless that at the end of the 1980 decade there will be a demand for a roughly 30 percent better fuel economy over the 1982 level.

This program explored the feasibility of achieving the 30 percent fuel economy improvement without sacrificing performance. It was assumed in the study that mass production technologies needed to achieve that goal will have been available by 1990.

2 NOTATION

be : brake specific fuel consumption [BSFC]
bi : indicated specific fuel consumption [ISFC]
Pe : brake mean effective pressure [BMEP]
Pi : indicated mean effective pressure [IMEP]
Pf : friction mean effective pressure [FMEP]
Pft : pumping loss caused by throttle
Pfc : compression loss caused by gas leak and cooling
Pfv : valve train drive loss
Pfm : friction losses except Pft, Pfc, and Pfv

3 OBJECTIVES OF DEVELOPMENT

3.1 Development Plan

The development was aimed at:

(1) attaining a 30 percent better fuel economy over the base '82 model on the Japanese 10-mode city cycle for emissions testing, while complying with the Japanese 1978 emissions regulation.

(2) guaranteeing an engine quietness and performance at least equivalent to those of the base '82 model, thus maintaining or upgrading the driveability of the current production car.

(3) assuring that the fuel economy improvement goal be realized in the 10-mode city cycle which represents a primarily low-speed operation in Japan.

It has been confirmed that engine modification is effective for fuel economy improvement in the low speed dominant mode (1). In this program, therefore, the vehicle modification was confined to such weightsaving and vehicle resistance reduction as will not affect the car's safety and reliability, and the emphasis was put on the engine modification to improve fuel economy.

3.2 Development procedure

The following three steps were taken for development:

Step 1: The vehicle improvement items were specified according to the development plan. Then a fuel economy improvement from vehicle modification (10 percent) was estimated from a simulation based on data in Table 2, and the improvement goal for engine modification was set at 20 percent. The planned vehicle modification consisted of material substitution for vehicle weight reduction from the 1000 kg rank to the 875 kg rank and the use of tires with low rolling resistance.

Step 2: To achieve the fuel economy improvement goal by the engine modification, study was made on the technologies shown in Table 3. Because the compatibility of each technique to the vehicle was unknown, there was a possibility that a built-up concept car might end up with a less fuel economy gain than expected. Therefore, selection was made of those technologies presumed to provide more than a 30 percent gain in total. The selected technologies are outlined below.

1) Improvement of mechanical efficiency --- The fuel economy improvement effect $\Delta be/be$ by the friction loss reduction ΔPf can be expressed as:

$$\Delta be/be = bi \cdot \Delta Pf/(Pe + Pf).$$

Consequently, $\Delta be/be$ increases as the engine load decreases. Therefore, an attempt was made to reduce the piston system friction, which is the greatest of the losses by the main engine components.

2) Improvement of combustion efficiency --- A swirl intensification system called "Dual Induction System" (DIS) was used to ensure a wide open throttle (WOT) output while offering a rapid combustion in the part load region.

3) Reduction in pumping loss --- A variable displacement system (VDS) was incorporated in which the engine can run on two cylinders under light load and switches to four cylinders for sufficient output under heavy load.

4) Electronic control system --- sophisticated controls of fuel flow, EGR ratio, and ignition timing would be required with the use of the variable systems mentioned in 2) and 3) above. To cope with this, an electronic control system was used. Figure 2 outlines the fuel economy improving systems mentioned.

Step 3: The development was pushed forward with those fuel saving technologies integrated in the vehicle; gear ratio optimization was attempted to match engine output with vehicle load characteristics, while at the same time the potential of fuel economy was pursued along with evaluation of performance and driveability.

4 MAJOR TECHNOLOGIES USED IN THE FUEL-EFFICIENT CONCEPT CAR AND THEIR EFFECTS

4.1 Engine Modification

(1) Improvement of Mechanical Efficiency

The relationship between BMEP (Pe), IMEP (Pi), and FMEP (Pf) can be described as:

$$Pe = Pi - Pf.$$

The engine friction factors were analyzed through motoring tests at 25 rev/sec, which is a typical engine speed on the 10-mode procedure. The results are shown in Table 4. The friction of the piston accounts for approximately 50 percent of main engine components' friction loss. The piston friction loss (Pfp) was analyzed by varying the total tension of the piston rings (Wr), i.e., top ring, second ring, and oil ring. Presuming that the piston friction loss consists of the piston ring friction loss (Pfr), which increases as the piston rings' total tension increases, and the piston skirt friction loss (Pfs), which is independent of Wr, as shown in Figure 3a, the following equations were formulated from experiments:

$$Pfp = Pfs + Pfr,$$
$$Pfs = 0.040 + 0.15 \cdot Cm,$$
$$Pfr = (0.023 + 0.03 \cdot Cm) \cdot Wr$$

where Cm = mean piston speed (m/sec)

It was found as a result that the piston friction loss is 40 percent on the skirt, and 60 percent on the rings. Thus, reduction of the piston skirt friction loss was attempted first. The piston skirt friction is influenced by the clearance between the skirt and the cylinder wall. The effect of piston-to-bore clearance (C) on piston friction is expressed by experimental equation, $Pfs \propto C^{-0.25}$. The piston clearances shown were measured at 20°C (room temperature). In the firing condtion, as piston skirt temperature rises, the clearance diminishes due to the difference in thermal expansion between the aluminium piston and the cast iron cylinder. Piston skirt's force resulting from thermal expansion toward the sliding surface can be reduced by decreasing piston skirt rigidity. In decreasing piston skirt rigidity, attention must be paid to the following matters. Because of its function of properly holding the piston posture, the piston skirt must not develop permanent deformations and is required to have favorable surface for lubrication. Consequently, an optimum combination of piston skirt rigidity and piston profile is required to assure that under the compression/combustion pressures, the piston skirt undergoes a proper elastic deformation and maintains good lubrication surface. In this program, a new profiling method using a composite coating (5) was employed, which permitted an optimal combination of rigidity and profile to be determined effectively, and this made possible some 50 percent reduction of skirt rigidity. As a result, at an average 90°C piston skirt temperature measured in the 10-mode cycle operation, the piston skirt friction loss was cut in half, as shown in Figure 3b.

As to the piston ring friction loss, investigation was made on the rings' influence by reducing the tensions of the top, second, oil rings, but it turned out that reduction of oil spring tension greatly affects the lubricant oil consumption. Therefore, tension of the top and second rings only was reduced by 20 percent. However, this measure reduced Pf by 3 percent, and together with the skirt rigidity reduction, Pf was cut by 10.8 percent.

(2) Control on Swirl Intensity --- Dual Induction System

EGR system is a simple and effective means to reduce NOx emission, but it has a problem of destabilizing the combustion. It is well known that intensifying the swirl increases the burning rate, thereby reducing the EGR-caused deterioration of combustion, with resultant benefits of reducing exhaust emissions and improving fuel economy (2)(3)(4).

However, the swirl-intensifying technologies to improve combustion generally tend, in WOT, to reduce output from increased intake air resistance and to increase engine noise from excessively fast combustion (2). To solve these problems, a "Dual Induction System" (DIS) was developed, which continuously adjusts swirl intensity according to the demand for the optimum swirl intensity on a given operation condition (4). DIS has a primary and a secondary passages (See Figure 2). The primary passage is so designed that the mixture is given an optimal direction to form a swirl in the combustion chamber. The secondary passage is designed to direct the mixture to the bore center straightly so as not to form a swirl. Therefore, by controlling the opening of the "Swirl Control Valve" (SCV) installed at the secondary passage, the ratio of the mixture coming into the two passages is controllable, and thus the swirl intensity is controllable. The engine performance and swirl intensity relative to SCV opening is shown in Figure 4a. The swirl intensity was measured on a flow test stand by varying the intake valve lift using the paddle wheel method. Data of MBT and BSFC were at an engine speed of 25 rev/sec and at a BMEP of 294 kPa. From these data, it is understood that fuel economy does not improve by simply intensifying the swirl even at part load; there exists an optimum swirl intensity, i.e., an optimum SCV opening. Thus the valve opening was set in such a way that at part load the swirl is intensified enough to maximize fuel economy. As is evident from this figure, change in the swirl intensity results in the change in MBT, making necessary a very sophisticated setting. In this program an electronic control system was employed to control both the swirl control valve opening and the ignition timing. That is, as shown in Figure 4b, those two factors were determined with relation to the engine speed and the manifold vacuum.

Then four other types of the engine shown in Figure 5 were compared with the DIS type to identify the performance improvement of DIS over other methods for swirl intensification. Type A in the figure does not employ any swirl intensification measure. Type B has an intake port, which gives the mixture an optimal direction so as to form a mixture swirl in the combustion chamber, and also has a cylinder head equipped with a guide wall as part of an intake valve seat to intensify the swirl further. Type C has a guide vane at the intake throat to spiral the mixture flow. Type D is the same as type C except that the intake port area is smaller.

Type B is used as Mazda Stabilized Combustion System (SCS) to comply with emissions standards in the U.S. and Japan. Type A is used in the other areas such as ECE. The base model in our program used the type B engine. Comparison with those four engines was made in maximum power at WOT and in fuel economy at an engine speed of 25 rev/sec and a BMEP of 294 kPa which are typical operational conditions in the city driving condition such as the 10 mode. The result is shown in Figure 5 with type A as a basis. Note the curves for type A, type B, type C, and type D. The swirl gets intensified in this order, but power loss also increases relative to fuel saving effects. On the other hand, DIS produces a maximum power equivalent to that obtainable by type A while improving fuel economy by 4 percent over type B, and by 8 percent over type A. This means, DIS produces approximately 8 percent more power than the case where the swirl is intensified so as to get a fuel economy equivalent to that of type D. When the output was allowed to decline to the level of type B by employing valve timing modification, DIS recorded a fuel economy gain of 9.3 percent over type A and 5.3 percent over type B. Then in order to check how DIS, capable of alleviating swirl intensity at WOT, is effective in engine noise reduction, comparison was made in the combustion pressure and engine noise between type A, B, and C engines under the conditions of an engine speed of 66.7 rev/sec, WOT, and an air-fuel ratio of 13:1, with the ignition timing at MBT. The engine noise was measured at a distance of 1 m from the cylinder block. Sound pressure level (SPL) in the DIS engine was less by 0.8 dB (A) than that in the type B engine. This is because of reduction in the high-frequency components above 500 Hz as shown in Figure 5; the DIS engine achieved the same level of quietness as the type A engine.

(3) Resetting Compression Ratio

Raising compression ratio is very effective in increasing thermal efficiency of a spark ignited engine, but its use is restricted because it induces knocking. However, it is known that an appropriate swirl accelerates combustion and thus is beneficial in suppressing knocking (2). Based on this fact, the feasibility of raising the compression ratio was studied on the DIS engine. The experiment was performed on the type B and DIS engines with the compression ratio varied from 9.0, the value employed in 1982 model, to 11.0, and comparison was made in WOT output at knock limit and in BSFC at an engine speed of 25 rev/sec and a BMEP of 294 kPa. Either type of engine showed an improved BSFC with a higher compression ratio but recorded a lower BMEP at the low speed region at WOT. With the DIS engine, however, an appropriate opening of SCV enables BMEP to increase at knock limit (2). Consequently, even when the compression ratio was raised to 10.0, BMEP equivalent to type B (1982 model) was obtained.

(4) Variable Displacement System (VDS)

In developing a variable displacement system, comparison was made between two methods of cutting out cylinders No.1, and No.4, namely, shutter valve method, and intake/exhaust valve deactivating method. In order to evaluate their effectiveness, the friction loss measured by motoring was analyzed using the following equation:

$$Pf = Pft + Pfc + Pfv + Pfm$$

The friction mean effective pressures (Pf) of both control systems measured at an engine speed of 25 rev/sec and at a throttle opening equivalent to 10 percent load are shown in Table 4, and load - BSFC characteristics at a 25 rev/sec engine speed are shown in Figure 6. By converting to a two-cylinder operation using the intake/exhaust valve deactivating method, Pft, Pfc, and Pfv were reduced with the result of Pf decreasing by 34.6 percent in total, and BSFC decreasing by 25 percent at a 25 rev/sec engine speed and a 10 percent load. With the shutter valve method, the exhaust gas was re-circulated to prevent dead cylinders from sucking in the charge, and because of this, Pf was reduced only by 18.4 percent resulting from Pft reduction and BSFC improvement was limited to 12 percent. Based on these observations, the intake/exhaust valve deactivation method was selected in this program to obtain the maximum improvement of BSFC, and the system was evaluated as installed in the vehicle.

However, as is shown in this load - BSFC curve, even the intake/exhaust valve deactivation method loses its benefits in improving fuel economy over the four-cylinder operation when load exceeds 30 percent load. The mechanism to control valves in the intake/exhaust valve deactivation method is shown in Figure 2. The intake and exhaust valves are deactivated by freeing the pivot of the rocker arm. It was authors' belief that if the two-cylinder operation system is to be adopted, it would have to offer adequate power performance combined with low vibration. As a consequence of evaluation tests on an engine dynamometer, it was tentatively decided to limit the two-cylinder operation to a steady-state running after warming up, and to finalize the conditions for two-cylinder operation by analyzing the running performance of a built up car. Shown in Figure 6 is a comparison in crank angular variation characteristic while idling between four-cylinder and two-cylinder operations. Idling with two active cylinders induced vehicle body vibrations lower than 10 Hz, to which the human being is most susceptible, so that it was judged intolerable in a car. Deactivation of valves was controlled by an electronic control system which senses the state of warmup and running condition. In the 10-mode cycle, the duration in which the aforesaid requirements for the two-cylinder operation --- adequate performance and low vibration are met --- accounts for 22 percent of the cycle. This ratio was calculated by excluding the time required to determine if the conditions are ready for deactivating.

(5) Optimization by Electronic Control System

As described in the foregoing, the swirl control valve needs to be controlled according to operating conditions to maximize the benefits of DIS, and the number of active cylinders in VDS has to be decided to suit an operating condition. Further, when such variable mechanisms work, demands on ignition timing and exhaust gas recirculation (EGR) ratio change. To provide such a sophisticated control at high degree of freedom, an electronic control system was adopted.

An electronically controlled single point injection system (SPI) was also incorporated into the fuel metering system to provide an air-fuel ratio suitable for a given operating condition. Through the monitoring of an operating condition from throttle opening, engine speed change, etc., the air-fuel ratio is set for the best fuel economy in a steady-state operation. In the acceleration phase, in order for a three way catalyst to make efficient reduction of nitrogen oxides (NOx) and to maintain acceleration performance, the air fuel mixture was controlled to the stoichiometric level. Using the electronic control system on the improved combustion engine made it possible to set a leaner air-fuel mixture at steady-state operation. This, combined with optimization of ignition timing and EGR ratio, permitted an 11 percent BSFC improvement at an engine speed of 25 rev/sec and a BMEP of 294 kPa (See Figure 7). In VDS when two cylinders are in operation at WOT, torque is approximately half that of four-cylinder operation, but conversely, torque in two-cylinder operation at light load at the same throttle opening is greater than that in four-cylinder operation because of the difference in pumping loss. Therefore, a system was adopted that is so designed to switch the two-cylinder and four-cylinder modes at the crossing point of torques of those modes, the point being determined by monitoring the engine speed and the manifold vacuum.

4.2 Vehicle Modification

The vehicle resistance has a grave influence on fuel economy and consists of aerodynamic drag, rolling resistance, acceleration resistance, and climbing resistance. When a vehicle is running on a level road the vehicle resistance mainly consists of the rolling resistance, aerodynamic drag, and acceleration resistance. In this project, however, to reduce the vehicle resistance efforts were focused on reduction in rolling resistance and vehicle weight.

(1) Reduction in Rolling Resistance

The principal factor in the rolling resistance is the internal loss by tire deformation, which is estimated to account for approximately 90 percent of the rolling resistance in straight running. Supressing the hysteresis of the material of the tread which causes a substantial loss is very effective in reducing the energy loss due to tire deformation. The material of tread rubber was modified to minimize friction loss --- an adverse effect of rolling loss reduction --- while simultaneously reducing energy loss. As a result, rolling resistance was reduced by approximately 30 percent from 1982 model tire.

(2) Reduction in Vehicle Weight

Since both rolling and acceleration resistances are functions of the vehicle weight, vehicle weight reduction should be very effective in fuel economy improvement. In selecting the items for weight reduction, evaluation was made on the feasibility of weight reduction throughout the vehicle. And the items were ranked in the order of importance:

Rank A : Already under development and technically feasible,
Rank B : Likely to increase cost but expected feasible,
Rank C : In need of considerable research and development.

In this project, in order to finish the development in a short period of time, efforts were directed to sufficiently feasible items of rank A and rank B. Chosen parts, materials used and weight reduction effects are shown in Table 5. The major method employed for weight reduction was material substitution with aluminum, plastics, magnesium, and high tensile steel for designing lighter-gauge components. This saved the vehicle weight by 70 kg, reducing the original 1982 model 840 kg to 770 kg. As a result, the test inertia weight at the 10-mode emissions measurement declined from the 1000 kg rank to the 875 kg rank.

(3) Determination of Gear Ratio

Gear ratio selection was made utilizing the 70 kg weight reduction and rolling resistance reduction for improving fuel economy. The target in determining the gear ratio was to retain a vehicle performance at least equal to that of the base model.

In determination of the gear ratio a simulation analysis was conducted to compare vehicle performance and fuel economy. Figure 8 shows the results of simulation when the axle ratio and vehicle weight were varied. The result indicates that by altering the gear ratio, the improvement in vehicle performance due to the vehicle weight reduction can be transformed to a fuel economy improvement of 7.5 percent.

5 PERFORMANCE OF CONCEPT CAR AND CONTRIBUTION OF FUEL ECONOMY IMPROVING TECHNOLOGIES

The engine and vehicle modifications discussed so far were incorporated in the concept car, and verification testing was performed on the completed car to identify achievements in fuel economy and performance. The improvements in fuel economy and performance on the proving ground are shown in Figure 9. The fuel economy data are measurements at 60 km/h. Cruising with two active cylinders cut fuel consumption by as much as 14 percent.

The 10-mode cycle fuel economy measured on a chassis dynamometer is shown in Figure 9. Test was performed adding each fuel saving measure in a cumulative manner to clarify its individual effectiveness. The effect of the VDS itself in fuel economy gain is 8.0 percent. When all these measures are combined, fuel economy improves 42.4 percent compared with the 1982 model. However, even in the full-time four-cylinder operation without the VDS, the fuel economy gain was better than the target of 30 percent in this program, and the emissions including NOx complied with the 1978 regulation as is shown in Table 6. Table 7 presents the conversion factors of the fuel economy saving measures taken by operating the concept car in the 10-mode cycle and the U.S. EPA test mode cycle combining the city mode and highway mode. In the U.S. test mode, which requires frequent operation at high speed, high load regions as compared with the 10-mode, VDS is less effective. Also, in the U.S. mode the gear ratio and the rolling resistance have greater effects on the fuel consumption, indicating that it is important to choose fuel saving measures suitable to local vehicle operating conditions.

6 SUMMARY

1) A concept car incorporating various fuel saving measures achieved a fuel economy gain of 31.9 percent, exceeding the 30 percent initial target, and exhibited a performance superior to the base 1982 model. Although there remain some tasks in respect to reliability confirmation and cost saving, part of the technologies used in this concept car will be applied to production models in the near future.

2) It was confirmed that employment of VDS further improves fuel economy by some 10 percent but has room for improvement, including vibrations at the two-cylinder operation and inadequate acceleration from a two-cylinder steady-state operation.

3) Benefits of various fuel saving measures change greatly with vehicle operating conditions, and improvement measures need to be carefully chosen to suit given operating conditions, with cost-effectiveness taken into consideration.

4) It was confirmed that variable mechanisms such as DIS and VDS were very effective measures for meeting conflicting needs. In view of the ever-increasing demands on the engine, there is a need for further efforts to develop variable technologies.

7 REFERENCES

(1) MUROKI, T., YAMAMOTO, H., IWAYA, M., and YOKOOKU, K. On the Development of a Lower Fuel Consumption Concept Car with High Performance. The Second International Pacific Conference on Automotive Engineering, Tokyo, 1983, 830904.

(2) NAGAO, A. and TANAKA, K. The Effect of Swirl Control on Combustion Improvement of Spark Ignition Engine. I Mech E Conf. Pub., 1983, 153-161.

(3) KOBAYASHI, K., TANAKA, K. and YOKOOKU, K. Mazda Stabilized Combustion System (MSCS) Emission Control System Development for Good Fuel Economy. Shell Automotive Technical Symposium, Tokyo, 1980, No.7.

(4) TANAKA, K., KOBAYASHI, K. and KAWASAKI, K. Combustion Improvement Technique in Mazda Engine. FISITA 18th International Congress, Hamburg, May 1980, 80.4.4.5.

(5) YAGI, T. and YAMAGATA, I. Experimental Method of Determining Piston Profile by Use of Composite Materials. SAE Paper 820769

(6) MAY, M. G. Lower Specific Fuel Consumption with High Compression Lean Burn Spark Ignited 4 Stroke Engines. SAE Paper 790386

Table 1 Specifications of base engine

Bore x Stroke	mm	77 x 80
Number of Cylinder		4
Displacement Volume	cm³	1490
Combustion Chamber		Hemispherical with Guide Wall
Compression Ratio		9.0

Table 2 Specifications of vehicle

		'82 Model	Fuel-Efficient Concept Car
Vehicle	Weight 10-mode Test Inertia Weight Tire Rolling Resistance (μr)	840kg 1000kg 0.014	770kg 875kg 0.009
Transmission Gear Ratio		1st 3.416 2nd 2.055 3rd 1.290 4th 0.918 5th 0.731	3.470 1.840 1.218 0.868 0.731
Axle Ratio		3.850	3.409

Table 3 Fuel saving technologies and technologies employed in fuel-efficient car

	Improvement Item	Improvement Factor	Improvement Technology
Engine	Mechanical Efficiency Increase	Mechanical Loss Reduction	[Piston & Rings], Valve Gearing, Crank System
		Auxiliaries Driving Loss Reduction	Optimum Control of Oil Pump, Water Pump, Alternator
	Combustion Efficiency Increase	Higher Compression Ratio	Compact Combustion Chamber (6)
		Rapid Combustion	[Swirl Intensification (DIS)]
	Pumping Loss Reduction		[VDS], Intake Air Preheating
	Radiation Loss Reduction		
	Exhaust Energy Recycling		
	Improved Utilization	Optimization by Electronic Control	[Ignition Timing Control], [Fuel Control], [EGR Control]
		Variable Mechanism	[DIS], [VDS], Valve Timing, Compression Ratio
Vehicle	Running Resistance Reduction		Aerodynamic Drag Reduction
			[Rolling Resistance Reduction]
	Weight Reduction		[Employment of Lighter Material]
	Improved Utilization		[Gear Ratio Optimization]

☐ = Employed Technologies

DIS : Dual Induction System
VDS : Variable Displacement System
EGR : Exhaust Gas Recirculation

Table 4 Factors of friction loss and comparison of effectiveness of displacement modulating methods

(Engine Speed 25 rev/sec, Throttle Opening Equiv. to 10% Load)

Components		4-Cyl.	2-Cyl. (Shutter Valve)	2-Cyl. (Valve De-activation)
Main Components	Piston	24.0 %	24.0 %	24.0 %
	Crankshaft	4.2	4.2	4.2
	Valve Train	13.2	13.2	9.4
Pumping, Compression, etc.	Throttle	25.7	7.3	6.5
	Compression, etc.	22.2	22.2	10.6
Auxiliary Components		10.7	10.7	10.7
Total ( ) = Reduction Rate		100.0	81.6 (-18.4)	65.4 (-34.6)

Table 5 Weight reduction

Lighter-Weight Material	Weight Reduction (kg)	Example of Parts A	Example of Parts B
Aluminium	36.29	Shift Fork, Engine Bracket	Differential Case, Brake Drum
Plastics	5.63	Air Cleaner, Cylinder Head Cover	Front Fender, Oil Pan, Bumper, Front Seat
High Tensile Steel	3.49	—	Front Frame Lower, Disk Wheel
Magnesium	3.53	—	Clutch Housing, Transmission Case
Lighter-Gauge, etc.	20.93	Dust Cover, Engine Block, Crankshaft	Battery Carrier, Coil Spring, Rear & Side Glass

A : Under development with good technical feasibility
B : Technologically feasible with cost increase

Table 6 10-mode emission

	Emission (g/km)			
	CO	HC	NOx	CO$_2$
'82 Model	0.31	0.14	0.21	142.6
Fuel-Efficient Concept Car	0.23	0.18	0.23	108.3
Emissions Standard	2.1	0.25	0.25	—

Catalyst used : Three-way catalyst
Durability-run for 50,000 km

Table 7 Conversion factor of individual fuel saving technologies

	Technology	Benefit (%)	10-mode Fuel Saving (%)	10-mode Conversion Factor	EPA Combined Fuel Saving (%)	EPA Combined Conversion Factor
Engine	Piston System Mechanical Loss Reduction	(Pf) 10.8	3.5	0.32	3.7	0.35
	Improved Combustion (DIS & Higher Comp. Ratio)	(BSFC) 7.0	6.2	0.89	5.0	0.71
	Optimization by Electronic Control	(BSFC) 11.0	8.0	0.72	6.6	0.60
	Pumping Loss Reduction (VDS)	(Pf) 34.6	8.0	0.23	3.8	0.11
Vehicle	Running Resistance Reduction (Tire)	(μr) 30.0	3.0	0.10	6.3	0.18
	Vehicle Weight Reduction	(Weight) 8.3	2.7	0.33	2.2	0.26
	Improved Utilization (Higher Ratio Gear)	(Gear Ratio) 16.3	5.0	0.31	10.1	0.62

Conversion factor : Fuel saving rate divided by benefit

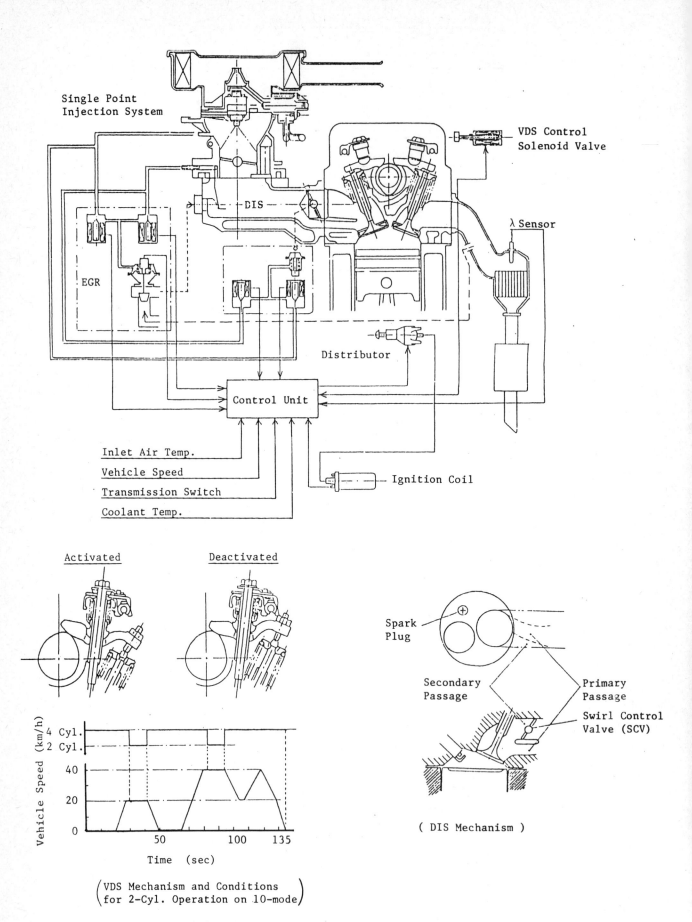

Fig 2 Schematic of fuel-efficient concept system

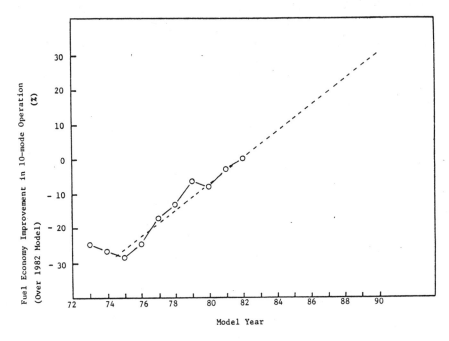

Fig 1 Trend of average 10-mode fuel economy of 1.2–1.6 litre cars in Japan

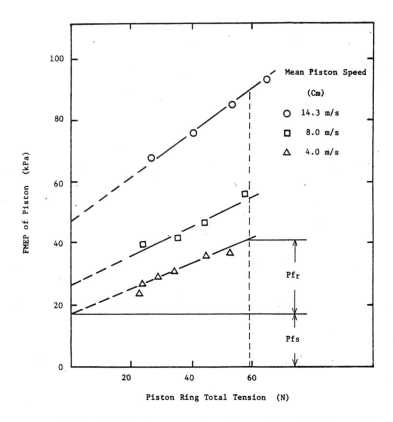

Fig 3a Relation between piston ring tension and friction loss (engine speed 25 r/sec, WOT)

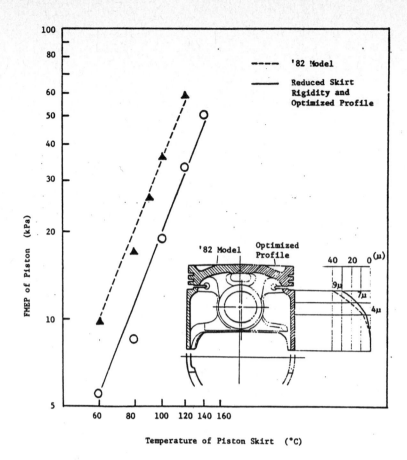

Fig 3b   The effect on friction loss reduction of skirt rigidity reduction and piston profile improvement

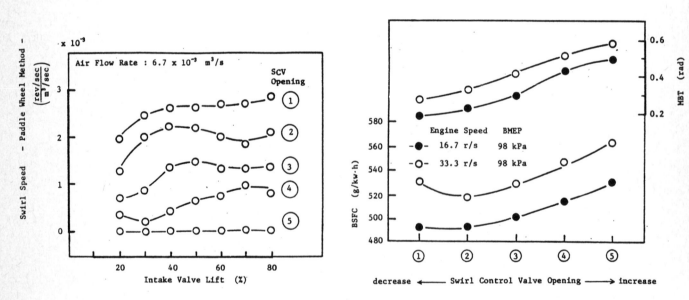

Fig 4a   Effects of SCV opening on swirl speed and engine performance

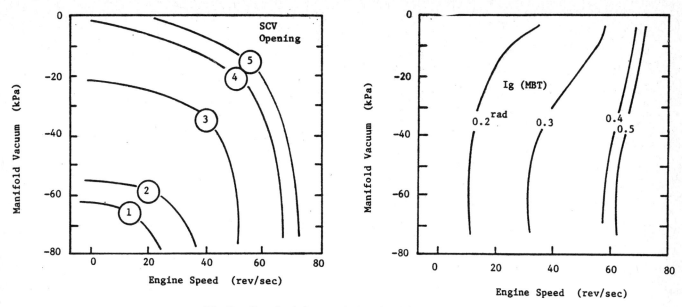

Fig 4b   Required characteristics of ignition timing and SCV opening

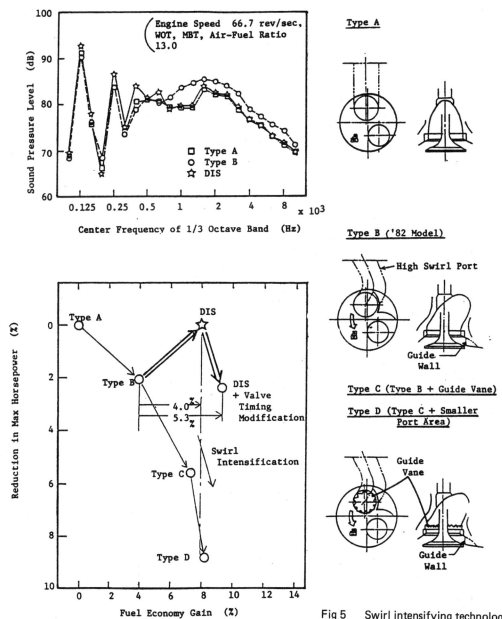

Fig 5   Swirl intensifying technologies and their effect on engine performance

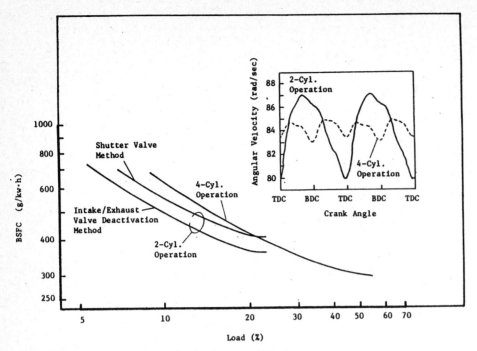

Fig 6 VDS's fuel saving benefits (engine speed 25 r/sec, air-fuel ratio and Ig. timing best set) and its effect on crank angular velocity fluctuation (engine speed 13.3 r/sec, idle)

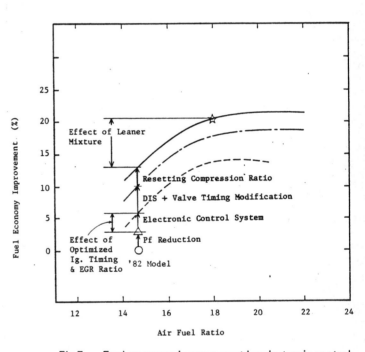

Fig 7 Fuel economy improvement by electronic control system (engine speed 25 r/sec, BMEP 294 kPa)

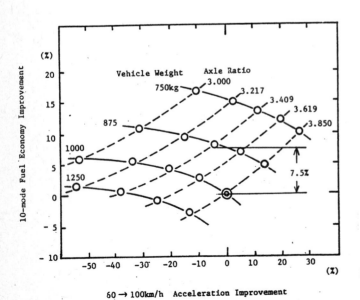

Fig 8 Effects of vehicle weight and axle ratio on 10-mode fuel economy and acceleration

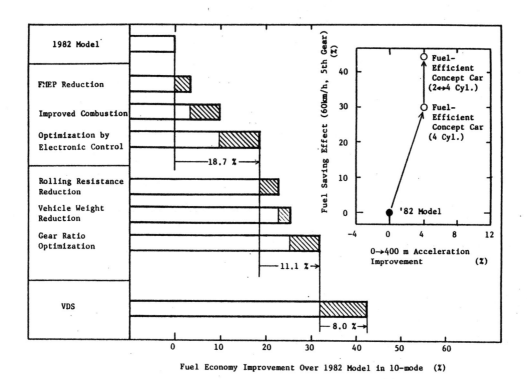

Fig 9 Fuel saving effect of a fuel-efficient concept car in 10-mode and 60 km/h

C438/84

SAE 841310

Methods of reaching vehicle fuel consumption and weight reduction

P E G BAUDOIN and H L LAGRANGE
Régie Nationale des Usines Renault, Rueil Malmaison, France

SYNOPSIS : The parameters influencing the fuel consumption of a vehicle are shown, and the different methods to reduce it, especially by reducing the weight, are explained, with their limits.

1 INTRODUCTION

The energy crisis obliges all motor car manufacturers to take action to reduce the consumption of vehicles.

RENAULT has a research programme, subsidized by the French State, which has as its object the production of prototypes of vehicles whose weighted fuel consumption is to attain 3 litres per 100 km, according to European Standards.

This programme has been named VESTA (Véhicule Econome de Systèmes et Technologies Avancés - Advanced Systems and Technologies Vehicles). It began in January 1981 and will end in 1985. The possible approaches to the design of such a vehicle will now be described.

2 FUEL CONSUMPTION PARAMETERS

The fuel consumption of a vehicle at any moment is directly proportional to the driving force at the wheel rim F_d and inversely proportional to the efficiency of the power train :

$$C_i = K \cdot \frac{F_d}{\eta_{PT}}$$

The driving force can be expressed by the following equation :

$$F_d = Mgf \cos\alpha + \tfrac{1}{2} \rho SC_d V^2 + Mg \sin\alpha + M^*\gamma + F_b$$

$Mgf \cos\alpha$ = rolling resistance

$\tfrac{1}{2} \rho SC_D \cdot V^2$ = air resistance

$Mg \sin\alpha$ = force due to gradient

$M^*\gamma$ = force due to acceleration

M^* = inertia of vehicle, i.e. weight + rotational inertias

F_b = braking force, if any.

Various devices which enable the fuel consumption at any moment to be read on the dashboard are on the market, but it is not always easy for the driver to make use of this information to drive more economically for three reasons :

- with engines that have a carburettor, this information is provided with a non-negligible delay ;

and more fundamentally

- no distinction is made between forces that dissipate energy and those that do not ;

- this does not make it possible to draw a distinction between losses that are inevitable and those that could be avoided.

On the one hand, the rolling resistance, the air resistance and the braking force, if any, are forces that dissipate energy, whereas the force of inertia or the weight component parallel to the road do not dissipate energy, but convert the energy supplied by the power train into kinetic or potential energy, which are not degraded thermodynamically.

On the other hand, the efficiency of the power train consists of several components :

- the inescapable efficiency due to the conversion of heat to work or CARNOT efficiency ;

- the generalized mechanical efficiency related to engine and transmission friction, the laminar flow of gases, and essential auxiliary equipment (water pump, oil pump), about which something can be done.

These internal energy-dissipating forces make the specific fuel consumption rise from its theoretical asymptotic value (about 230 g/kWh for a spark ignition engine) to its real value.

Finally, the last category of energy-dissipating forces concerns those driving the alternator, the power-assisted steering pump, the air-conditioning compressor, etc...

It is important to have an idea of the breakdown of the fuel consumption of a vehicle according to the main energy-dissipating forces. The table below refers to a RENAULT 9 GTL vehicle.

Weight = 840 kg + 140 kg

C_p = 0.38

Rolling: 140 N/ton

It will be noted:

- that the power train bears a large share of responsibility in all cases
- aerodynamic resistance comes second,
- that the share of responsibility of braking is slightly under estimated for this conventional fuel consumption case in relation to the real world consumption.

Table 1. Example of fuel consumption breakdown at a steady 90 km/h, at a stead 120 km/h and on the European urban cycle for a Renault 9 GTL.

Consumption (L/100 km)	90	120	ECE	3x1/3	%
Resistance to rolling	0.99	1.15	0.99	1.04	15.5
Aerodynamic resistance	2.10	3.75	0.37	2.07	30.9
Braking	0	0	0.85	0.28	4.2
Generalized friction of the power train (Carnot efficiency excluded)	2.31	2.40	5.19	3.30	49.3
TOTAL	5.40	7.30	7.40	6.70	

3 RESISTANCE TO ROLLING

The choice of wheels and tires assumes particular importance in the preliminary design of a vehicle, because it determines its interior roominess and external dimensions and therefore has an influence on its total weight. This choice also affects road behaviour, as well as resistance to rolling.

As was seen above, resistance to rolling is proportional to the weight and the rolling coefficient.

Methods enabling weight to be reduced will be examined later, but in the meantime consideration will be given to the parameters that have an effect on the coefficient f.

3.1. Pressure of tires - p

For a given tire, the relationship is essentially as follows:

$$\frac{\Delta f}{f} = -0.4 \frac{\Delta p}{p}$$

Since the usual pressure is about 2 bars, this tells us that an increase in pressure by 0.5 bar enables drag to be reduced by 10 %.

In addition, it is known that increasing the pressure tends to diminish the drift angle of the tires. There therefore remain the comfort and skid-resistance qualities to be considered.

It is undeniable that an increase in pressure has an adverse effect on comfort and special research would have to be carried out to discover means of mitigating this disadvantage.

3.2. External diameter

If only the resistance to rolling aspect is considered, all the studies carried out have shown that for a vehicle of given weight and at equal pressure, a tire with a large external diameter and small tire width is preferable to a tire of small external diameter and large tire width.

But other arguments militate in the opposite sense:

- the service life of the tire;
- skid resistance;
- the incidence on the vehicle weight, without taking aesthetics into account.

Finally, in the present state of knowledge, compromises are still very subjective. For the Vesta vehicle, we have chosen an external diameter which corresponds to an 11" wheel rim, which is not on the market at present.

3.3. Structure of tire

It is known that a radial structure is more favourable than a diagonal structure, so far as resistance to rolling is concerned. The quality of the rubber used also has an influence on this coefficient. Tire manufacturers are at present making great progress in this connection and it can be reasonably expected that resistance to rolling will in the near future drop from 140 N/T to 100 N/T or even lower, at normal speeds.

4 AERODYNAMIC RESISTANCE

It was seen above that it is possible to ascribe about a third of the standard average fuel consumption to aerodynamic resistance.

It will be remembered that the aerodynamic force equation is :

$$\frac{1}{2} \rho \, S \, C_D \, v^2$$

Although drivers have a great and sometimes unknown to them opportunity to reduce their fuel consumption by driving at reasonable speeds, the motor car manufacturer can only act on the frontal area S and the air drag coefficient C_D.

4.1. Frontal area - S

The cross-sections of the passenger compartment opposite the front and rear passenger seats depend much on the roominess that is desired for a given range of vehicles. Limit as closely as possible the space for passengers has been made possible in the last years by the utilization of curved side windows. Today; there is not much to be hoped for under this heading.

In view of the fact that the frontal area in aerodynamic shapes is usually located in the front third of the length of the vehicle, the designer has interest in paying particular attention to the cross-section at the rear seats. The cross-section corresponding to the front seats is then obtained quite easily during the aerodynamic design work.

4.2. Air drag coefficient C_D

This coefficient depends essentially on the external shape of a vehicle, which stylists and aerodynamics specialists must optimize, but it will be seen later that to obtain a low CD, account must be taken of the air circulation in the engine compartment and of the trim of the vehicle with respect to the ground.

In the seventies, the average C_D of European vehicles was about 0.45, but it will probably drop to 0.30 in the next few years. It is generally agreed that for land vehicles a quasi-asymptote is obtained when C_D = 0.16.

On the EVE vehicle (Figure 1), which has a total length of 4.45 m, RENAULT has obtained a C_D of 0.22. It is know that the greater the length to frontal area ratio, the easier it is to obtain a low C_D.

On the VESTA vehicle (Figure 2), we laid down a C_D figure of 0.25 as the objective for an overall length of 3.27 m. This short length led to the necessity of the rear part being truncated vertically. Although some general rules are known for sketching an aerodynamic shape, the optimization work of the final shape requires close collaboration between stylists and aerodynamics specialists. A large number of details, blend radii of the areas with respect to one another, trailing edges, position of window panes, rear-view mirrors etc, are just as important as the general shapes.

4.3. Cooling drag

By definition, this is proportional to the difference between the aerodynamic forces measured on the vehicle with and without circulation of air in the engine compartment. Expressed as ΔC_D, it varies greatly with the vehicle. In a series of tests carried out between 1977 and 1980 by the RENAULT Aerodynamics Laboratory on 95 vehicles of all makes, it was found that the extreme values were 0,01 and 0,07.

These figures show clearly the importance of the cooling air circuit to the engineer responsible for designing a low-fuel consumption vehicle.

Two notions can help him to make his initial choices :

- the power required to circulate air through the engine compartment so as to cool the latter is necessarily provided by the engine. This power is represented by :

$$P \text{ air} = Qv \times \Delta p$$

Qv = air flow

Δp = pressure losses in the whole of the piping system.

These two terms must be therefore minimized.

To reduce Qv, the following advice can be offered :

. make provision for all the air flow to be used in fact for cooling, which is possible by placing an airtight duct between the front face of the vehicle and the radiator and putting seals around the headlights or between the various metal sheet parts ;

. choose the radiator and its dimensions so that it does not require a high air flow rate. A radiator that is too small requires a high air flow rate to remove heat. The total pressure losses Δp can be broken down into three parts:

$$\Delta p = Pe + Pf + ps$$

pe = pressure losses upstream of the radiator ; they are due essentially to the radiator grille and the inlet deflector.

pf = pressure losses due to the radiator core itself ; these depend of course on its technology, but especially on its design base on a constant calorific power (Figure 3).

ps = losses downstream of the radiator ; these generally consist largely of those due to the motor-driven fan and its support.

It should also be pointed out that if it is desired to avoid using the electric motor-driven fan as much as possible, the total pressure loss must remain compatible with the upstream dynamic pressure $1/2 \rho V_o^2$ supplied by the travelling speed V_o of the vehicle.

The air passing through the radiator core can be regulated by means of a variable geometry device placed, for example, at the inlet of the radiator grille.

- The other important notion to have in mind is the disturbance of the aerodynamic field of the vehicle by the circulation of the cooling air. Unfortunately, the present state of knowledge does not enable this to be considered in any way other than experimentally. A priori, it can be taken that the largest negative effects are due to the air outlets of the engine compartment. It should be possible to choose the locations of these outlets, as well as the orientation of the air velocity vectors so as to lower the value of C_D.

An illustration of this is given in Figure 4.

It shows the schematic layout of a set-up comprising a sleeve downstream of the radiator opening out onto the front part of the bonnet. This has enabled the cooling C_D to be reduced to 0.007.

5 THE BRAKING FORCE

The braking force is obviously intermittent and very closely connected to the driver's way of driving and the road itself. In cities, according to the ECE cycle, this represents 11% of the fuel consumption. It is therefore not negligible.

A manufacturer has only two possibilities open to him :

- reducing the weight of the vehicle ;

- recovering some of the energy usually dissipated in the brakes,

which imposes the problem to transform and store that energy.

6 GENERALIZED POWER TRAIN FRICTION

It was seen that generalized power train friction is responsible for about half of the average fuel consumption, which is considerable. It is more or less proportional to the size of the power train, which itself is linked to the performance levels stipulated in the specification of the vehicle.

It will be seen later how the specification can be used to determine the respective influences of the weight, air and rolling resistances on fuel consumption.

6.1. Choice of engine and transmission

It will be recalled that heat engines can be classified according to their thermodynamic cycle : spark ignition, Diesel, Rankine, Stirling, Brayton and others.

From the point of view of consumption, they differ in their efficiencies in accordance with the speed and power level.

In addition, they differ by their specific output and by the maximum torque curves in accordance with the speed.

This paper will be limited to two of the most common types : the spark-ignition engine and the Diesel engine.

As regards transmission, this can be classified under three categories : stepped transmission, manual or automatic control, continuous transmission, and energy storage transmission.

These latter have a special theoretical interest, especially in town use, but they seem to be of limited application, since their extra cost is considerable and they will therefore not be considered here.

To reduce fuel consumption, it is obvious that an engine and a transmission with low generalized friction must be chosen. A detailed analysis of th precautions to be taken in the design of these two components would largely exceed the scope of this paper, which is specifically directed towards the design of a vehicle.

It is possible simply to give the choices that have been made for the VESTA vehicle ;

- 4 stroke spark ignition engine with 3 cylinders with a crank shaft and a conrod assembly on bearings ;

- 2 stroke direct injection Diesel engine with 2 cylinders ;

- a 5 speed gear box ;

- a belt variator.

6.2. Adaptation of engine and transmission

Whatever type of engine or transmission chosen, it will be found that consumption at a given power level is very much influenced by the transmission ratio. The table below gives the consumption figures for a RENAULT 18 with spark-ignition and a manual gearbox, at 90 km/h in various gears :

Table 2. R18 fuel consumption et 90 km/h.

Gear	3rd	4th	"6"th
Engine speed (RPM)	4760	3125	1650
Consumption (1/100 km)	8.88	6.40	5.04
Difference %	+ 76%	+ 27%	−

The following ratio will be called the "transmission adaptation coefficient" :

Least possible consumption for the required power

Real consumption at the operating point obtained.

A good transmission adaptation coefficient often saves more fuel than can be hoped from any progress in engine and transmission design.

An improvement of this coefficient requires the transmission ratio range to be extended towards overdrive ratios, which raises two kinds of problems according to whether transmission is manual or automatic.

In the case of a manual gear box, increasing the number of gears beyond 5 is difficult to accept for reasons of driving ease and extension of the fifth gear must remain within reasonable limits so that the driver does not have to change gear too often.

In the case of an automatic stepped gear transmission, a similar reasoning can be advanced, taking the discomfort due to the changes being made automatically into account. In that case, the best compromise for a private car would seem to be 4 gears combined with a torque converter.

In the case of a continuous transmission, it is theoretically possible to provide for very high overdrives, but these are limited in practice by technology.

In the case of a variable pulley and belt or metal chain variator, variation ranges from 1 to 6 can be obtained without too many difficulties, which enables the adaptation coefficient to be improved appreciably.

6.4. Consumption due to auxiliary equipment

6.4.1. Oil pump

Although an oil pump is an integral part of the engine since the latter requires to be lubricated a few words may be said about this component. In the present engines, a fixed displacement pump is generally found driven at a rate proportional to that of the engine, and discharge valve calibrated at 2 to 4 bars. The power absorbed by this component in accordance with the speed of the engine is shown in Figure 5.

The hatched part represents the energy dissipated in the discharge valve.

At maximum speed, the power absorbed by the pump can reach 4% of the maximum engine power.

A careful investigation of the lubrication needs of the engine and a variable displacement pump would enable both the generalized friction of the engine and its size to be reduced while giving the same performance.

6.4.2. Water pump

Its energy consumption is generally much less than that of the oil pump, and practically negligible in the overall energy budget. It may nevertheless be pointed out that the adoption of an electric water pump enables a cooling strategy to be implemented, increasing the rate at which the temperature rises and thus reducing consumption when the engine is cold.

6.4.3. Alternator

The generation of electric current on board vehicles is appreciably different from that which has been seen so far.

Electrical energy needs vary very greatly in accordance with weather conditions (rain, outside temperature) and light conditions (day or night, fog). The size of the alternator is proportional to the size of the equipment more than to the dimensions of the vehicle.

Finally, the efforts made by manufacturers to reduce fuel and the natural tendency to increase the equipment levels of vehicles will in the future give added weight to the responsibility of the alternator in the overall fuel consumption.

For example, we have calculated that on the VESTA with an average alternator and drive efficiency of 40% (as against 25 to 30% on present vehicles), and with a marginal cost of energy supplied by the heat engine equal to 190 g/kWh, the fuel consumption related to the generation of electricity would be as follows :

0.5 l/100 in towns,

0.15 l/100 on roads

0.1 l/100 on motorways

6.4.4. Other auxiliary equipment

It is quite evident the designer must also take precautions to minimize fuel consumption due to other auxiliary equipment : power-steering pump, air-conditioning compressor, etc. It must also not be forgotten that the driving belts can dissipate non-negligible amounts of energy as compared with that absorbed by the auxiliary equipment.

7 RESPECTIVE INFLUENCES OF WEIGHT, DRAG COEFFICIENT AND ROLLING RESISTANCE COEFFICIENT

The three external parameters that influence the fuel consumption are
 the weight
 the drag coefficient or the frontal area
 and the rolling resistance coefficient.

How to calculate their respective influences ?

Two cases are to be considered, after one changed

- the power train is not modified, which means that the performances are changed.

- the power train is either downsized or upsized in order to maintain the level of performance in accordance with the general specification of the vehicle.

This latter case is more complex, but leads to a general solution and will be considered now.

We will assume that a power train has a generalized friction consumption that is proportionnal to its size.

We will also assume that the principles and technology of the power train are defined elsewhere.

How to calculate the new size of the power train?

Starting from the basic design of the vehicle and considering the various performance criteria (maximum speed, time to accelerate from o to 100 km/h, and so on) it is possible to determine the most unfavourable one : it is the one leading to the largest power train. It is now possible to calculate the amount of energy consumed by each of the forces that constitute the driving force, to achieve the corresponding performance and to attribute to each of them a share of responsibility in the power train sizing, proportional to that amount of energy.

As those forces are well known functions of the three parameters, it will be possible to calculate the new size of the power train and the new fuel consumption of the vehicle.

If the parameters are subject to large variation it becomes necessary to check that the sizing performance criterion remains the same.

In the case of the Vesta, such a calculation gave the following results :

basic definition : weight : 520 + 140 = 660 kg
C_D = 0,25
f = 100 N/T

new definitions: 1. weight : 660 - 66 = 594 kg
2. C_D = 0,25 - 0,025 = 0,225
3. f = 100 - 10 = 90 N/T

fuel consumption savings:

	Same performance		Same powertrain	
Weight - 66 kg	.176 l/100	5.7%	.070 l/100	2.3 %
C_D - .025	.115 l/100	3.8%	.085 l/100	2.8 %
f - 10N/T	.61 l/100	2.0%	.046 l/100	1.5 %

It will be noticed that the differences in consumption for the same variation of a parameter are much larger when the drive train is modified to retain the basic performances, this observation being particularly true for the weight.

8 WEIGHT REDUCTION

8.1. Importance of weight reduction

Even before the oil crisis, manufacturers attempted to contain the natural growth in the weight of vehicles. The reason for this is evident for industrial vehicles, in view of the regulations on the Gross Vehicle Weight Rating and axle loads. Any reduction in unladen weight results in an increase in the useful load.

For manufacturers of private cars, the motivation is less evident. It corresponds to the observation that the price per kilogram of vehicle at a given equipment level varies little. To a certain extent, a judicious reduction in weight is a means of obtaining a reduction in the cost price.

With the advent of oil crisis, weight reduction has become essential and it has been seen that weight is involved :

- in the resistance to rolling,
- in the forces due to gradients,
- in the forces due to acceleration,
- as well as in the braking forces

But even if the forces due to gradients and acceleration are not energy-dissipating forces, the foregoing has shown that the weight has a very great effect on consumption at equal performance levels.

As regards aerodynamics, the rate of development is very rapid so that within some ten years, the average Cx of 0.45 will drop to 0.30, i.e. a saving of 30%.

The possibilities of weight reduction are more limited, particularly since manufacturers can influence only the unladen weight, which is only a fraction of the weight involved in the equations.

In passing, mention should be made that the notion of weight reduction has brought that all experts consider that an isolated but sufficiently large reduction in weight opens up possibilities of further weight savings. It is generally agreed that there is a coefficient of 1.3 to 1.4 between total weight reduction and initial weight reduction.

8.2. Acceptable cost of weight reduction

This is a very controversial question, and it is true that this cost depends very largely on the point of view from which it is considered.

Symbols:

P_o = price per litre of fuel at moment of vehicle purchase

X = its supposedly constant annual increase

d = annual number of kilometres travelled

t = vehicle utilization duration in years

y = interest on the non-invested capital

c = reduction in consumption in l/100 km

S = Corresponding extra cost at the vehicle purchase moment.

We assumed that the increase in the cost of fuel is constant and the amount not invested or not spent is invested at an interest of y. The calculation gave :

$$S = \frac{d}{100} \cdot c \cdot P_o \cdot (1 + y)^{-1/2} \cdot \frac{u^t - 1}{u - 1}$$

in which $u = \frac{1 + x}{1 + y}$

Numerical assumptions

P_o = 5 FF

x = 20%

d = 12,000

t = 3 and 10

y = 8,5% for a private person and 18% for the community.

c = 0,00334 (Vesta)

Results

Private person

Amortization time	3 years	10 years
Acceptable extra cost S	6.4. FF/kg	31.6 FF/kg

	Community	
Amortization time	3 years	10 years
Acceptable extra cost S	5.6 FF/kg	19.9 FF/Kg

It is seen that the private customer, who would find it difficult to plan for more than 2 or 3 years ahead is not prepared to pay more than 5 to 6 F per kilogram saved; he will only pay this if he perceives the fuel saving of the lighter vehicle, understands the economies and is willing to wait for a later pay-off. A community reasoning on a national scale, may conclude that an extra cost of 20 F per kilogram saved, is acceptable or even more, as in that case it would probably be possible to take into account a different tax policy applying on the extra cost to increase, for instance, energy independency.

8.3. Weight reduction methods usable by the designer

The person who is in charge of the overall design of a vehicle has to take various approaches in accordance with the state of progress of a project.

8.3.1. Downsizing of the power train through reduction of the drag coefficient :

When we assumed that the design of a power train was determined, for example, to an extent of :

- 70% by the force due to acceleration
- 20% by the air resistance
- 10% by the rolling resistance

we implicitly stated that a reduction in the C_D would enable the size of the engine to be reduced. This provides one way of reducing weight.

8.3.2. Stress travel

The design of any new vehicle begins by preliminary designs wich consist in placing the various components in relation to one another in a layout. It could be considered to be premature to think of weight reduction at this stage ; in fact, this is the key stage in the search of weight reduction. The designer will seek to obtain a compact vehicle by combining the various components so as not to waste any space. He must also pay great attention to the travel of stresses. Figure 6 illustrates this principle.

In a motor vehicle, rigidity has often to be provided where it is essential for the proper operation of the assembly, for example, twisting and flexural rigidity of a structure. But it is not sufficient to reason statistically. The vibrations and acoustics disadvantages due to a weakness in design are subsequently difficult to overcome without appreciably increasing the weight.

By this type of reasoning, it is possible to forecast that the positioning of the engine in the front, with its driving wheels at the back will be heavier than a front drive solution. Figures 7 and 8 are two variants of design for a front wheel drive vehicle with a transverse engine.

In Figure 7, the drive train suspension dampers are very far from the rigid part of the passenger compartment, as also the bearings on the Mac-Pherson axle stabilizing bar. In Figure 8, the various stress introduction points have been brought as close as possible to the rigid areas. The routing of the vibration stresses is shorter.

8.3.3. Systems study

A systems study consists in analysing a system or function as scientifically as possible, making if possible a model of these so that it can be used to optimize the system in accordance with the parameters that can be chosen. A systems study enables the overall-operation to be understood and it is useful also for purposes other than weight reduction.

Using this method, it is possible, for example :

- to minimize the quantity of water to be introduced into the cooling circuit
- to reduce the weight of a starter.

But its interest resides especially in the oppertunity it gives to determine if the weight reduction is really the main objective for the considered component : As has already been stated, a radiator that is too small leads ultimately to an increase in drag or in extra consumption of the motor fan. In that case, it is possible to optimize the area of the radiator. Similarly, the conclusion would be reached that too large a reduction in the weight of the alternator would reduce its efficiency and increase the total fuel consumption.

8.3.4. Value analysis

This method, which has greatly developed in the last decade, has cost-reduction as its priority objective, and is very readily applicable to weight reduction. It will be recalled that in the methodical sequence involved there are three stages :

- drafting of functional specification ;
- evaluation of the necessary minimum ;
- search for solutions.

The notion of "necessary minimum" is very important since it determines the asymptote which must be approached. The evaluation of the necessary weight minimum is an essential stage prior to obtaining the necessary minimum price.

Although value analysis has been applied for a long time almost exclusively to mass-produced parts or components, it is being used increasingly frequently in the design of new components. In that case, the preparation of the functional specification often enables the existence of gaps in knowledge to be revealed and consequently a more scientific approach to be adopted, and even a systems study to be undertaken.

Two examples illustrate the application of this method to weight reduction :

Reduction in weight of R20/R30 rear axle

Alternative	1	2	3
	Series	New Design	Improved series
Weight kg	49.5	37.7 to 38.2	37.2 to 40.1
Saving %	Base	26 to 28 %	19 to 25%

Reduction in weight of R9 front seat

WEIGHT	Passengers		Driver	
	Series	FVA	Series	FVA
Frame	5.7	3.5	6.1	4.4
Head-rest	0.9	0.6	0.9	0.6
Padding	3.0	3.0	3.0	3.0
Underframe	4.8	3.8	4.9	3.9
TOTAL	14.4	10.7	14.9	11.9
Saving %	Base	25.7 %	Base	20.1 %

8.3.5. Calculation by finite elements

This method of calculation is now in current use thanks to the existence today of powerful computers. It is already used in static and dynamic calculations and is beginning to be applied to anisotropic materials and non-linear strain.

But these calculations do not give weight optimization information directly, but only the data, which requires to be analyzed by all the art and skill of the engineer.

The utilization of these programmes is sometime a very onerous and slow task due to the time required to prepare the data. If care is not taken the result of the calculations will become available when the drawings have already been sent to the production shop.

It is believed that in the future attention should be directed to two types of programme :

- Rudimentary grids, which require the intuition of the operators ; this programme is not costly to prepare and use, and proves to be quite didactic.

- More accurate, finer, but more elaborate grids, obtained automatically in computer-assisted design. This type of programme is particularly important in designing the car body structure, which should therefore be obtained more rapidly.

8.3.6. Combination of functions

The combination of functions in a same part has been sought for a very long time with the object of saving weight and cost. The most outstanding examples of this since the beginning of the motor car have been :

- the integration of the chassis and the body, which has given rise to the self-supporting body shell ;
- the assembly of the front wheel hubs on ball joints that has enabled the degrees of freedom of the vertical displacement and steering motor to be combined, and assembly on two crossed axles, which still exists on lorries, to be replaced ;

- the integration of the headlights with the mudguards into the general shape of the vehicle has enabled performance to be improved and many intermediate parts to be eliminated ;

- placing the gear box and drive axle in the same housing has enabled small, light weight vehicles to be made, and is becoming very common today in front wheel drive, transverse engine cars in the low and middle price car range.

- the MacPerson axle is a very fine example of function integration, since the shock absorber participates in the vertical guidance of the wheel and serves as a support for the spring and, in addition, the stabilizer bar participates in the steering of the wheel.

Two recent examples can also be mentioned, which may become common very rapidly :

- Rear axles in the form of a deformable axle combined in a single unit and two attachment points, which not very long ago was done with two arms and four attachment points ;

- Integration of the expansion chamber with the radiator water tank enables various parts to be eliminated (hose, collars, fasteners,...) and weight and costs to be reduced.

8.3.7. The new materials

These are often the means that spring to mind when weight saving is considered. Yet they are not always easy to use, especially if there are cost limitations.

The vehicle designer is generally not a materials specialist, but it is well that he should know to tackle this problem.

Why chose one material rather than another ?

Above all, the choice criteria, which depend on the functions to be fulfilled, must be defined.

For example :

If it is desired to obtain a required rigidity with an assembly of stamped and welded parts, the specific moduli of elasticity must be compared (ratio of modulus of elasticity to specific weight).

It is found that steel and aluminium are on the same level. This means that for a given rigidity, an aluminium structure will be lighter than an equivalent steel structure only if it is possible to save weight on parts that contribute only little to rigidity. If it is desired to store energy in an elastic form, it is found that orientated glass fibre resin composites are better than spring steel.

If a fairly complex housing is to be made, aluminium and, even more, magnesium are better than cast iron, essentially because the material can be distributed better.

Sandwich materials

The aeronautical industry makes great use of honeycomb structures, which have an extraordinary rigidity for their weight. One could be tempted to use this construction principle with two strong plates with a light core between them.

There are, however, some traps to be avoided.

Honeycombs and sandwiches are very effective for flat floors, but can be disappointing for more complex shapes.

Figure 9 compares a ribbed trunk floor made of 8/10 steel with a 2/10 steel, 6/10 polypropylene, 2/10 steel laminate, which has the same plane flexural rigidity. By making the same stamped ribs with this material, the flexular rigidity obtained is only 4/10th of that of the initial solution. This is due to the fact that in a flat test specimen, the neutral fibre is inside the material, whereas with an omega rib, the neutral fibre is on the exterior of the material.

A second example relates curved panels such as roof panels, floor wells, door panels, etc. These panels are subjected to complex stresses :

- shearing in the plane : the core does not participate very much ; only one plate is loaded if glued on one side ;
- membrane vibration : the interest of the sandwich increases with vibration frequency ;
- acoustic transparency : the sandwich is efficient ;
- denting : the sandwich performs well locally compared with metal sheet, but it is weaker in the peripheral area.

In general, the less the panel is curved, the more interesting the sandwich structure is.

8.3.8. Processing means

To reduce vehicle weight, the designer must also know about processing means and methods and developments in this field. For the processing method influences both the grade of materials (and therefore their mechanical characteristics) and the shape of the part : minimum thickness, drafts, variable thickness possibilities, etc.

Depending on the processing method, there will be more or less waste material. New methods or methods that are still little used in the motor car industry may enable the weight of parts to be reduced (for example, fluoturning, magneto-forming, electron beam welding, plasma spraying, etc...).

A good illustration is provided by the so-called lost foam casting process, which was used for a long time by toolmakers for casting large frames, but which has just started to be used in series production. A model of the part is made of expanded polystyrene in on or several parts glued to one another. It is then covered with a fine coating of refractory material and placed inside a container containing unbonded sand, which fills all the free spaces left by the model. This process enables thickness excesses to be avoided and the weight of parts such as manifolds, cylinder heads, crankcases, crankshafts, etc. to be reduced.

9 CONCLUSIONS

Designing a motor car that uses energy very economically is a fascinating task and involves re-examining all the habits of design and engineering offices and familiarizing oneself with new technologies. This also reveals that there are quite a number of gaps in knowledge today in the motor car field. This is encouraging for the new generation of engineers. They will have much to discover.

Finally, this type of design work sometimes leads to surprises like those we had in the VESTA programme :

- we had made style models in the form of single units with a very forward and very inclined windscreen ; we obtained a better drag coefficient only with the form finally adopted.

- an extensive study using the finite elements calculation method enabled us to save more than 40% on the weight of the body rigidity structure, while retaining the use of steel plate,

- The first design of the doors and bonnet made use of composite materials. On examining the various alternatives, it proved that an equivalent or better result could be obtained by combining steel plate with composite materials.

Fig 1   Renault Eve+ prototype

Fig 2   Renault Vesta prototype

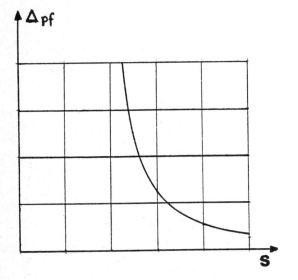

Fig 3   Pressure losses due to the radiator itself in relation to its frontal area at constant calorific power

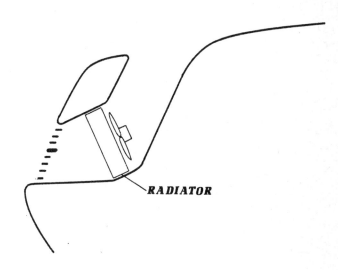

Fig 4   A sleeve downstream of the radiator opening out on to the front part of the bonnet reduces the cooling drag

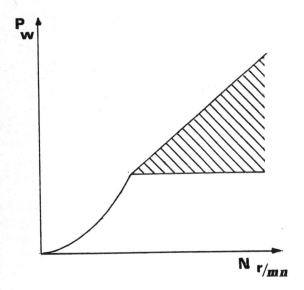

Fig 5   Power absorbed by the oil pump

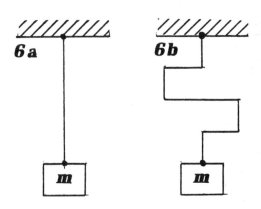

Fig 6   Stress travel
(a) minimum weight support
(b) longer travel leads to increase weight particularly if equivalent stiffness is needed

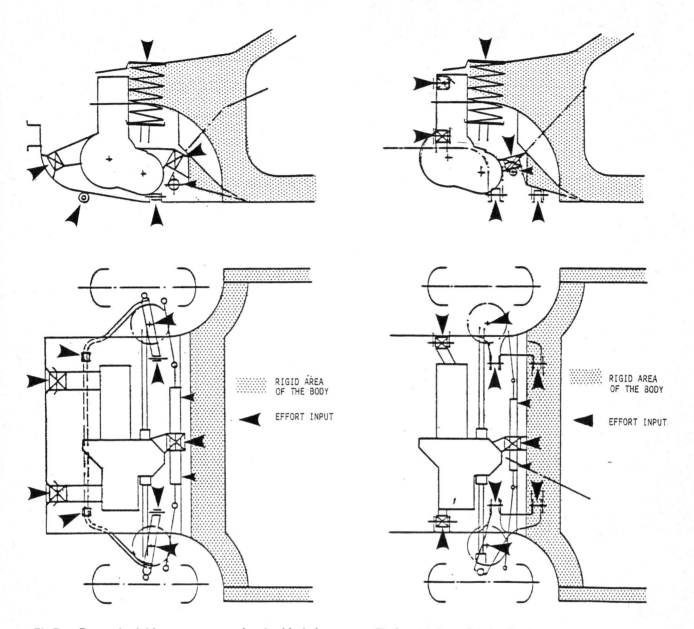

Fig 7 Front wheel drive, transverse engine. In this design the different effort inputs are all located in the engine compartment

Fig 8 Variant of design for the same vehicle as fig 7 but with effort inputs located as close as possible to the rigid area

SANDWICH MATERIALS
Trunk floor

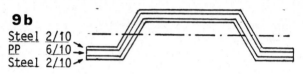

Curved panels like roof - floor wells

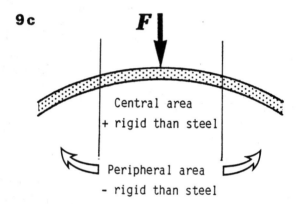

Fig 9   Sandwich materials
   (a) 8/10 steel trunk floor
   (b) the same design with steel-PP laminated offers only 40% of the all-steel rigidity
   (c) curved panels with a foam core

C448/84

SAE 841311

Fuel economy and performance – the case for lightweight vehicles

A ANCLIFF, BA, MSc, CEng, MIMechE
BL Technology Limited, Lighthorne, Warwickshire

SYNOPSIS The influence of weight changes on fuel consumption is examined on lightweight, functionally equivalent vehicles and a relationship established. This is used to predict the effects of weight changes on future vehicles when engine efficiency is improved and drag losses reduced well below current standards.

1 INTRODUCTION

The fuel crises of the 1970s started the energy conservation movement and a natural target was the private passenger car. This led either to direct or indirect interference by governments, with the requirements for corporate average fuel economy in the USA, and in the UK, the publication of the fuel consumption of all the new vehicles being offered for sale. These published figures are not an accurate representation of actual consumption experienced by users for a number of reasons and these have been well documented elsewhere (1).

In the USA, large cars were downsized to achieve improvements in economy through weight saving while in Europe and Japan efforts were concentrated on reducing drag and increasing the efficiency of the power trains. Weight saving on small cars generally means material substitution, with plastic and aluminium replacing steel. Both of these materials are considerably more expensive in terms of cost per kg, and suppliers have used the concept of Lifetime Energy Use (2), (3) to show that the energy saved over an assumed life of 160 000 km (100 000 miles) is both energy and cost effective.

Now that the more obvious savings have been made what further benefits do weight reductions offer in terms of fuel saving?

There has been a wide range of values published for the fuel saved per kg of weight saved and much of this has been related to vehicles that are typical of the US market. It was postulated (2) that weight reduction, without corresponding changes to engine and drive train, could account for about one half of the total fuel economy improvements achievable. Magee (4) pointed out that it is useful to realise that the fuel consumed per km travelled should increase linearly with increase in vehicle weight. The major problem was to obtain the appropriate data to identify this change, because functionally equivalent vehicles were required. A value of fuel saved of 0.57×10^{-2} m^3 per kg of weight saved over a vehicle life of 160 000 km was derived for manual transmission cars, and estimates from various sources were tabulated which showed both the shorter term or 'application to existing vehicle' estimates and long run estimates. These are repeated in Table 1 and, in view of the wide variations in these results, it was felt that a closer study of the effects of weight saving would be valuable to provide data for very lightweight vehicles.

2 METHODOLOGY

2.1 Vehicle specification

A computer simulation was used to predict the acceleration performance, maximum speed and fuel consumption of a range of hypothetical vehicles with kerb weights varying from 550 kg to 850 kg. Predictions are required to show the relationship between weight reduction and reduced fuel consumption when the vehicle is also subjected to changes in other parameters that effect fuel consumption. The other variables considered in this paper, whose influence needs to be separated from the effects of weight reduction, are engine efficiency, aerodynamic drag, rolling resistance and acceleration performance.

In order to arrive at functionally similar vehicles, the time to accelerate from 0 to 96.5 km/h (60 mile/h) was used with the vehicle at kerb weight plus 180 kg. The major features of the vehicle which remained constant were:-

Frontal area   1.8 m^2

CG height above ground   0.51 m

Wheel base   2.4 m

Weight distribution   60:40 Front:Rear

Front wheel drive

Five speed manual transmission

The gear ratios were specified to meet the following criteria:-

Low ratio first gear was selected so that a fully laden vehicle would start from rest on a 25 per cent gradient. 4th gear was selected to give maximum vehicle speed at peak power. 5th

Table 1 Estimates of change in fuel consumption with weight (after Magee)

	Year	Litres per 160 000 km per kg	Method
1	1970	5.8	Weight and aerodynamic considered
2	1973	11.6	Sales weighted average (SWA)
3	1975	11.6	(SWA)
4	1976	11.6	(SWA)
5	1978	4.1 Short Term 9.0 Long Term	GM - No technical discussion
6	1979	18.2	All cars
7	1979	4.5 Short Term 9.0 Long Term	Aerodynamic correction
8	1979	4.1 Short Term 9.0 Long Term	Aero and dyno corrosion
9	1980	1.7 Short Term 3.9 Short Term 5.7 Long Term	Experiments on Ford cars with manual transmissions
10	1980	2.9 Short Term 6.3 - 7.6 Long Term	Extension of (9) range due to uncertainties in various factors
11		2.2 Short Term 4.5 - 8.5 Long Term	Additional factors taken into account

gear was chosen to be 25 per cent higher than 4th while 2nd and 3rd gears formed a geometric progression between 1st and 4th gear ratios.

The study consisted of analysing a number of vehicle configurations over the 550 kg to 850 kg kerb weight range that typified the best attributes of current practice. These attributes were:-

(i) Aerodynamic Drag C_d = 0.35

(ii) Coefficient of Rolling Resistance C_{rr} = 0.012

(iii) Engine b.s.f.c. map of Fig 1a

The analysis was then repeated and the attributes changed to the following values:-

(i) C_d = 0.2

(ii) C_{rr} = 0.007

(iii) B.s.f.c. map of Fig 1b

It is judged that all these values are achievable over the next decade and it is conceivable that current benefits arising from weight reduction may be reduced to a level that would not warrant any on-cost brought about by material substitution.

No attempt was made to optimise the intermediate gear ratios for minimum fuel consumption. This would have been an added refinement but it was felt that, for the purpose of this exercise, the relative results would not be materially affected. Reference 5 gives one method for optimising gear ratios for minimum fuel consumption but this would have necessitated many more computer runs if adopted here.

2.2 Computer simulation

Computer simulation offers repeatability and can predict small changes in performance and fuel economy based on changes made to vehicle specification. In the case of vehicles which are currently not available, simulation is the effective means of obtaining information about their performance characteristics. While it is not the intention to predict absolute values for fuel economy and performance, some measure of accuracy of the simulation procedure used can be obtained from Fig 2. These results were part of the validation exercise carried out on the BL Technology Energy Conservation Vehicle programme. The accuracy of the fuel consumption predictions was very sensitive to the values of the b.s.f.c. of the actual engine used in the vehicle in which the test measurements were made. The experimental values were obtained by driving the vehicles at constant speeds and makin accurate measurements of speed, distance and fuel used. A number of runs were made in each direction and an average taken to give the fuel consumption at the particular speed. The urban consumption measurements were made in a similar way, with the driver following a 'driver's aid' installed in the vehicle. The experimental values were sensitive to the experience of the driver and his ability to maintain the correct speeds and accelerations without making unnecessary throttle adjustments. The results reflect, more accurately than rolling road results, realistic values of consumption. The computer simulation is intended to predict the road performance and fuel consumptions as measured by these experimental methods. The predictions should not only produce the trend, but also give a reasonable indication of the absolute values of fuel consumption. The program used can simulate fuel consumption for any driving cycle, and in this exercise, the ECE 15 urban driving cycle was used together with constant speeds of 90 km/h and 120 km/h in 5th gear to derive an overall consumption. The vehicles were assumed to be on a level road and in zero wind velocity conditions.

3 RESULTS

3.1 Performance calculations

Fig 3 shows the results applicable to a 750 kg kerb weight vehicle using current technology. The variables of acceleration performance and maximum speed are plotted against engine capacity. The maximum speed is not considered to be a design goal but a product of the effort to reduce energy losses. The range of acceleration times to reach 96.5 km/h span from 18 seconds to 8 seconds, and the engine capacity to achieve these times varies from 0.9 to 2.5 litres. Fig 4 shows the same information for vehicles using the advanced technologies. Engine capacity is now 0.66 to 1.97 litres for the same span of acceleration times and maximum speed is predicted at about 254 km/h as against 210 km/h for a vehicle using current technology.

3.2 Fuel consumption predictions

The fuel consumption for each vehicle weight and acceleration criterion can be calculated now that the engine capacities have been established. These calculated values are shown in Tables 2 and 3. Values are expected to vary linearly with vehicle weight for each level of acceleration performance. Figures 5 and 6 are plots of the overall fuel consumption and the constituent parts from which the overall consumption is derived. The graphs indeed confirm that, for the range of weights being considered, the fuel consumption does vary in an approximately linear manner and the slopes are listed in Table 5. The intercept is a measure of the energy loss from aerodynamics and other resistances, and also the inefficiency of the engine.

4 DISCUSSION OF PREDICTED RESULTS

4.1 Acceleration and maximum speed

Maximum acceleration performance is a function of the power to weight ratio of the vehicle and is not influenced by the choice of gear ratios. The various drag losses are not significant either up to speeds of 100 km/h. Obviously lighter weight vehicles needs smaller power units for the same acceleration performances. Even with the smaller power outputs, the low aerodynamic drag enables the vehicles to exceed current maximum speeds by a large margin. Fig 7 shows the available tractive effort in each gear for a 750 kg vehicle using advanced technology plotted against vehicle speed together with the road load line. With the high gearing in 4th and 5th gears, it can be seen that small changes in external resistances, such as gradient and wind speed, will have a marked effect on maximum speeds. This could have far reaching effects on brake and tyre equipment and may require the fitting of speed limiting devices. There is a limitation on the tractive effort in both 1st and 2nd gears over approximately the first five seconds with the choice of front wheel drive. A four wheel drive layout would reduce the acceleration time of this particular combination of weight and power to reach a given speed.

4.2 Fuel consumption

Generally for a given acceleration performance, a smaller engine operates in the area of lower b.s.f.c. in the constant speed modes and the ECE 15 urban cycle. Another benefit of smaller engines on the urban cycle is that less fuel is used in the idle and overrun conditions. However, examination of Figures 3, 4, 5 and 6 show that there is not a great deal of benefit to be derived in terms of fuel consumption in producing light weight low drag vehicles with a 0 - 96.5 km/h acceleration time greater than 12 seconds. In some instances, fuel consumption actually starts to increase as acceleration performance and engine size are reduced. This is most pronounced on the 120 km/h constant speed mode for the current technology vehicle. This is because the engine size has been reduced to the point where it is operating in areas of the fuel map that are beyond the minimum b.s.f.c. and are increasing towards the full throttle condition. When fuel consumption is plotted against vehicle mass for functionally similar vehicles as in Figures 5 and 6, there is the linear variation as expected from the review of previous work. The slopes of the lines vary from one operating mode to another with the steepest slope being obtained on the urban cycle as would be expected, and the smallest gradient on the constant 120 kph mode where aerodynamic forces are predominant.

The values of the slopes and intercept for the overall consumption are listed in Table 4. In order to derive the overall fuel consumption of a vehicle with a kerbweight of 750 kg and an acceleration time to 96.5 km/h in 12 seconds, it is now a matter of simple arithmetic. As an example of a current technology vehicle, the fuel consumption is given by

$$FC = (.0328+(750 \times 3.37 \times 10^{-5}))100 = 5.8 \text{ L/100 km}$$

It is evident from these values that the benefits of weight reduction are less with the future technology vehicle than with current technology. If a vehicle is considered with an acceleration time of 12 seconds to 96.5 km/h, the lifetime (160 000 km) fuel savings for the current and future technology vehicles respectives are predicted to be:-

5.39 litres per kg of weight saved

3.68 litres per kg of weight saved

The 5.39 litres is at the bottom end of the values quoted in (4) while the 3.68 litres is more pessimistic than any long term predictions previously quoted.

The intercept value which is a measure of engine efficiency and other energy consuming parameters, shows that bigger gains in economy can be obtained by concentrating on these areas. This can be demonstrated by considering the fuel consumed over a distance of 20 000 km (an aver-

Table 2 Current technology vehicle
        Fuel consumption litres/100 km

MODE	KERB WEIGHT kg	ACCELERATION TIME 0 - 96.5 km/h (sec)			
		8.5	12	15	18
ECE 15 URBAN CYCLE	550	7.74	5.79	5.26	5.11
	650	8.43	6.28	5.80	5.52
	750	8.33	6.80	6.23	5.95
	850	8.88	7.37	6.69	6.36
CONSTANT 90 km/h	550	5.19	4.38	4.25	4.35
	650	5.44	4.56	4.43	4.43
	750	5.79	4.77	4.61	4.56
	850	6.00	5.00	4.79	4.70
CONSTANT 120 km/h	550	6.79	6.23	6.33	6.55
	650	7.00	6.39	6.42	6.58
	750	7.32	6.61	6.55	6.63
	850	7.51	6.96	6.66	6.72
OVERALL	550	6.38	5.13	4.86	4.87
	650	6.80	5.43	5.17	5.09
	750	6.96	5.76	5.45	5.32
	850	7.30	6.14	5.74	5.57

Table 3 Future technology vehicle
        Fuel consumption litres/100 km

MODE	KERB WEIGHT kg	ACCELERATION TIME 0 - 96.5 km/h (sec)			
		8.5	12	15	18
ECE 15 URBAN CYCLE	550	4.96	3.93	3.62	3.42
	650	5.56	4.32	3.97	3.70
	750	6.14	4.73	4.27	4.04
	850	6.82	5.1	4.57	4.36
CONSTANT 90 km/h	550	2.86	2.34	2.23	2.09
	650	3.04	2.47	2.36	2.26
	750	3.20	2.58	2.49	2.38
	850	3.38	2.73	2.60	2.50
CONSTANT 120 km/h	550	3.58	3.12	3.03	2.97
	650	3.74	3.25	3.16	3.07
	750	3.89	3.37	3.27	3.19
	850	4.11	3.51	3.37	3.30
OVERALL	550	3.78	3.06	2.87	2.71
	650	4.12	3.28	3.08	2.92
	750	4.44	3.52	3.28	3.13
	850	4.83	3.75	3.46	3.32

Table 4 Values derived from figures 5d and 6d

Acceleration Time to 96.5 km/h	Slope $\frac{1/km}{kg}$	Intercept 1/km	Technology
8.5	3.07×10^{-5}	.0469	Current
12	3.37×10^{-5}	.0328	Current
15	2.93×10^{-5}	.0359	Current
18	2.33×10^{-5}	.0359	Current
8	3.35×10^{-5}	.0186	Future
12	2.30×10^{-5}	.0179	Future
15	1.97×10^{-5}	.0179	Future
18	2.03×10^{-5}	.0160	Future

age years motoring) by a vehicle capable of accelerating to 96.5 km/h in 12 seconds.

A saving of 458 litres could be made by changing the technology and further saving of 92 litres made by reducing the weight from 750 kg to 550 kg.

Table 5 Litres of fuel used per 20 000 km

Vehicle Weight	750 kg	550 kg	Difference
Current Technology	1161	1027	134
Future Technology	703	611	92
Difference	458	416	

5 CONCLUDING REMARKS

No attempt has been made to specify how the vehicle weight would be reduced, but to quantify the effects of weight reduction when other vehicle parameters changed. 550 kg is a weight target that can be achieved but material substitution on a large scale would be necessary. Readers are referred to (6) for a review of material usage in the European automobile industry where it is stated that no material is adopted for a mass production model unless it offers one of three things: An overall cost saving; some benefit without a cost increase; or a major benefit with an acceptable cost increase. With a future technology vehicle, the projected fuel saving for a weight reduction of 200 kg is about 740 litres over the life of the vehicle. Assuming a cost per litre of £0.4 this indicates a lifetime saving of about £300.

It should also be noted that there are other technical problems that have to be addressed when very lightweight high speed vehicles are being considered. The maximum speed fully laden condition has major impacts on brake and tyre equipment. There may be stability problems at high speeds in cross wind conditions on vehicles with low aerodynamic drag coefficients and refinement, in terms of ride, noise and vibrations may deteriorate in lightweight vehicles. Finally, in the UK at least, there are limitations imposed on towing capacity so it may be that these technical factors will put a limit on the lightest practical vehicle.

REFERENCES

1 HELLMAN, K. H., MURREL, J. D., 'Why vehicles don't achieve the EPA MPG on the road and how that shortfall can be accounted for'. Passenger Car Meeting, Troy 1982, SAE paper 820791.

2 COCHRAN, C. N., MCGLURE, R. G., 'Weight saving materials, energy and the automobile', 1980. Proceedings of the First International Automotive Fuel Economy Research Conference, US Department of Transportation, pp 254.

3 WHEELER, M. A., 'Lightweight materials and life cycle energy use'. International Congress and Exposition Detroit, Michigan, 1982, SAE paper 820148.

4 MAGEE, C. L., 'The role of weight reducing materials in automotive fuel savings'. International Congress and Exposition Detroit, Michigan, 1982, SAE Paper 820147.

5 CHANA, H. E., FEDEWA, W. L., MAHONEY, J. E., 'An analytical study of transmission modifications as related to vehicle performance and economy'. International Automotive Engineering Congress and Exposition, Detroit, 1977, SAE Paper 770418.

6 WHITBREAD, C., 'The car of the future in Western Europe'. The Economist Intelligence Unit Special Report 155 Chp. 7 1984.

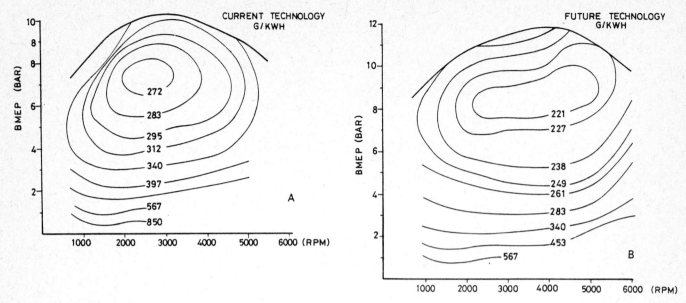

Fig 1   Fuel maps used in calculations

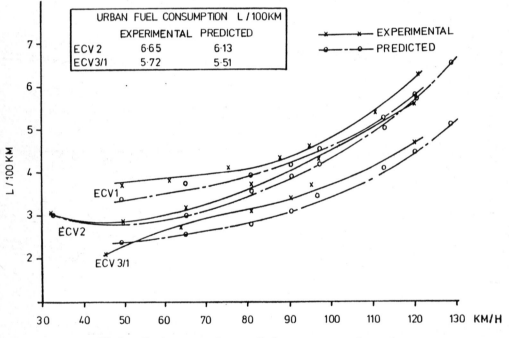

Fig 2   Fuel consumption; predictions versus experimental results from the energy conservation vehicle programme

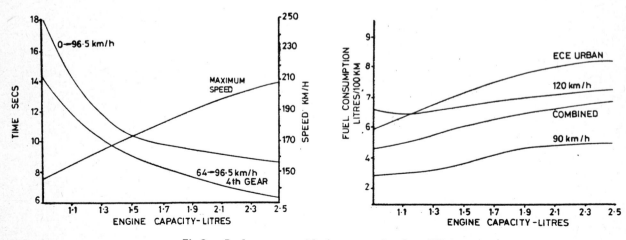

Fig 3   Performance and fuel consumption for a 750 kg vehicle using current technology

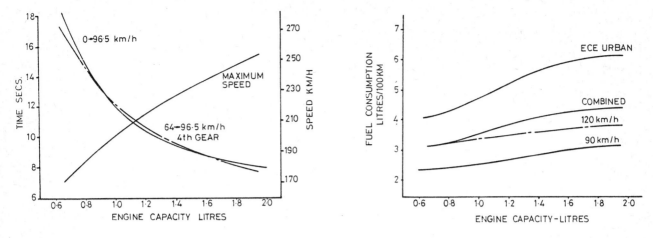

Fig 4　Performance and fuel consumption for a 750 kg vehicle using future technology

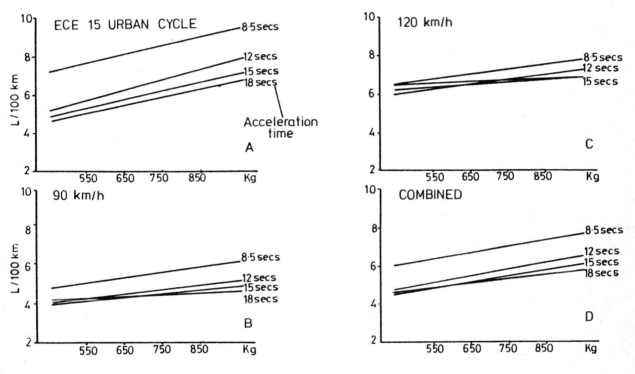

Fig 5　Fuel consumption versus weight current technology

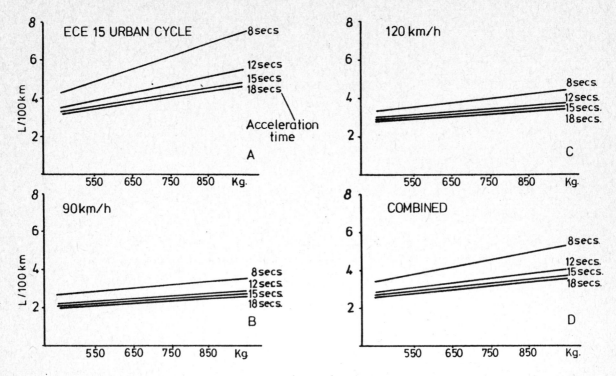

Fig 6 Fuel consumption versus weight future technology

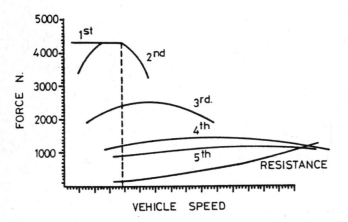

Fig 7 Maximum transferable tractive force for a 750 kg vehicle